Turbulent Shear Flows 3

Selected Papers from the Third International
Symposium on Turbulent Shear Flows,
The University of California, Davis, September 9–11, 1981

Editors:

L. J. S. Bradbury F. Durst B. E. Launder
F. W. Schmidt J. H. Whitelaw

With 244 Figures

Springer-Verlag Berlin Heidelberg New York 1982

Leslie J. S. Bradbury
Department of Mechanical Engineering, University of Surrey,
Guildford, Surrey GU2 5XH, England

Franz Durst
Sonderforschungsbereich 80 der Universität Karlsruhe, Kaiserstraße 12,
D-7500 Karlsruhe 1, Fed. Rep. of Germany

Brian E. Launder
Department of Mechanical Engineering, University of Manchester,
Institute of Science and Technology, PO Box 88,
Manchester M60 1QD, England

Frank W. Schmidt
Mechnical Engineering Department, The Pennsylvania State University,
University Park, PA 16802, USA

James H. Whitelaw
Department of Mechanical Engineering, Imperial College of Science and
Technology, Exhibition Road, London SW7 2BX, England

ISBN-13:978-3-642-95412-2 e-ISBN-13:978-3-642-95410-8
DOI: 10.1007/978-3-642-95410-8

Library of Congress Cataloging in Publication Data. International Symposium on Turbulent Shear Flows (3rd: 1981: University of California, Davis). Turbulent shear flows 3. Bibliography: p. Includes index.
1. Shear flow - Congresses. 2. Turbulence- Congresses. I. Bradbury, L. J. S. (Leslie John Stanley), 1936–.
II. Title TA357.I59 1981 620.1'064 82-16916

Softcover reprint of the hardcover 1st edition 1982

Preface

In spite of intensive efforts over many decades, the problem of turbulence remains as challenging as ever and the number of papers, books and conferences on this topic continues to grow. As experimental techniques and computing power have developed, the breadth of investigations into the structure and development of turbulent flows has increased to encompass many diverse fields of application in engineering, physics, biology and so on. As a consequence, it is now very difficult for a single research worker to keep in touch with the many developments that are taking place in turbulence. One of the few opportunities for obtaining some overall view of the subject arises from large international symposia on turbulence and, although they have some drawbacks, it is this opportunity that is one of their main merits.

The International Symposium on Turbulent Shear Flows has now been held on three occasions and they seem to be established as a major opportunity for papers on a very diverse range of topics to be presented at a single meeting. This volume is a collection of papers from the third symposium that was held at the University of California, Davis from 9–11 September 1981. The papers are divided into four sections entitled Wall Flows, Scalar Transport, Recirculating Flows and Fundamentals. This collection represents about a third of the total number of papers presented. Inevitably, there is some uneveness in the coverage of various sections but, nevertheless, the selection is reasonably representative of the range of papers presented. As with previous volumes, each section is preceded by a brief introductory article whose purpose is to make some general observations about the various sections and to fit the individual papers into the context of the general topic.

As with the earlier symposia, we would like to acknowledge the financial support of the Research Offices of the United States Army, Navy, and Air Force and the National Science Foundation. We would also like to thank the many individuals at the University of California, Davis who helped with both organising and running the conference and express again our appreciation to the Fluids Engineering and Heat Transfer Divisions of the American Society of Mechanical Engineers for their assistance with publicity.

The task of reviewing and selecting papers was carried out by a papers committee and an advisory committee many of whose members later performed sterling work as chairmen of technical sessions at the conference and who also have been a valuable source of helpful criticism. Their work has been greatly appreciated. The committees consisted of:

R.J. Adrian, University of Urbana, Urbana, USA

J.C. André, EERN/GMD, 92100 Boulogne, France

L.H. Back, Jet Propulsion Lab., Pasadena, Calif., USA

H.A. Becker, Queen's University Kingston, Ont., Canada

R. Borghi, O.N.E.R.A., France
S. Corrsin, The John Hopkins University, Baltimore, Md., USA
J.J. Domingos, University of Lisbon, Lisbon, Portugal
C. du Pont Donaldson, ARAP, Princeton, N.J., USA
R. Dumas, Institut de Mécanique Statistique de la Turbulence, Marseilles, France
H. Fiedler, Technische Universität Berlin, Berlin, Fed. Rep. of Germany
I.S. Gartshore, University of British Columbia, Canada
V.W. Goldschmidt, Purdue University, W. Lafayette, Ind., USA
A.D. Gosman, Imperial College, London, U.K.
R. Günther, Universität Karlsruhe, Karlsruhe, Fed. Rep. of Germany
K. Hanjalić, Masinski Fakultet, Sarajevo, Yugoslavia
T.J. Hanratty, University of Illinois, Urbana-Champaign, Ill., USA
R.J. Herring, NCAR, Boulder, Colo., USA
A.K.M.F. Hussain, University of Houston, Texas, USA
W.P. Jones, Imperial College London, U.K.
P.N. Joubert, University of Melbourne, Melbourne, Australia
J. Laufer, University of Southern California, Calif., USA
A. Libby, University of California, San Diego, Calif., USA
J.L. Lumley, Cornell University, Ithaca, N.Y., USA
O. Martynenko, Heat and Mass Transfer Institute, Minsk, USSR
J. Mathieu, Ecole Centrale de Lyon, Ecully, France
H. McDonald, Scientific Research Associates Inc., Glastonbury, Conn., USA
Y. Mori, Tokyo Institute of Technology, Tokyo, Japan
K. Owen, Owen International, Palo Alto, Calif., USA
W.C. Reynolds, Stanford University, Stanford, Calif., USA
W. Rodi, Universität Karlsruhe, Karlsruhe, Fed. Rep. of Germany
M.W. Rubesin, NASA Ames Research Center, Calif., USA
A.K. Runchal, Dames & Moore, Los Angeles, Calif., USA
I. Wygnanski, Tel-Aviv, Israel
J.C. Wyngaard, CIRES, Boulder, Colo., USA

Finally, it is a pleasure to record our thanks to the authors for meeting the various deadlines we set and to Springer Verlag for their help in producing this third volume in the series.

Karlsruhe, June 1982 *The Editors*

Contents

Part I Wall Flows

Part II Scalar Transport

Part III Recirculating Flows

Part IV Fundamentals

Part I

Wall Flows

Introductory Remarks

Helmut Eckelmann

Max-Planck-Institut für Strömungsforschung, Böttingerstraße 4–8
D–3400 Göttingen, Fed. Rep. of Germany

Wall flows, which mean in this context wall bounded turbulent flows, can generally be classified into two main groups; external flows which arise around bodies (boundary layers) and internal flows which arise through a space confined by walls (pipes and channels). Although both groups exhibit a difference – the boundary layer thickness increases in streamwise direction whereas in pipes and channels the flow remains restricted to the space confined by the walls – both flows show a common behaviour in the vicinity of the wall. They both have a viscous sublayer, a buffer layer and for both the law of the wall is valid in the inner region. The key for the understanding of wall bounded turbulent flow should therefore be searched in the region close to the wall. The viscous sublayer is very thin and experimentally not accessible in most of the flows. The buffer layer, which is in most practical cases thick enough for experimental investigations, is in the opinion of the writer, the region which one should look at.

This chapter contains a selection of papers on wall flows which were presented at the 3rd Symposium on Turbulent Shear Flows held at the University of California at Davis. The major parts of the papers describe experimental work and one of them deals with a theoretical model of the turbulent boundary layer and calculations on this model which lead to longitudinal vortices occuring in the buffer laver. Such vortices were first observed experimental by *Bakewell* and *Lumley* (1967). One of the papers is both numerical and experimental and deals with calculations and experimental investigations obtained in a pipe flow.

The first four papers in this chapter deal with unsteady wall bounded turbulent flows, a subject which is very important for many engineering applications. Recently *Carr* (1981a) reviewed this flow type. The many laminar, transitional and turbulent unsteady flow experiments that have been performed are fully referenced in a AGARDograph by *Carr* (1981b).

In a turbulent channel flow with forced velocity oscillations of small amplitude Binder and Kueny made measurements over a wide range of frequencies. They found that both mean flow and mean turbulent intensity are not affected by the forced oscillations. The phase averaged streamwise turbulent intensity is not simply proportional to the amplitude of the velocity oscillations but depends on the wall distance and the oscillation frequency. The ratio of Stokes thickness $\sqrt{2\nu/\omega}$ (where $\omega = 2\pi f$) to viscous length ν/u_τ (ν being the kinematic viscosity of the fluid and u_τ the friction velocity) becomes an important parameter for this problem.

In a large oscillating water tunnel which is in principle a water filled U-tube Hayashi and Ohashi investigated the unsteady turbulent boundary layer with a hot-film probe and by a flow visualization technique that uses a thin milk layer on the wall. The hot film measurements yielded that the velocity phase leads that of the wave amplitude by about 15°;

in addition, both Reynolds stress and turbulent energy were found to be larger in the deceleration than in the acceleration period of the free stream period. The visual studies showed that the disturbances at the wall lag behind the free stream velocity maximum.

The response of a turbulent boundary layer to oscillations of the free stream velocity was studied by Parikh, Jayaraman and Reynolds and by Cousteix, Javelle and Houdeville. The first authors confirmed the results of Binder and Kueny that both mean velocity and mean turbulence intensity were unaffected by the imposed oscillations. In addition they found that the boundary layer thickness and Reynolds stress distribution across the boundary layer becomes frozen over the oscillation cycle at their mean values. The frequencies in this experiment ranged from zero to approximately the bursting frequency f_B which corresponds due to measurements of *Rao* et al. (1971) to a Strouhal number $f_B \delta / U_\infty$ of about 0.2. Cousteix et al. also found that the mean flow field was not affected by the unsteady effects. They interpreted this by assuming that the boundary layer responds to the perturbations induced by the pulsation of the external flow as a mechanical system with a small damping.

Zhang and Lilley present a theoretical model of the coherent structure of the turbulent boundary layer with zero pressure gradient. They show that a self-generating coherent structure arises in the calculation with their model. For a given initial disturbance amplitude Zhang and Lilley obtained strong streamwise vortices occurring periodically with opposite sign in the spanwise direction of the flow. The non-dimensional wave length of a counter-rotating vortex pair was found to be of the order of 100 ν/u_τ over a wide range of Reynolds numbers, a result which is in good agreement with the experimental investigations of *Lee* et al. (1974), *Blackwelder* and *Eckelmann* (1979) and *Kreplin* and *Eckelmann* (1979) who all had evidence for counter-rotating streamwise vortices occuring frequently in the wall region wall of bounded turbulent flow. The distance between two counter-rotating vortices was experimentally determined to be of the order of 50 ν/u_τ. As early as 1959 *Kline* and *Rundstadler* speculated that the low speed streaks having a spacing of about 100 ν/u_τ are related to an organized vortical structure in the wall region.

The mechanism of turbulent mass transfer at a solid surface was investigated by Campbell and Hanratty. From the mass balance equation the fluctuating concentration field was calculated by using measured values of the fluctuating velocity field. They found that at large Schmidt numbers the concentration boundary layer close to the surface acts as a filter in such a way that only velocity fluctuations of much lower frequency than the most energetic velocity fluctuations are effective in transporting mass. With increasing Schmidt number smaller and smaller fractions of turbulence energy are effective in determining the magnitude of the mass transfer coefficient.

The third paper in this chapter by Gibson, Verriopoulos and Nagano is concerned with the effects of longitudinal curvature on turbulent heat transfer, a problem which had been given little attention in the past. More is known about effects of such curvature on the turbulent boundary layer. The AGARDograph by *Bradshaw* (1973) fully reviewed this subject up to that time. The paper of Gibson et al. reports measurements of mean velocity, mean temperature and surface heat flux in a boundary layer growing on a mildly curved convex plate for which the radius of curvature R is about 100 times the boundary layer thickness δ. The main result of this investigation is that the curvature depresses the surface heat flux more than it depresses the skin friction, i.e., with increasing length Reynolds number the Stanton number falls more rapidly from flat plate values than does the skin friction coefficient.

References

Bakewell, H.P., Lumley, J.L.: Phys. Fluids **10**, 1880 (1967)
Blackwelder, R.E., Eckelmann, H.J.: Fluid Mech. **94**, 577 (1979)
Bradshaw, P.: AGARDograph **169** (1973)
Carr, L.W. (1981a): Proc. of the IUTAM Syp. on Unsteady Turbulent Shear Flows. (Springer Verlag, Berlin, Heidelberg, New York 1981), p. 3–34
Carr, L.W.: AGARDograph **265** (1981b)
Kline, S.J., Rundstadler, P.W.: J. Appl. Mech. E **26**, 166 (1959)
Kreplin, H.-P., Eckelmann, H.: J. Fluid Mech. **95**, 305 (1979)
Lee, M.K., Eckelman, L.D., Hanratty, T.J.: J. Fluid Mech. **66**, 17 (1974)
Rao, K.N., Narasimha, R., Badri Narayanan, M.A.: J. Fluid Mech. **48**, 399 (1971)

Measurements of the Periodic Velocity Oscillations Near the Wall in Unsteady Turbulent Channel Flow

G. Binder and J.L. Kueny

Institut de Mécanique, Université Scientifique et Médicale de Grenoble,
Institut National Polytechnique, B.P. 53 X,
F–38041, Grenoble-Cedex, France

Abstract

Measurements in turbulent channel flow with forced velocity oscillations of small amplitude have been performed over a wide range of frequencies. The results show that the mean flow and the mean turbulent intensity are not affected by the forced oscillations. The amplitude and the phase shift of the periodic velocity fluctuation follow the laminar Stokes solution at high frequency. At low frequencies near the wall the gradient of the amplitude becomes steeper than in the Stokes flow and the phase shift decreases to slightly negative values. Further from the wall a phase lead is again observed. The results also show that the phase averaged longitudinal turbulent intensity is not simply proportional to the velocity oscillations; the amplitude ratio is position and frequency dependent. It is shown that the Stokes thickness l_s^+ non-dimensionalized with the mean viscous sublayer thickness is the important parameter.

Nomenclature

A_{nq}	amplitude of nth mode of q	x, y	longitudinal and transverse coordinates
f	frequency of forced oscillations	ϕ_{nq}	phase lead of nth mode of q
f_t	characteristic frequency of turbulence	ν	kinematic viscosity
		τ	wall shear stress
f_v	frequency of validation of the laser Doppler signal	$\omega =$	$2\pi f$
		$y_s =$	y/l_s
		$(\ )^+ =$	$(\ )/l_\nu$
h	channel half-height	$(\ )_c$	centerline
$l_s = \sqrt{2\nu/\omega}$	Stokes length	$(\bar{\ })$	time mean
$l_\nu = \nu/u_\tau$	viscous wall-length scale	$(\tilde{\ })$	periodic fluctuation
		$(\ )'$	turbulent fluctuation
t	time	$\lvert\ \rvert$	amplitude of periodic fluctuation
T	period of forced oscillations	$\langle\langle\ \rangle\rangle = (\bar{\ }) + (\tilde{\ })$	ensemble average
u	longitudinal velocity	$q = \bar{q} + \tilde{q} + q'$	for any quantity q
u_τ	shear velocity	$\langle q \rangle =$	$\bar{q} + \sum_{n=1}^{\infty} A_{nq} \cos(n\omega t + \phi_{nq})$
$\Omega_h =$	$\omega h/\bar{u}_c$		

Introduction

Many flows of practical importance in aerodynamics, hydrodynamics, turbomachines, bio-fluid dynamics are unsteady. Yet, in the turbulent regime these flows are not well understood if at all even in the simplest case of the flow over a flat plate.

Experimental investigations of unsteady turbulent boundary layers have so far been rather scarce and have yielded little information on the wall layer. For instance, such vital information as the amplitude and phase shift of the wall shear stress is almost completely lacking. On the other hand, several turbulent closures developed for steady flows have been applied to unsteady flows but their predictions of the wall shear stress differ considerably from one another giving even different trends for its phase-shift towards large non-dimensional frequencies [Ref. 6, Fig. 6]. Since there is little experimental data available with which the various predictions can be compared, it is difficult to evaluate the merits and defects of the various models.

The need for measurements in unsteady turbulent boundary layers is also apparent and has been stressed many times in recent years. The purpose of the present investigation is to provide some data in this field and to contribute to the understanding of unsteady turbulent flows.

Apparatus. Instrumentation. Data Acquisition

The measurements are performed in a water channel: length: 2600 mm, height: 2 h = 100 mm, span: 1000 mm, $\bar{u}_c$ = 0–50 cm/s. The test section is 1000 mm long and located at the down stream end. A divergence, adjustable up to about 20°, can be imposed on the section walls in order to subject the flow to a mean unfavourable pressure gradient. The oscillating flow is driven by a reciprocating piston with adjustable speed and stroke mounted on the caisson upstream of the grids and honeycomb. The return flow is via a free surface in order to minimize the effects of unsteady pressure gradients.

The velocities were measured with a laser-Doppler velocimeter operating in the fringe mode. The period of the Doppler signal was determined with a counter. Since one of the goals was to perform measurements close to wall it was necessary to minimize the size of the probe volume. The waist of the focused beam was reduced to 0.08 mm by making use of a beam expander (5×). The half angle between the beams was about 6° and the resulting length of the probe volume was 0.8 mm. With this optical arrangement it was possible to make measurements as close as 0.25 mm from the wall corresponding to $y^+ \cong 2$ for $\bar{u}_c$ = 17.5 cm/s. No correction for probe volume truncation was necessary. Owing to the small size of the probe volume the shear noise contributed only 3% to the measured turbulent intensity at $y^+ = 3$.

Great flexibility in the characteristics of the transmission optics was achieved by use of standard optical components mounted on benches fixed onto a rigid plate.

The Doppler counter is intrinsically a digital period-meter delivering a digital word for every validated Doppler period. This word is fed into a microprocessor where the phase locked ensemble averages are performed. It is also programed to perform simple time integrations. The averaged data is transferred to a computer (NORD 10) where it is stored and analysed.

Averaging on the microprocessor is performed directly on the Doppler periods. It may be shown that the mean velocity in turbulent flow obtained from the mean Doppler period is not biased towards the larger velocities as would be the case if the mean frequency were used (*Tiederman* effect).

Averaging Procedure

Ensemble averaging is done at 100 points equally spaced over the period called channels hereafter for the sake of clarity. A problem arises because the phase averaging has to be performed on digital values arriving randomly in time and not from a continuous analog signal as from a hot wire. For this reason, the average in the ith channel is calculated by taking all the values arriving at the microprocessor during the time interval $iT/100$ to $(i+1)T/100$ from the beginning of each cycle. The maximum difference in phase angle between the values averaged in any channel is therefore $\pm 1.8°$. If f_v is the average frequency at which the validated values occur, then the average number of samples contributed to the ensemble in one channel during one cycle is $f_v/100f$. In particular, if $f_v < 100f$ a sample is not contributed to this average at every cycle.

It should be pointed out that a systematic error is introduced if the phase average is obtained from the analog output of the Doppler-counter by taking systematically one sample per channel and per cycle as is usually done with analog signals. Indeed this output is provided with a hold-circuit which prevents the signal to drop to zero between successive validations and this produces a phase lag and an amplitude reduction of the periodic oscillation $\tilde{u}$. For instance for a pure sine-wave these would respectively be

$$\Delta\varphi = \pi f/f_v \quad \text{and} \quad \sin\Delta\varphi/\Delta\varphi.$$

The validation frequency f_v depends upon many factors in particular upon the particle crossing rate through the probe volume and upon the fringe visibility. Each of these two factors causes f_v to decrease as the wall is approached. Indeed the particle crossing rate is proportional to $\langle u\rangle(y)$ and the fringe visibility decreases rapidly when the intersections of the direct beams with the solid boundary come into the field of view limited by the aperture of the receiving optics. At $y^+ \cong 3, f_v$ was typically less than 10 Hz. The systematic error $\Delta\varphi$ varies also rapidly with distance from the wall precisely in the region where the true phase changes take place. The phase lags published previously [8] may be accounted for by systematic error.

Such errors are avoided by the phase averaging procedure outlined above. This procedure has the additional advantage of accelerating the statistical convergence (especially slow near the wall where the turbulence intensity is high) when $f_v > 100f$, as long as

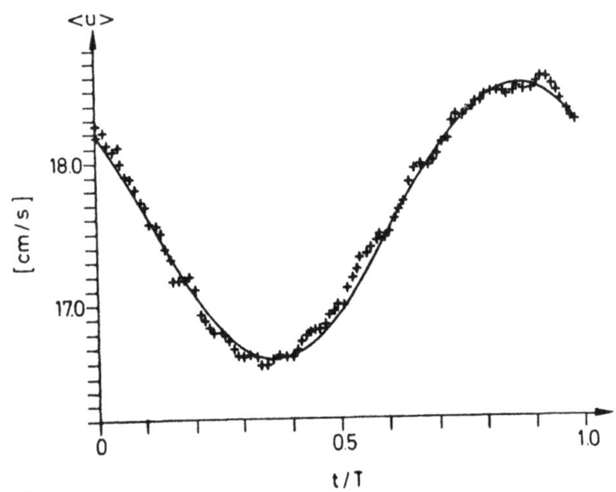

Fig. 1. Ensemble averaged velocity on channel centerline. $T = 7.32$ s; 500 samples/channel; $A_0 = 17.6$ cm/s; $A_1 = 0.96$ cm/s; $A_2/A_1 = 5.8\%$; $A_3/A_1 = 0.8\%$

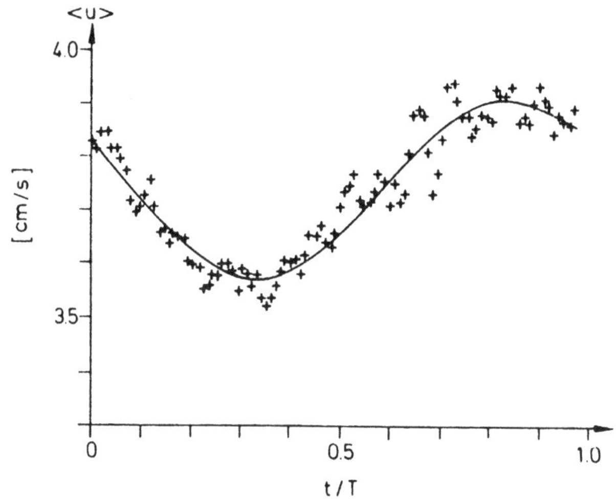

Fig. 2. Ensemble averaged velocity at $y^+ = 4.6$; $T = 7.32$ s; 5000 samples/channel; $A_0 = 3.74$ cm/s; $A_1 = 0.166$ cm/s; $A_2/A_1 = 11.3\%$; $A_3/A_1 = 1.7\%$

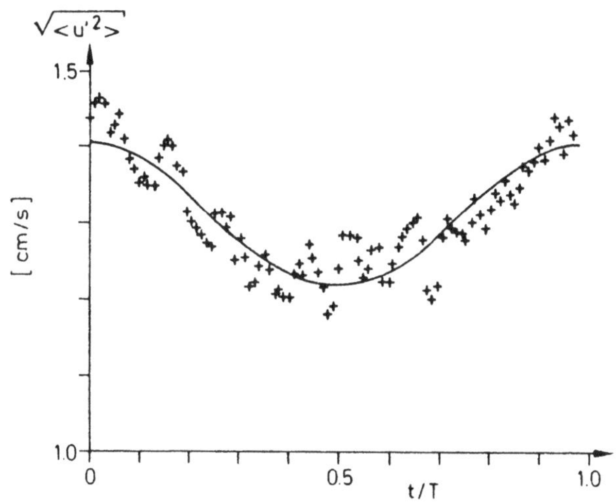

Fig. 3. Ensemble averaged turbulent intensity at $y^+ = 4.6$; $T = 7.32$ s; 5000 samples/channel; $A_0 = 1.31$ cm/s; $A_1 = 0.094$ cm/s

roughly $f_v \leqslant 10 f_t$, where f_t is the characteristic frequency of the turbulent energy containing eddies. This is especially useful at low forcing frequencies which in the present experiments were as small as 0.07 Hz. If in this case only one sample were taken in each channel per cycle, than it would take almost four hours to complete one phase average with a sample size of 1000 values per channel. Actually such a sample size proved to be insufficient near the wall even for $\langle u \rangle$ and quite inadequate for $\langle u'^2 \rangle$. The uncertainty on the phase due to insufficient statistical convergence is especially large.

The measurements reported here correspond therefore to an average of 5000 samples per channel or an integration over one hour. This proved to be adequate for u as shown on Figs. 1 and 2 corresponding to measurements on the centerline and at $y^+ = 4.6$ for $\bar{u}_c = 17.5$ cm/s. It may be observed that there remains appreciable scatter about the fundamental Fourier mode near the wall. $\langle u'^2 \rangle$ at the same point is shown on Fig. 3; the periodic variations of the turbulent intensity are clearly distinguishable from the random scatter. It should, finally, be remarked that the present measurements correspond to amplitudes of the forced oscillations on the centerline of 5% or less. For larger amplitudes the statistical scatter would be relatively smaller.

Experimental Checks

The two-dimensionality of the flow was checked at three stations 10 h apart from each other. The mean flow is established at least 10 h upstream of the section X = 2400 mm where the measurements were taken for the velocity u_c = 17,5 cm/s.

The instrumentation and data acquisition were checked in two ways: first in steady turbulent flow in the channel where good agreement was found for both u/u_τ and $\sqrt{u'^2}/u_\tau$ vs y^+ with previous measurements (Figs. 4,5) and secondly in laminar oscillating flow in an oil channel where good agreement was found for the amplitude A_{lu} and the phase shift ϕ_{lu} vs y_s with the theoretical values of the Stokes flow.

Vibrations of the wall generated by the oscillating pressure could produce large systematic errors on $\tilde{u}$ especially close to the wall where $\partial\bar{u}/\partial y$ is large. It should first be remarked that the oscillating pressure in a given section is proportional to $\partial\tilde{u}/\partial t$ and hence to $\omega|\tilde{u}|$. The problem of wall vibrations is therefore most severe with high frequency and large amplitude oscillations. These vibrations were measured with an ultrasonic depth gauge having a sensitivity of 10 μm equivalent to $l_\nu/12$. At the most frequent case. For the highest frequency f = 0.66 Hz and an amplitude $|\tilde{u}|$ = 1.8 cm/s – almost twice as large as the one used in the experiments reported here – the wall displacement was too small to be measured. At the same frequency but with an amplitude of 9 cm/s, the maximum displacement was 15 μm. If it is assumed that the displacement is proportional to the amplitude at a given frequency then it is about 1.5 μm also $l_\nu/80$ in the

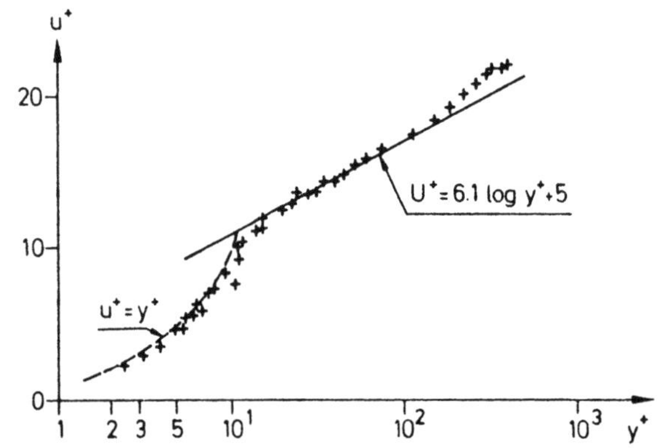

Fig. 4. Mean velocity in steady and unsteady flow

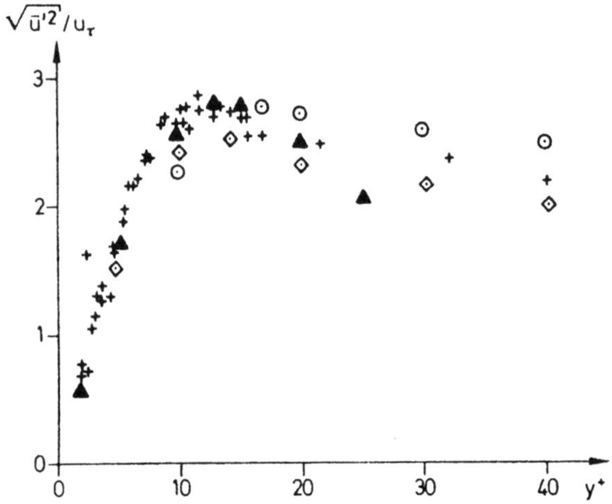

Fig. 5. Turbulent intensity near wall in channel flow. Steady flow: (○) Comte-Bellot, (▲) Eckelmann. Steady and unsteady flow: (◇) Acharia, (+) present investigation

most severe conditions of these experiments. Since the maximum mean velocity gradient is equal to u_τ/l_ν at the wall the resulting velocity oscillation is $(\partial\bar{u}/\partial y) \times l_\nu/80 = u_\tau/80$. The smallest amplitude measured is about $|\tilde{u}_c|/10$ at $y^+ = 3$. Since $|\tilde{u}_c| \cong 1$ cm/s and $u_\tau = 0.8$ cm/s in most cases, the largest relative error on the amplitude is about $(u_\tau/80)/(|\tilde{u}_c|/10) \cong 10\%$. This error decreases with the distance from the wall because $\partial\bar{u}/\partial y$ decreases and $|\tilde{u}|$ increases.

An indirect check of the negligible effect of the wall vibrations may be found on Fig. 5, where the extrapolations to $y = 0$ of the amplitude data points fall reasonably close to the origin. If the wall vibrations were important the measured amplitudes would not tend to zero with the distance.

The preceeding orders of magnitude i.e. a 10% error on the amplitude at $y^+ \cong 3$–4 by a wall deflection of only 1.5 μm in the present case of a rather thick viscous sublayer ($l_\nu \cong 120$ μm), must be stressed. Experimenters in the field of unsteady wall flows should be aware of the fact that very small wall deflections may cause large errors on the measurements of the periodic velocity oscillations near the wall.

Experimental Conditions

All measurements reported here except for one set were taken for the following conditions:

$\bar{u}_c = 17.5$ cm/s $\quad u_\tau = 0.80$ cm/s $\quad R_{eh} = 8.8 \times 10^3$

$A_{1u_c}/\bar{u}_c = 5\%$ $\quad T = 0.7$–13.5 s $\quad \Omega_h = 2.5$–0.1

The corresponding range of the non-dimensional Stokes length is $l_s^+ = 3.8$–16.6.

In order to obtain smaller reduced frequencies, the centerline velocity had to be increased. The lowest frequency case corresponds to the following conditions:

$\bar{u}_c = 50.2$ cm/s $\quad u_\tau = 1.9$ cm/s $\quad R_{eh} = 25 \times 10^3$

$A_{1u_c}/\bar{u}_c = 2\%$ $\quad T = 13.5$ s $\quad \Omega_h = 0.046$

$l_s^+ = 39.4$

Because the piston producing the oscillatory motion has a limited stroke, the amplitude could at most be 2% with this value of $\bar{u}_c$ and f. The amplitude of the second harmonic of the forced velocity oscillations on the centerline was less than 6% of the fundamental and the amplitude of the higher harmonics was less than 1%.

The reproducibility of the oscillating flow on the channel centerline was ± 2.5% r.m.s. for the amplitude and ± 1.1% r.m.s. for the phase of the fundamental mode.

Results and Discussion

Mean Velocity and Mean Turbulent Intensity

Mean velocity profiles measured in oscillating flow are indistinguishable from the steady flow curve at all frequencies tested (Fig. 4). This confirms the previously published results [4, 6, 9] but is at variance with the latest data of *Ramaprian* and *Shuen-Wei* [7] whose profile seems to show a point of inflexion near the wall. These authors argue that the mean flow in their experiment has been affected by the periodic oscillations because the forcing frequency f is of the same order as the characteristic frequency f_t of the turbulence

($f \cong f_t$) and that an interaction between the two types of fluctuations takes place. Since our highest Strouhal number based on the half channel width $\Omega_h \cong 1.2$ is even larger than the one of *RS* which was about 0.5 ($R\omega/u_m$ in their case; pipe radius: R), this argument alone cannot explain the differences between these observations. It should be added that the highest frequency in *Karlsson's* [5] experiment and the two frequencies used by *Acharia* [4] were also such that $f \cong f_t$ and no effect on the mean flow was observed.

The other parameter which, together with the frequency, would affect the turbulence and therefore the mean flow is the relative amplitude of the oscillation. This parameter is quite different in [7] and the present experiments, being respectively about 35% and 5%. In *Karlsson's* measurements at high amplitude (33%) and at a medium or high frequency, there seems to be a slight effect on the mean velocity, but from the plotted data it is difficult to conclude whether this effect is significant or not. In the present work the amplitude had to be kept small in order to avoid flow reversals near the wall because the laser velocimeter was not equiped with a frequency shifting device.

The mean turbulent intensity (Fig. 5) is also not affected by the forced oscillations as observed in other experiments [4, 6, 9]. The data of the turbulent shear stress in [4, 6, 9] point towards the same conclusion but the measurements could of course not be performed as close to the wall as those of $\sqrt{\overline{u'^2}}$. The insensitivity of the turbulence to the forced oscillations explains that of the mean velocity, which can only be affected by $\overline{u'v'}$ in channel flow where the shear stress of the periodic motion $\overline{\tilde{u}\tilde{v}}$ is zero because $\partial\tilde{u}/\partial x = 0$ and hence from continuity $\tilde{v} = 0$. Actually it is shown in [9] that $\overline{\tilde{u}\tilde{v}}$ is small even in an unsteady boundary layer in a mean unfavorable pressure gradient.

Periodic Velocity Oscillations

The evolutions of the amplitude and of the phase of the fundamental mode of the forced oscillations vs the non-dimensional distance y_s ($y_s = y/l_s$) for five cases are shown on Figs. 6 and 7. The parameter $l_s^+ = l_s/l_\nu$ measures how far the viscous Stokes layer extends into the inner layer of the steady turbulent wall flow. It is clear that small values of l_s^+ correspond to high frequencies.

Despite some differences, it is quite remarkable how close the amplitudes fall to the laminar Stokes layer in all cases for $y_s < 1$. To within about ± 50% the slope $(\partial A_{1u}/\partial y)_o$ at the wall is that of the laminar oscillating flow. This behaviour contrasts sharply with that of the mean flow where the gradient at the wall in the turbulent case u_τ/l_ν is at least an order of magnitude steeper than in the laminar regime $\bar{u}_c/h$, the ratio being $(u_\tau/\bar{u}_c)\,(u_\tau h/\nu) = (u_\tau/\bar{u}_c)\,\mathcal{R}_e^*$; where typically $20 < \bar{u}_c/u_\tau < 40$. The diffusion of the oscillating wall shear stress into the fluid seems also little affected by the presence of the turbulence produced by the mean flow.

In the two high frequency runs $l_s^+ = 5.5$ and 7 the amplitude and the phase follow the laminar Stokes curve quite closely. In particular there is a large phase lead near the wall. The scatter of the data points is sufficiently small that the extrapolation to $y = 0$ is significant, yielding a 45°-phase lead at the wall as in laminar flow. *Karlsson* [5] mentions that the largest phase lead obtained was 35° at $y = 0.01''$ for $l_s^+ = 11$ inferred from his data. The corresponding Stokes flow at the same distance gives a phase lead of 36° which confirms the present high frequency results despite the difference in l_s^+. *Parikh* et al. [9] also find a steep raise of the phase with a lead up to 30° near the wall at their highest frequency. An appreciable undershoot of the amplitude for $l_s^+ = 5.5$ is observed beyond $y_s = 2$ which does not exist in laminar flow.

Acharya's [4] experimental results for $l_s^+ = 9$ ($f = 24$ Hz) in an air channel are in qualitative agreement with the present measurements. But his results for the $l_s^+ = 7$ case

(f = 40 Hz) are completely different from ours. Actually the present $l_s^+ = 7$ case was run precisely in order to duplicate *Acharya*'s experimental conditions. He measured a phase difference at the wall of about 180° and the amplitude goes through a sharp maximum equal to A_{1u_c} at $y \cong 0.5$ and then through a minimum equal to $A_{1u_c}/8$ at $y_s = 1.9$. Wall vibrations could explain a large part of these discrepancies since *Acharya* measured deflections of about $l_v/6$, despite the loading of the tunnel walls with lead bricks. These deflections are about 10 times larger than in the present experiment and since the amplitudes are somewhat smaller than here (3.6% compared to 5%) the resulting error should be about ten times larger, also of the order of 100%, according to the argument exposed above. It seems that *Acharya* drew the wrong conclusions from his wall vibration measurements because he argued that since the deflections where of the order of the hot-wire diameter that the hot-wire would not see them.

The main features of the high frequency behaviour may be understood if it is recalled that in the steady turbulent wall flow, the turbulent shear-stress is smaller than the viscous one for $y^+ < 12$ and if it is observed that the thickness of the viscous Stokes layer is about $2y_s$. Hence, below $y^+ = 12$, $\tilde{u}$ (as $\overline{u}$) is mainly subjected to viscous diffusion and, if $2l_s^+ < 12$, the amplitude and phase of $\tilde{u}$ reach the outer values before the turbulence can actively affect them. It is seen that the present high frequency cases $l_s^+ = 5.5$ and 7 approximately satisfy this requirement.

The evolution of $\tilde{u}$ in the *low frequency* cases is more complex. In the proximity of the wall a steepending of the amplitude gradient with respect to the Stokes flow is clearly noticeable for $l_s^+ = 16.6$ and 39.4. This is to be expected since most of the variations of $\tilde{u}$ are in the fully turbulent part of the inner layer. Yet, it is surprising that there is so little difference between these two cases though the values differ by a factor of 2.5. For large values of l_s^+ one would expect $\partial\tilde{u}/\partial y$ near the wall to become independent of l_s^+. This does not seem to happen or then the value of 39.4 is still too small. It should be remembered that because of the physical limitations of the pulsating device the amplitude on the centerline was only 2% in the latter case. But it is doubtful that at such low levels the amplitude itself influences the flow.

The evolutions of the amplitude for $y_s > 0{,}5$ at the two lowest frequencies are strikingly different. For $l_s^+ = 16.6$, $\partial|\tilde{u}|/\partial y$ keeps roughly the same value until the amplitude reaches the outer level as in the high frequency cases, while for $l_s^+ = 39.4$ the slope of $|\tilde{u}|$ decreases abruptly, a behaviour reminiscent of the mean velocity's.

The intermediate case $y^+ = 12.2$ is somewhat puzzling since near the boundary, $|\tilde{u}|$ varies as in the high frequency cases whereas further away it evolves nearly as in the lowest frequency case.

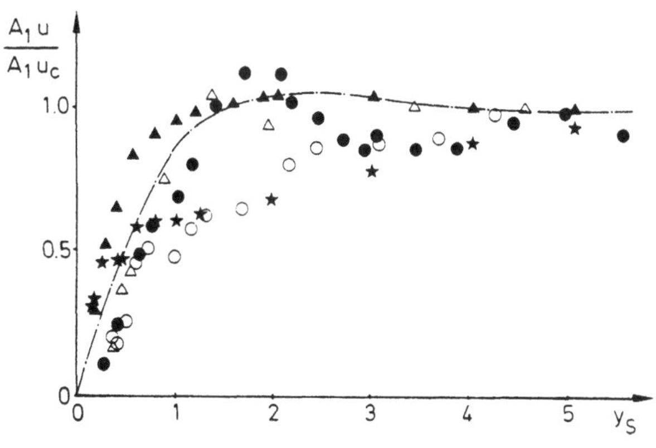

Fig. 6. Relative amplitude of fundamental mode of the periodic velocity oscillations vs non-dimensional distance y_s from the wall. (●) $l_s^+ = 5.5$; (△) 7.0; (○) 12.2; (▲) 16.6; (★) 39.4; (—·—) laminar Stokes flow

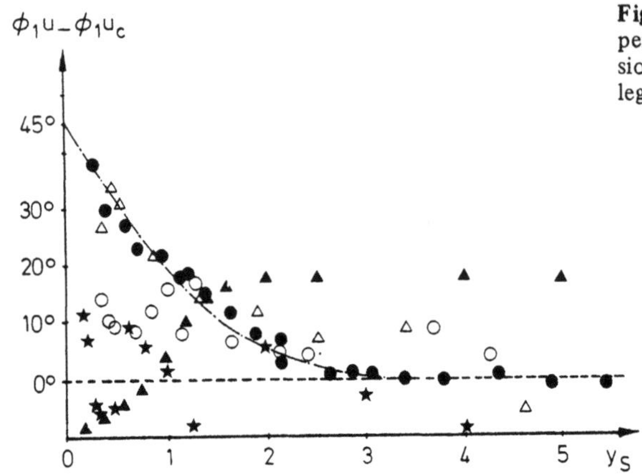

Fig. 7. Phase of fundamental mode of the periodic velocity oscillations vs non-dimensional distance y_s from the wall. (For legend see Fig. 6)

As l_s^+ increases the phase shifts away from the laminar Stokes curve. In particular, the phase lead at the wall tends to zero and becomes negative. Extrapolation to wall except for $l_s^+ = 16.6$ is hazardeous because of the scatter of the data especially for $l_s^+ =$ 39.4. In this latter case the eduction of $\tilde{u}$ from the turbulence is much slower because of the small value of the centerline amplitude and it has been observed that the phase is quite sensitive to the scatter of the points around the fundamental mode (see Fig. 2). There is quite a bit of disagreement on the phase among the various experiments but it is felt that an appreciable part may just come from measurement inaccuracies.

Oscillating Wall Shear-Stress

Since the amplitude of $\tilde{u}$ follows approximately the laminar Stokes solution, the amplitude of the oscillating shear stress is

$$A_{1\tau} \cong \mu \left(\frac{\partial A_{1u}}{\partial y}\right)_{y \to 0} = \sqrt{2}\ \mu \left(\frac{A_{1u_c}}{l_s}\right)$$

so that

$$\frac{A_{1\tau}}{\overline{\tau}} \cong \sqrt{2}\ \frac{A_{1u_c}}{u_\tau}\ \frac{1}{l_s^+} \tag{1}$$

or

$$(A_{1\tau}/\overline{\tau})/(A_{1u_c}/\overline{u}_c) \cong \sqrt{2}\ (\overline{u}_c/u_\tau)/l_s^+ \tag{2}$$

Let us recall that in a laminar boundary layer this relative amplitude ratio equals $3\sqrt{\Omega_x}$ in the high frequency limit and $h/\sqrt{2}\, l_s$ in laminar channel flow. The above relation shows the importance of the non-dimensional Stokes thickness expressed in wall parameters l_s^+. It seems that as long as $l_s << \vartheta$ or h/l_s^+ is a more relevant parameter than ϑ/l_s or h/l_s or $\omega x/\overline{u}_c$.

The above relations also show the combined influence of the frequency and of the forcing amplitude on the oscillating shear stress. They further point towards the importance of a mean unfavorable pressure gradient due to the presence of u_τ in the denominator since then $u_\tau \to 0$.

Periodic flow reversal requires $|\tilde{\tau}|/\bar{\tau} > 1$, also:

$$|\tilde{u}_c|/\bar{u}_c > \frac{1}{\sqrt{2}} \frac{u_\tau}{\bar{u}_c} l_s^+ \tag{3}$$

which shows that the amplitude must be large unless the frequency is very large or u_τ is very small as in a mean unfavorable pressure gradient.

Turbulence is only affected by the strain rate i.e. velocity gradients and not directly by velocity oscillations whatever the frequency, so that there can be no effect over the whole layer where $\partial\tilde{u}/\partial y = 0$. The oscillating flow can therefore only modify the turbulence and hence the mean flow if the oscillating gradient $\partial|\tilde{u}|/\partial y$ contributes substantially to the total gradient $\partial < u > \partial y$ in the wall layer, also if $|\tilde{\tau}|/\bar{\tau} \sim 1$. This requires Eq. (3) to hold. But this is not a sufficient condition for if the $\partial|\tilde{u}|/\partial y$-gradient is mainly located between $y^+ = 0$ and $y^+ \cong 5$ where $\langle u'v'\rangle/u_\tau^2 << 1$, this gradient cannot substantially contribute to the turbulence production. It must also be required, in addition, that $l_s^+ > 5$. These two conditions are not simultaneously satisfied in the present experiments. It is easily seen that if $l_s^+ = 10$, the forcing amplitude must already be larger than 30% since $u_\tau/\bar{u}_c \cong 1/22$.

It may, therefore, be concluded that the mean flow and the mean turbulence characteristics can neither be affected if the frequency is low because condition (3) cannot be satisfied unless the amplitude is very large or u_τ very small nor if the frequency is high because l_s^+ is then too small.

Periodic Variations of the Turbulent Intensity

The amplitude of the fundamental mode of the periodic variations of the turbulent intensity $\sqrt{\overline{u'^2}}$ for four cases are shown on Fig. 8 and 9 respectively with respect to u_τ and A_{1u}. As $\sqrt{\overline{u'^2}}/u_\tau, A_{1u'}/u_\tau$ tends to zero with y^+, reaches a maximum and slowly decreases to zero as y^+ increases.

Since the amplitude of the run $l_s^+ = 39.4$ is 2.5 times smaller than in the others, it may be seen that for $l_s^+ > 12$ the maximum value of the ratio

$$\left(\frac{A_{1u'}}{u_\tau}\right) \Big/ \left(\frac{|\tilde{u}_c|}{\bar{u}_c}\right) \sim 4 \ .$$

The oscillations of the turbulent intensity are also larger than the forcing of the mean flow. It is clear also that these oscillations are much smaller at high frequencies as shown

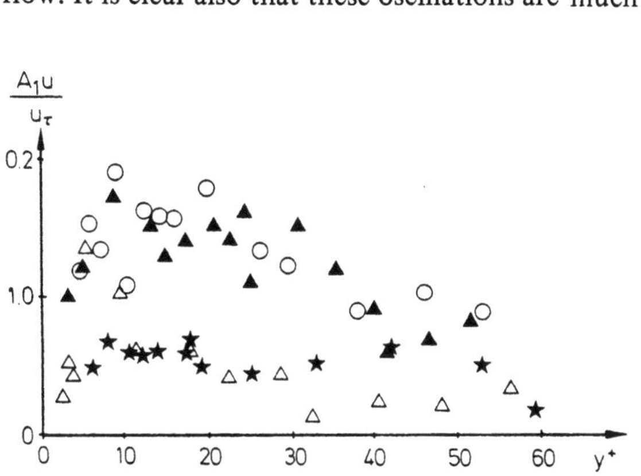

Fig. 8. Amplitude of fundamental mode of the periodic variations of the turbulent intensity vs y^+. (For legend see Fig. 6)

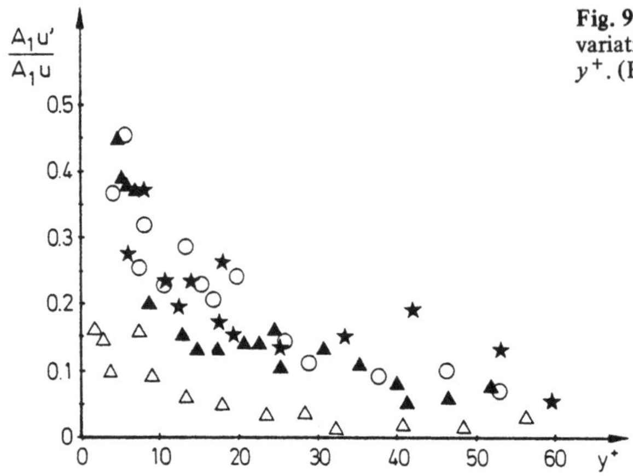

Fig. 9. Relative amplitude of the periodic variations of the turbulent intensity vs y^+. (For legend see Fig. 6)

by the values for $l_s^+ = 7$. This seems to confirm partially the argument above concerning the requirement $l_s^+ > 5$ for the turbulence to be affected by the forced oscillations. These observations are confirmed by Fig. 9 where the turbulent intensity variations are referred to the local amplitude of the velocity. It is also seen that the relative intensity near the wall reaches about 50% a value which is twice as large as that of the ratio $\sqrt{\overline{u}'^2}/\overline{u}$ of the mean, which tends to .24 according to *Eckelmann* [3].

There also appears to be an amplification of the turbulent intensity with respect to the imposed velocity oscillations. This should lead to a somewhat weaker requirement than (3) for a modification of the mean turbulence and mean flow.

The level of the u'-oscillations remains, however, low in these experiments as compared to the mean turbulence $\sqrt{\overline{u}'^2}/u_\tau$ (see Fig. 5 and 8), especially if the comparison is made on $(\langle u'^2 \rangle - \overline{u'^2})$ and $\overline{u'^2}$ rather than on the r.m.s. values. It is seen that the oscillations of the normal shear-stress are only about 1% of the mean value. This draws the attention to the fact that the u'^2-oscillations were buried here in a random noise one hundred times larger in mean amplitude. Proper eduction requires also a very large sample size. These results on the turbulence oscillations should also be viewed with caution and only be taken in a qualitative sense. The problem would certainly be less severe with large amplitude forcing.

Concluding Remarks

Velocity measurements in turbulent channel flow subjected to low amplitude periodic oscillations have been performed. They show that at high frequencies the amplitude and the phase of the velocity oscillations near the wall behave as in laminar Stokes flow. At lower frequencies this conclusion remains approximately valid for the amplitude very close to the wall but not for the phase-lead which is smaller and even changes sign. On the other hand, the mean characteristics of the flow are not affected by the presence of the forced oscillation.

From these observations it is inferred that the Stokes thickness $l_s = \sqrt{2\nu/\omega}$ and the viscous length $l_\nu = \nu/u_\tau$ of the steady turbulent wall flow are the relevant length scales in the wall region and hence that the relevant parameter governing the evolution of the velocity oscillations is $l_s^+ = l_s/l_\nu$. This parameter shows how far the laminar Stokes layer would reach into the inner layer of the steady turbulent flow if there were no interactions between the two flows. Since below $y^+ = 12$ the mean flow is dominated by viscous effects and since the total thickness of the Stokes layer is approximately $2l_s$ it is argued that these interactions are weak if $l_s^+ < 6$. Furthermore, at lower frequencies $l_s^+ > 6$ these interactions become only important if the forcing amplitude on the channel centerline $|\tilde{u}_c|/\overline{u}_c$ is at least of the order of $(u_\tau/\overline{u}_c)\, l_s^+$. Since the inner layer is the same in channels and boundary layers with zero or mild mean pressure gradients these conclusions are expected to hold in the latter flows. In flows with small mean pressure gradients the above requirements lead to large forcing amplitudes and this may explain why in nearly all published experiments no effect of the forced oscillations on the mean flow was observed.

It would, of course, be interesting to test the preceeding conclusions with measurements in flows subjected to large amplitude oscillations. In such conditions the eduction of the periodic characteristics would also be more accurate and, therefore, the measurements of the velocity oscillations in the immediate proximity of the wall and especially of the periodic variations of the Reynolds stresses would be more reliable.

The non-dimensional frequency $\omega X/\overline{u}_c$ has so far often been used to correlate unsteady turbulent boundary layer results. This parameter has a clear physical meaning only in the unsteady laminar flat plate flow where it is proportional to ϑ^2/l_s^2. In the turbulent case even for the flat plate this ratio depends also upon R_{ex}. The use of $\omega X/\overline{u}_c$, supposes that at a given station X the boundary layer history has been the same from one experiment to another. This is not a serious objection for the flat-plate case since a virtual origin can easily be defined but in the presence of pressure gradients $\omega X/\overline{u}_c$ has little significance. The non-dimensional Stokes length l_s^+ introduced here seems, therefore, more appropriate especially to correlate data in the wall layer. It is, however, not claimed here that l_s^+ is the only parameter relevant to oscillating turbulent wall flows. It may, in particular, be expected that l_s^+ will only play a minor role when its value is large, say $l_s^+ > 100$, at very low frequencies and that a parameter like $\vartheta/l_s = \mathcal{R}_e^*/l_s^+$ will be more relevant.

Acknowledgement. Financial support from the Direction des Recherches, Etudes et Techniques from the Ministère des Armées is greatfully acknowledged.

References

1. Tellonis, D.P.: Unsteady boundary layers separated and attached. J. Fluids Eng., **101**, 29–43 (1979)
2. Comte-Bellot, G.: *Ecoulement turbulent entre deux parois parallèles.* (Publications Scientifiques et Techniques du Ministère de l'Air, 1965) 160 p.
3. Eckelmann, H.: The structure of the viscous sublayer and the adjacent wall region in a turbulent channel flow. J. Fluid Mech. **65**, 439–459 (1974)
4. Acharya, M.: "Measurements and Predictions of a Fully Developed Turbulent Channel Flow with Imposed Controlled Oscillations", Ph.D. thesis, Stanford Univ., (1975) Ref. 75–25, 492
5. Karlsson, S.K.: An unsteady turbulent boundary layer. J. Fluid Mech. **5**, 622–636 (1958) Ph.D. Dissertation, The John Hopkins University (1969)
6. Cousteix, J., Desopper, A., Houdeville, R.: "Structure and Development of a Turbulent Boundary Layer in Oscillatory External Flow", Proc. Symp. on Turbulent Shear Flows, Penn. State Univ., Univ. Park, Pa. (1977)
7. Ramaprian, B.R., Shven-Wei Tu: An experimental study of oscillatory pipe flow at transitional Reynolds numbers. J. Fluid Mech., **100**, 513–544, (1980)
8. Binder, G., Favre-Marinet, M.: *The Inner Layer in Unsteady Turbulent Boundary Layers* (EUROVISC III, Liverpool 1979)
9. Parihk, P.G., Reynolds, W.C., Jayaraman, R., Carr, L.W.: "Dynamic Behaviour of an Unsteady Turbulent Boundary Layer", Proc. IUTAM Symp. on "Unsteady Turbulent Shear Flows" Toulouse, May 5–8, 1981

A Dynamical and Visual Study on the Oscillatory Turbulent Boundary Layer

T. Hayashi and M. Ohashi

Hydraulic Laboratory, Department of Civil Engineering,
Chuo University, Kasuga 1-chome, Bunkyo-ku,
Tokyo 112, Japan

Abstract

Measurements of velocities in the purely oscillatory turbulent boundary layer over a horizontal wall have been made using a large oscillating water tunnel. The procedure of periodic ensemble averaging, modified by application of a finite Fourier series for the estimation of the most probable values of the periodic averages, has permitted the separation of the periodic components and the turbulent fluctuations. From the data decomposed the vertical profiles and the time variations, of mean velocities, Reynolds stresses and turbulent energy, are obtained.

The turbulent structure of the oscillatory boundary layer has been studied utilizing the flow visualization technique of thin-layered milk method developed by the authors. Visual studies elucidate the measurement results of turbulent quantities and explain the mechanism of the generation, development and decay of turbulence occurring during each cycle of the oscillatory motion. The large-scale features of turbulence which cause negative Reynolds stresses are examined.

Nomenclature

$\hat{A}_s$	Fourier coefficient of cosine part
$\hat{B}_s$	Fourier coefficient of sine part
a	Free stream amplitude
h	Harmonic number
k	Turbulent energy
T	Period of oscillatory motion
t	Time
U_0	Free stream velocity amplitude
$\bar{u}$	Mean velocity in longitudinal direction
u'	Velocity fluctuation in longitudinal direction
Var	Cumulative variance
w'	Velocity fluctuation in vertical direction
$X_{i,j}$	A measured datum
$\bar{X}_j$	Periodic ensemble mean of data
$\hat{X}_j$	Estimated periodic ensemble mean of data
$\bar{\bar{X}}$	Overall-time-averaged value of all data
z	Probe height measured from bed
σ^2	Variance
$\hat{\sigma}^2$	Estimated variance
δ	Stokes-layer thickness = $(2\nu/\omega)^{1/2}$
λ	Stokes parameter = $(U_0/2)(2\nu/\omega)^{1/2}/\nu$
ν	Kinematic viscosity of fluid
ω	radian frequency

Subscripts

i,j jth sample obtained during the ith cycle
m Number of samples during a cycle
n Number of sampling cycles

Introduction

Experimental studies of the purely oscillatory turbulent boundary layer date back to the work of *Jonsson* [1], who determined shear stress profiles on the basis of the velocities measured with a micropropeller current meter only. Later, in a similar procedure *Jonsson* and *Carlsen* [2] determined turbulent characteristics such as shear stresses, eddy viscosities, energy loss and boundary layer thicknesses. But since more profound experimental and data processing techniques to study the structure of turbulence in such boundary layers have become available, attention has been focussed to more refined measurements of oscillatory turbulent boundary layers. The experiments of *Hino* et al.[3] revealed the fact that turbulence is generated suddenly in the decelerating phase of the free stream velocity and that the profile of the velocity distribution changes drastically. Similar observations were reported by *Merkli* and *Thomann* [4] ,who performed experiments on the oscillatory boundary layer in a circular pipe. Detailed measurements of turbulent characteristics such as mean velocity profiles, Reynolds stresses and turbulent energy were made by *Cousteix* et al. [5]. Measurements of turbulent intensities and Reynolds stresses were also made by *Kobashi* and *Hayakawa* [6] for the purpose of studying the effects of unsteadiness of flow on the transition mechanism and development of turbulences in the boundary layer. Experiments by *Ramaprian* and *Mueller* [7] showed entire profiles across the boundary layer, of periodic velocity and ensemble-averaged turbulence intensity in longitudinal direction. Measurements of Reynolds stresses were made by *Anwar* and *Atkins* [8], who were in concern primarily concerned with the turbulence characteristics in the tidal flow, the period of oscillation being very large, and the effect of non-steadiness of the flow on the structure of turbulence being not remarkable.

Recently, the writers [9] presented the results of detailed measurement of turbulent characteristics and also visualization of the transition to tubulence, and supplied new information on the structure of turbulence. In another recent study, *Hino* et al. [10] reported the spatial correlations determined for the purpose of elucidating the coherent structure of turbulence in the boundary layer.

In spite of these significant previous studies, a question remains whether the behavior of the generation of turbulence observed at the end of the decelerating phase of the free stream velocity is similar to the behavior of the generation of turbulence in the steady uni-directional flow, i.e., the behavior of bursting, or not.

Another question to be elucidated is the large-scale features of turbulence which are supposed to exist immediately after the direction of free stream velocity is reversed, i.e., at the beginning of the accelerating phase of the absolute value of the free stream velocity.

The main purpose of this paper is to elucidate these questions performing detailed measurement of turbulence and also visualization of disturbances.

Experimental Apparatus and Data Processing Procedure

The experiment was conducted in a large water tunnel of Lundgren-Sørensen-type. This simulated, on a prototype scale, conditions due to the wave motion near the sea bed. The apparatus forms a U-shaped tube consisting of two vertical risers, each with a height of 3.5 m, and a horizontal tunnel which is 17.5 m long, 0.75 m wide and 0.2 m deep. By means of a piston the water is made to oscillate in the U-shaped tube with periods and amplitudes as desired. The general view of the apparatus is shown in Fig. 1.

The movement of the water surface in the open riser was recorded by the use of a wave amplitude meter. The phase angle of the oscillatory motion was defined as shown in Fig. 2. Velocities were measured by a X-type hot-film probe. This probe was inserted in the water tunnel, and the height of the probe above the bed was changed in distances of 2 mm between z = 0 and 20 mm, 3 mm between z = 20 mm and 35 mm, 5 mm be-

Fig. 1. General view of oscillating water tunnel

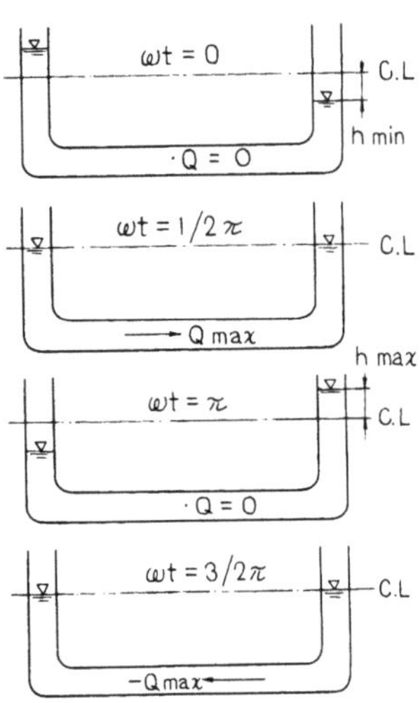

Fig. 2. Definition of the phase angle of oscillating motion

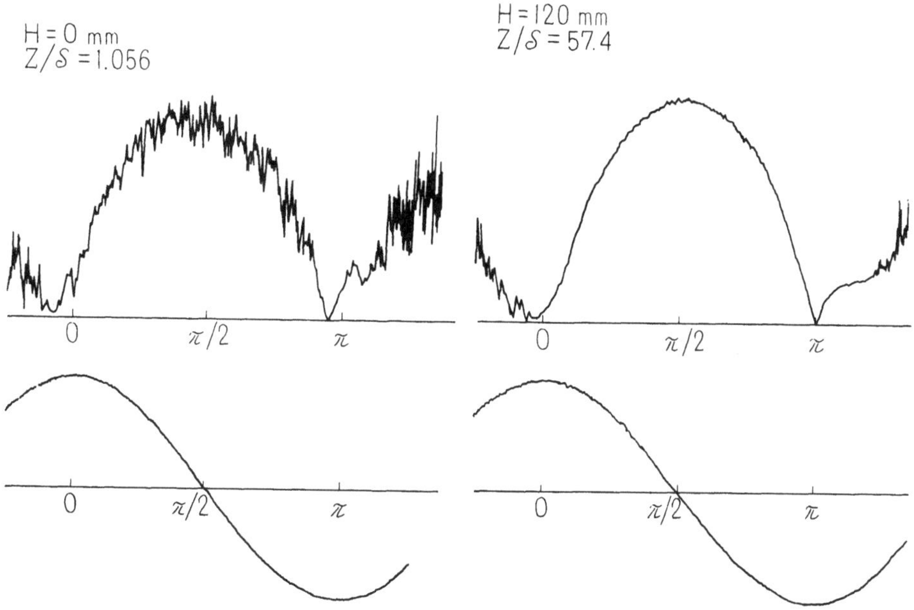

Fig. 3. Record of velocity measurement (*above*: Output of X-type hot-film probe, *below*: Water level in riser)

tween z = 35 mm and 50 mm, and 10 mm between 50 mm and 120 mm. A typical time record of velocity measurement is shown in Fig. 3.

In order to estimate the most probable values of phase averages of turbulent quantities of the oscillatory flow out of the data obtained the finite Fourier series described below was adopted to the periodic ensemble means of the data.

Let $X_{i,j}$ be a datum, in which i = cycle ($i = 1, \ldots, n$) and j = number of a datum in each cycle ($j = 1, \ldots, m$). All data obtained are arranged in a matrix form as

$$\begin{bmatrix} X_{1,1} & X_{1,2} & \cdots\cdots\cdots & X_{1,m} \\ X_{2,1} & X_{2,2} & \cdots\cdots\cdots & X_{2,m} \\ \cdot & \cdot & & \\ \cdot & \cdot & & \\ \cdot & \cdot & & \\ X_{n,1} & X_{n,2} & \cdots\cdots\cdots & X_{n,m} \end{bmatrix}$$

Then, the periodic ensemble mean is given as

$$\overline{X}_j = \frac{1}{n} \sum_{i=1}^{n} X_{i,j} \qquad (j = 1, \ldots, m) \ . \tag{1}$$

We fit a finite Fourier series with respect to j to the periodic ensemble means obtained by Eq. (1), the finite Fourier series being expressed as

$$\hat{X}_j = \bar{X} + \sum_{s=1}^{h} \hat{A}_s \cos \frac{2\pi sj}{m} + \hat{B}_s \sin \frac{2\pi sj}{m} \quad , \tag{2}$$

where

$$\bar{X} = \frac{1}{m} \sum_{j=1}^{m} \bar{X}_j$$

$$\hat{A}_s = \frac{2}{m} \sum_{j=1}^{m} (\bar{X}_j - \bar{X}) \cos \frac{2\pi sj}{m}$$

$$\hat{B}_s = \frac{2}{m} \sum_{j=1}^{m} (\bar{X}_j - \bar{X}) \sin \frac{2\pi sj}{m} \quad .$$

Now, the problem is to select an optimal number for this h. For this purpose we first calculate the variance of $\bar{X}_j$ from $\bar{X}$ as Eq. (3) and the variance of the sth harmonic adopted in Eq. (2) designated by $\bar{X}_{js}$ as in Eq. (4).

$$\sigma^2 (\bar{X}_j) = \frac{1}{m} \sum_{j=1}^{m} (\bar{X}_j - \bar{X})^2 \tag{3}$$

$$\hat{\sigma}^2 (\bar{X}_j)_s = (\hat{A}_s^2 + \hat{B}_s^2) / 2 \qquad (s = 1, ..., h) \quad . \tag{4}$$

With these variances thus obtained we calculate the cumulative variance with respect to s as

$$\text{Var}\ \{\bar{X}_j\} = [\hat{\sigma}^2 (\bar{X}_j)_s / \sigma^2 (\bar{X}_j)] \times 100 \qquad (s = 1, ..., h) \quad . \tag{5}$$

A typical cumulative variance is shown in Fig. 4. The number for h was determined in such a way as to make the value of this cumulative variance 99% [11]. Furthermore,

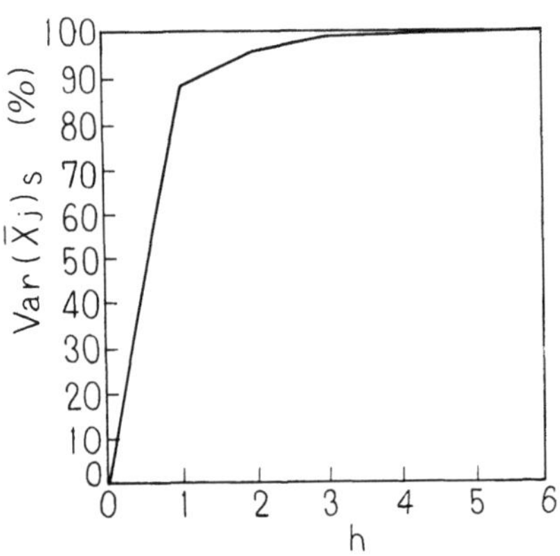

Fig. 4. Cumulative variance of X_j

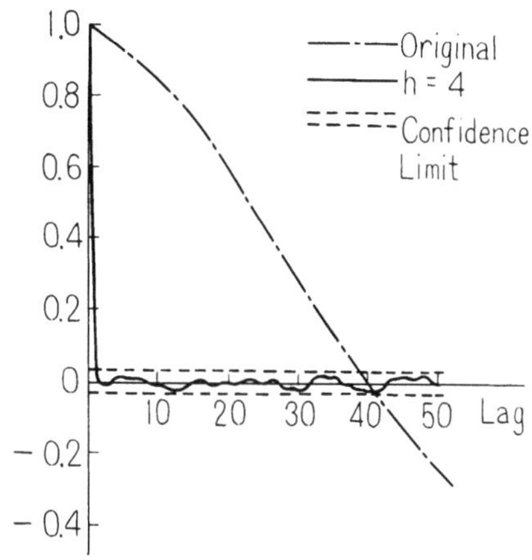

Fig. 5. Correlogram of $\bar{X}_j - X_j$

the correlogram was calculated from the autocorrelation of $\bar{X}_j - X_j$, and from this correlogram it was checked whether the periodic components had properly been eliminated (Fig. 5).

Experimental Results

The experimental conditions adopted in this study are listed in Table 1.

Table 1. Experimental conditions

	Run-1	Run-2	Run-3	Run-4
T [s]	9.8	9.7	11.5	13.5
U_0[cm/s]	59.0	67.3	58.8	49.5
ω [1/s]	0.63	0.65	0.55	0.47
λ	439	498	473	433
δ [cm]	0.213	0.208	0.227	0.246
a [cm]	93.7	103.5	106.9	105.3

Figs. 6 and 7 show the dimensionless velocity $\bar{u}/U_0$ of Run-1 plotted against dimensionless height from the wall. In these figures, Fig. 6 a, b are for the acceleration period and deceleration period, respectively, of the free stream, and Fig. 7 a,b are for the early stage of the acceleration period and the finishing stage of the deceleration period, respectively. It is seen from Fig. 3 and also Fig. 6 that the phase of velocity advances that of wave amplitude meter by about $\pi/12$.

Figs. 8 and 9 show the vertical profiles and the time variation, respectively, of turbulent shear stresses in Run-1, and Figs. 10 and 11 show the vertical profiles and the time variation, respectively, of turbulent energy of the same run. Since only two components of turbulent velocity were measured in this study, the turbulent energy k was calculated assuming the formula written below, which was used by *Eckelmann* in his study on low-Reynolds-number turbulence [12].

$$k = (3/4)\,(\overline{u'^2} + \overline{w'^2}) \qquad (6)$$

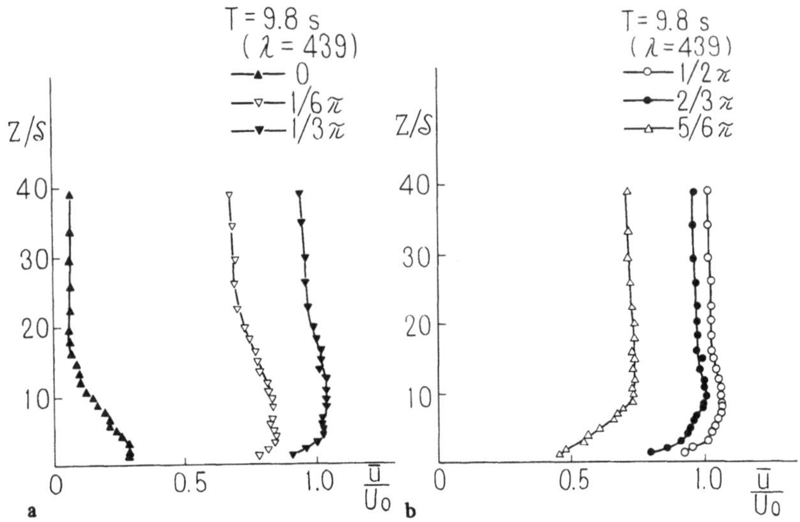

Fig. 6 a, b. Mean velocity profiles. **(a)** During the period of acceleration. **(b)** During the period of deceleration

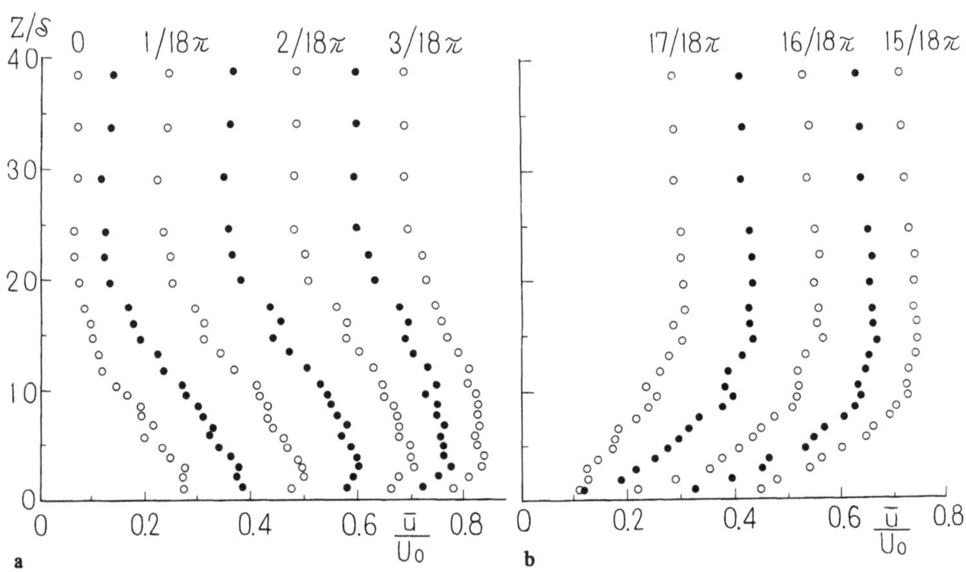

Fig. 7 a, b. Mean velocity profiles. **(a)** At the early stage of acceleration period. **(b)** At the finishing stage of deceleration period

Figures 12 and 13 show the time variations of the Reynolds stresses and turbulent energy, respectively, of Runs-2, 3 and 4.

It is seen from Figs. 8–13 that both the turbulent shear stress and the turbulent energy are larger in the period of deceleration than in the period of acceleration, of the absolute value of free stream velocity. Also, it is to be noted in Figs. 9 and 11

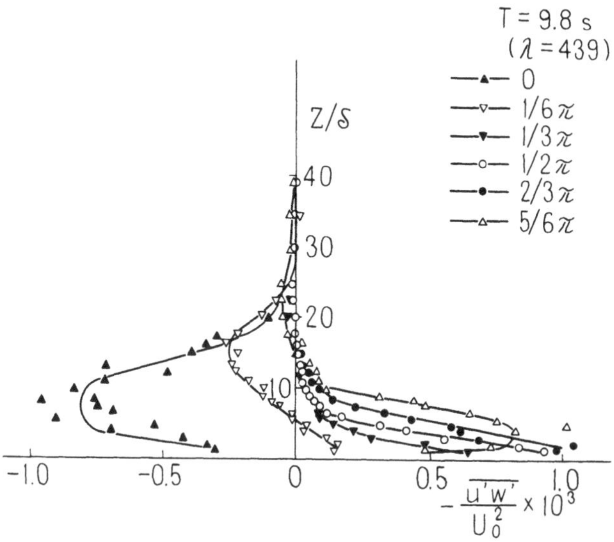

Fig. 8. Turbulent shear stress profiles (Run-1)

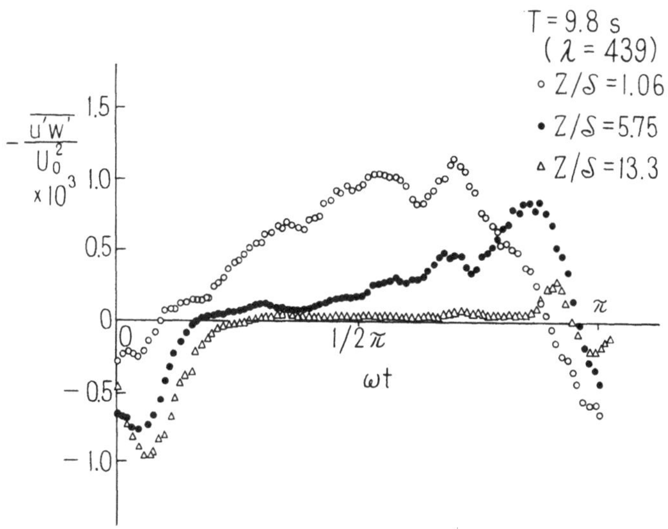

Fig. 9. Time variations of turbulent shear stress (Run-1)

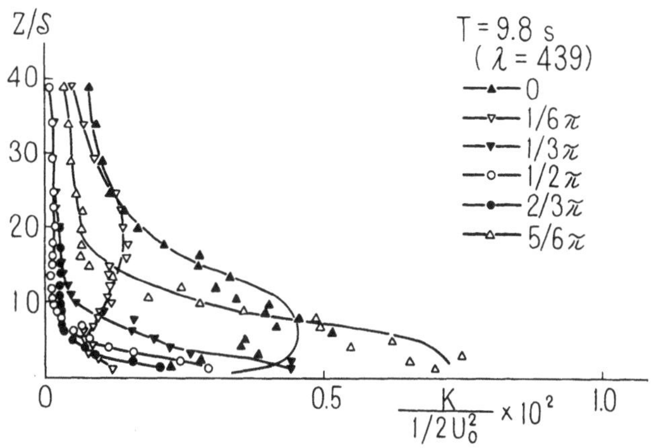

Fig. 10. Turbulent energy profiles (Run-1)

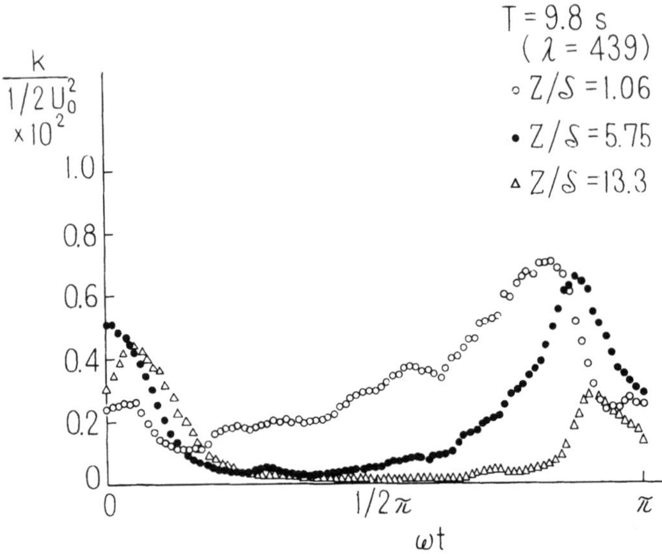

Fig. 11. Time variations of turbulent energy (Run-1)

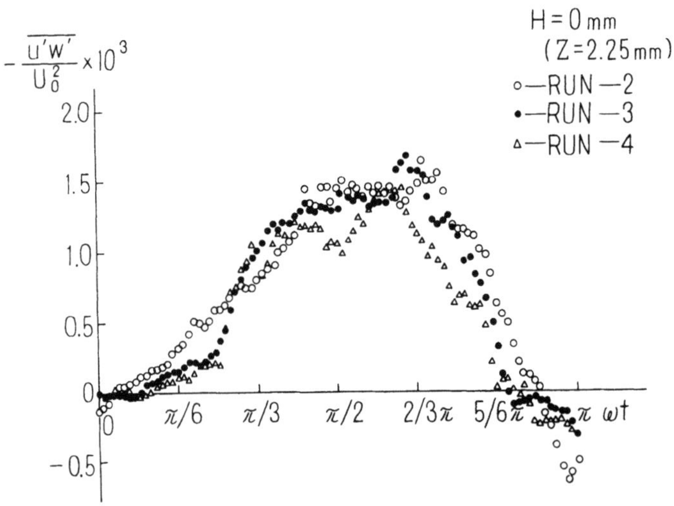

Fig. 12. Time variations of turbulent shear stress

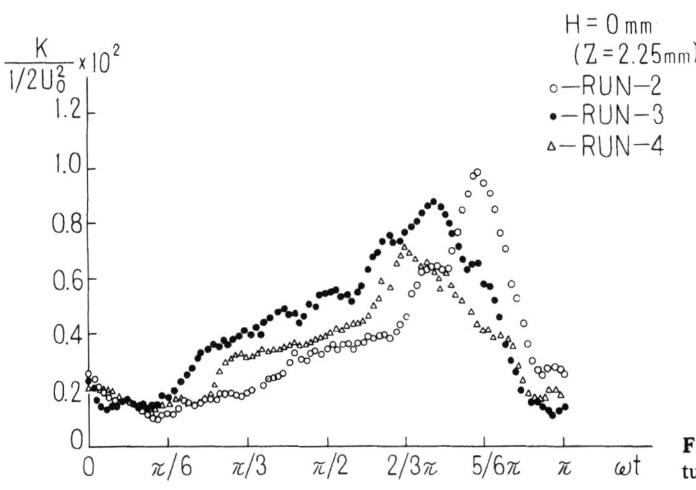

Fig. 13. Time variations of turbulent energy

that time of occurrence of the maxima of turbulent shear stress and turbulent energy tend to be delayed as the height of the point of measurement increases.

Runs–2, 3 and 4 have a same free stream amplitude, but different radian frequencies. It is seen from Figs. 12 and 13 that the turbulent shear stresses do not show distinct difference between these runs, but time of occurrence of the maximum of turbulent energy tends to be delayed as radian frequency increases.

Flow Visualization and Observation

Flow visualization provides efficient means for the investigation on the generation, development and decay of turbulence in oscillatory boundary layers. The visualization was made by two methods. The first was "thin-layered milk method" developed by the authors and the other was the hydrogen-bubble method. In the thin-layered milk method, prior to an experiment the bed of water tunnel was covered with a layer less than 1 mm of powdered milk dissolved in water, for more than 2 m of length of the bed, the specific gravity of the solution of powdered milk having been adjusted to be in the range 1.002–1.008.

In order to lay solution of powdered milk uniformly on the bed, first the solution was cooled by several degrees centigrade and then it was gently poured on the bed through a flexible vinyl pipe having a guide at the tip utilizing the difference of the density of this solution relative to the density of the water. Furthermore, in order to make this layer more uniform a slight unidirectional flow of water was produced by adding water slowly in the riser. After having filled water up to the proper height in the riser another 15 min were waited before the start of experiment, so that the temperature of the milk layer became equal to that of water in the tunnel.

After this preparation the piston-stroke of water tunnel was started from rest to sinusoidal movement (Fig. 14). The upper wall of the water tunnel had a lucite part (Fig. 15). The movement of milk was recorded photographically with a motor-driven camera set stationary above the tunnel, and the velocity of flow at the point located 12 cm high from bottom on the symmetrical axis of the tunnel was measured by a KANOMAX 15 mW He-Ne laser-Doppler anemometer and recorded in a data recorder.

A series of pictures of Fig. 16 show patterns of disturbances occurring on the bed of the tunnel. When the speed of the water flow was a maximum, i.e. when $\omega t = \pi/2$, no disturbance was observed on the bed. Disturbances occur after $\omega t = \pi/2$. A series of pictures of Fig. 17 are for nearly a quarter period of oscillation having the instant of no free stream velocity $\omega t = \pi$ in it.

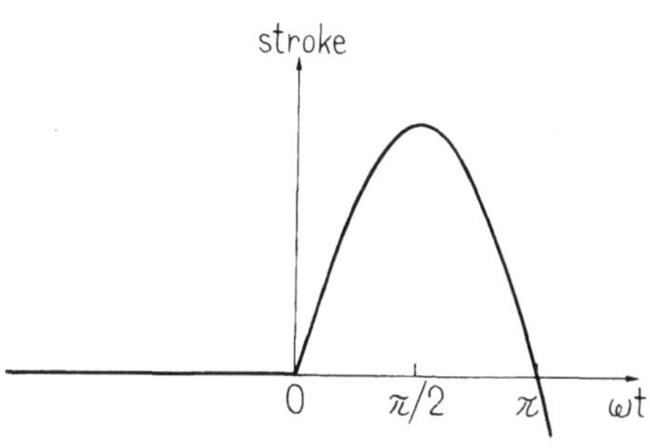

Fig. 14. Piston-stroke

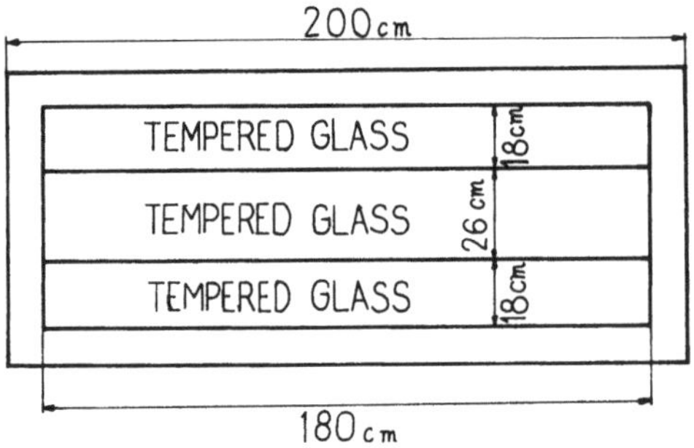

Fig. 15. Dimensions of the lucite part installed in the upper wall of water tunnel

The process of generation and development of turbulence which has been observed from these series of pictures is as follows:

Turbulence at the Late Stage of the Deceleration Period

1) Little turbulence was generated in the acceleration period preceding the period of deceleration (Fig. 16 a, b).
2) At the beginning of the period of deceleration flow disturbances appear on the bed and are deformed to U-shaped loops (Fig. 16 c).
3) The tips of the U-shaped loops move away from the bed owing to the self-induction effect of the loops (Fig. 16 d).
4) When ωt approaches π, i.e. when the free stream velocity is decelerated to almost zero, the U-shaped loops eject turbulent bursts (Fig. 16 e, f).

Turbulence at the Early Stage of the Acceleration Period

1) The disturbances formed at first on the bed are shown in (Fig. 17 a).
2) The amplitude of these disturbances increases, and the disturbances are deformed to U-shaped loops (Fig. 17 b).
3) The tips of the U-shaped loops move away from the bed owing to the self-induction effect of the loops (Fig. 17 c).
4) However, when ωt proceeds to $11/12\ \pi$, i.e. when the velocity near the bed is decelerated to almost zero, the U-shaped loops stop their developing (Fig. 17 d).
5) At an early stage of acceleration period a large-scale structure of eddies is formed around the fringe of the boundary layer (Fig. 17 e, f).
6) These large eddies are decomposed into turbulence having higher frequencies (Fig. 17 g).
7) The turbulence tends to be relaminarized during the acceleration period. The clouds of milk already formed diffuse little and are simply carried back as a whole by the flow (Fig. 17 h, i).

In order to visualize the profiles of mean velocity and also the turbulence in a longitudinally vertical section a water tunnel experiment was made by means of hydrogen-bubble technique. Figure 18 shows the occurrence of bursting phenomenon during the late stage of the deceleration period. Figures 16 and 18 indicate that generation of turbulence is made during a limited range of angular phase. Figure 19 shows a large-scale structure of an eddy formed at an early stage of the acceleration period. The time-lines of hydrogen-bubbles show the phase lead of velocity near the bottom. The eddy

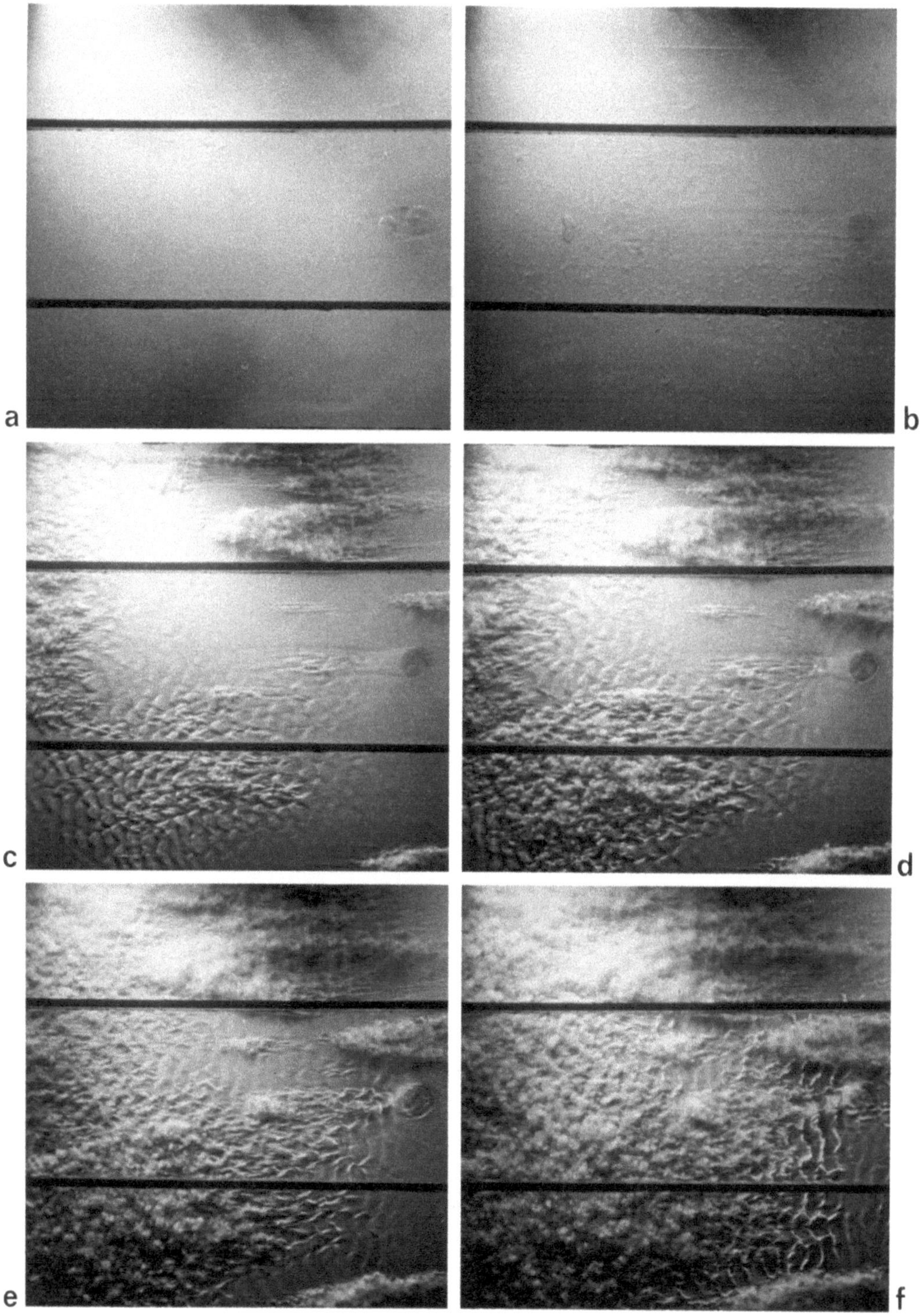

Fig. 16 a–f. Patterns of disturbances formed (Scope in actual size:64 cm × 69 cm) (T = 10.0 s)
Directions of the free stream velocity and the velocity near the bed, in (**a–f**) from right to left
(**a**) $\omega t = 0$ (**b**) $\omega t = 1/2\pi$ (**c**) $\omega t = 12/18\ \pi$
(**d**) $\omega t = 14/18\ \pi$ (**e**) $\omega t = 15/18\ \pi$ (**f**) $\omega t = 16/18\ \pi$

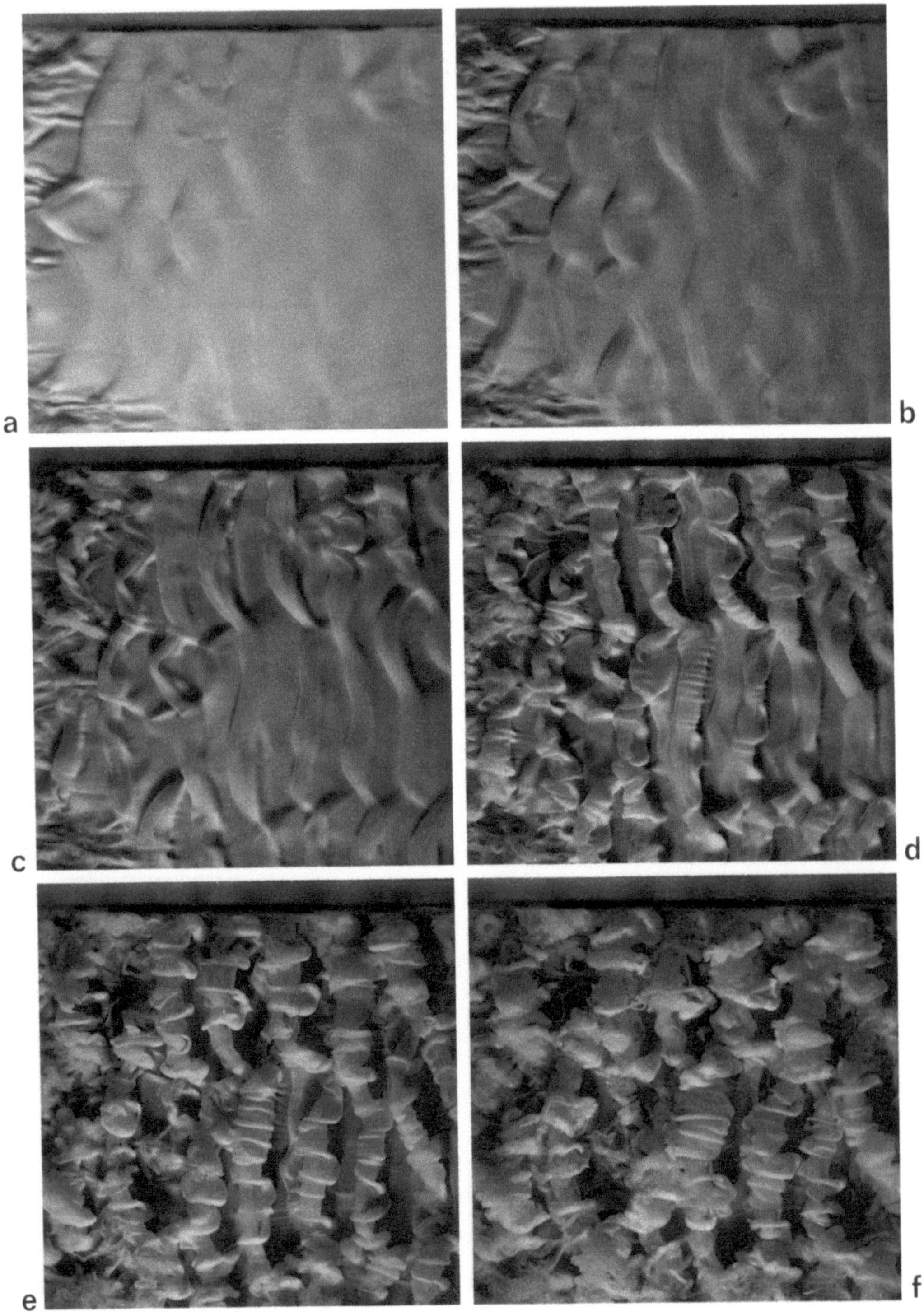

Fig. 17a-f. Figure caption see opposite page

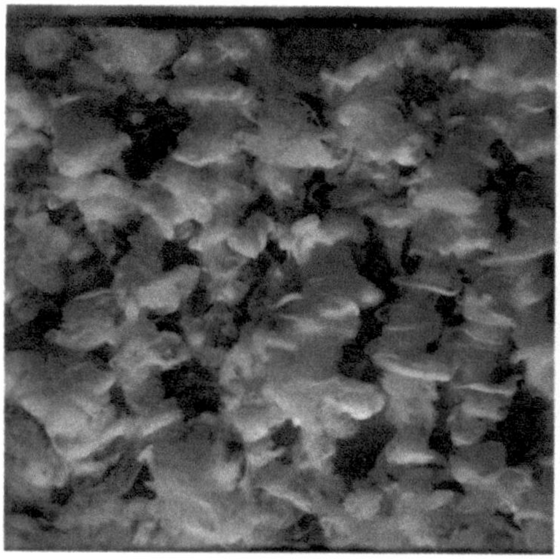

g

h

i

Fig. 17a–h. Large-scale eddies (Scope in actual size: 18 cm x 21 cm) (U_0 = 49.2 cm/s, T = 13.0 s, λ = 437) Directions of the velocity near the bed, in (**a–c**): from right to left, in (**d–i**): from left to right. Directions of the free stream velocity, in (**a–d**): from right to left, in (**e**): velocity 0, in (**f–i**): from left to right

(**a**) $\omega t = 14/18\ \pi$ (**b**) $\omega t = 15/18\ \pi$
(**c**) $\omega t = 16/18\ \pi$ (**d**) $\omega t = 17/18\ \pi$
(**e**) $\omega t = \pi$ (**f**) $\omega t = 19/18\ \pi$
(**g**) $\omega t = 20/18\ \pi$ (**h**) $\omega t = 21/18\ \pi$
(**i**) $\omega t = 22/18\ \pi$

Fig. 18. Turbulence at the late stage of deceleration period (Hydrogen-bubble technique, Time interval: 0.1 s, Direction of flow: from right to left. Scope in actual size from the bed: 9 cm)

Fig. 19. Large-scale eddy structure (Hydrogen-bubble technique. Time interval: 0.2 s, Direction of flow: from left to right. Scope in actual size from the bed: 8 cm)

formed around the fringe of boundary layer develops as the free stream velocity overtakes the forerunner velocity near bed.

Figures 17 and 19 show a structure of generation of turbulence different from that shown in Figs. 16 and 18, i.e. the structure of bursting.

Comparison between Fig. 9, Fig. 17 and Fig. 19 indicate that the features responsible to the occurrence of negative values of turbulent shear stress are such large-scale eddies as shown in Figs. 17 d, e, f and 19. It is noticed in Fig. 9 that the absolute value of the negative turbulent shear stress increases as the distance from the bed increases up to a certain height, although the value of the positive turbulent shear stress decreases as the distance from the bed increases. This result of dynamical measurement can be interpreted also as the consequence of the formation of the large-scale eddies (Figs. 17d–f, and 19).

Conclusions

New details on the nature of turbulence in purely oscillatory boundary layers were observed by dynamical measurement of turbulence and also from flow visualization technique. This study leads to the following conclusions.

1) There are two mechanisms responsible for the generation of turbulence in purely oscillatory boundary layers. One is the phenomenon of bursting which explosively occurs at the late stage of deceleration period of free stream velocity, this phenomenon being very similar to that of bursting in steady uni-directional flows. The other is a large-scale eddy structure causing negative turbulent shear stress, formed at the early stage of the acceleration period of free stream velocity.

2) Large-scale eddies occur at a distance around the fringe of the forerunner velocity layer which appears owing to the phase lead of the velocity near the bed, at the very early stage of acceleration period of free stream velocity when the free stream velocity is still very small. These large-scale eddies constitute a structure of turbulence peculiar to purely oscillatory boundary layers.

3) After the generation of large-scale eddies turbulence in the boundary layer decreases rapidly owing to relaminarization of turbulence during the period of acceleration. When the free stream velocity comes to be decelerated, turbulence in the deceleration period is generated and developed. In this way the process of generation, development and decay of turbulence is repeated in every half period of oscillation of flow.

References

1. Jonsson, I.G.: "Measurements in the Turbulent Wave Boundary Layer", Proc. 10th IAHR Congress, London, Vol. 1 (1963) pp. 85–92
2. Jonsson, I.G., Carlsen, N.A.: Experimental and theoretical investigations in an oscillatory turbulent boundary layer. J. Hydraul. Res. **14**, 45–60 (1976)
3. Hino, M., Sawamoto, M., Takasu, S.: Experiments on transition to turbulence in an oscillatory flow. J. Fluid. Mech. **75**, 193–207 (1976)
4. Merkli, P., Thomann, H.: Transition to turbulence in oscillating pipe flow. J. Fluid Mech. **68**, 567–575 (1975)
5. Cousteix, J., Desopper, A., Houdeville, R.: "Structure and Development of a Turbulent Boundary Layer in an Oscillatory External Flow", in *Turbulent Shear Flows I*, ed. by F. Durst, B.E. Launder F., W. Schmidt, J.H. Whitelaw (Springer, Berlin, Heidelberg, New York 1979) pp. 154–171
6. Kobashi, Y., Hayakawa, M.: "The Transition Mechanism of an Oscillating Boundary Layer", in *Laminar-Turbulent Transition*, ed. by R. Eppler, H.Fasel (Springer,Berlin, Heidelberg, New York 1980) pp. 102–109
7. Ramaprian, B.R., Müller, A.: Transitional periodic boundary layer study. J. Hydraul. Div. Am. Soc Civ. Eng. **106**, 1959–1971 (1980)
8. Anwar, H.O., Atkins, R.: Turbulent measurements in simulated tidal flow. J. Hydraul. Am. Soc. Civ. Eng. **106**, 1273–1289 (1980)
9. Hayashi, T., Ohashi, M., Takeyasu, S.: "Experimental Study on the Oscillatory Boundary Layer", Proc. 12th Symposium on Turbulence, Institute of Space and Aeronautical Science, Univ. of Tokyo, 1980, pp. 83–90 (in Japanese)
10. Hino, M., Kashiyanagi, M., Nakayama, T., Hara, T.: "On the Time-Space Correlation of the Turbulence of the Oscillatory Boundary Layer", Proc. 12th Symposium on Turbulence, Institute of Space and Aeronautical Science, Univ. of Tokyo, 1980, pp. 91–97 (in Japanese)
11. Fisher, R.A.: Test of significance in harmonic analysis. Proc. R. Soc. London. **125**, 54–59 (1929)
12. Eckelmann, H.: "Experimentelle Untersuchungen in einer turbulenten Kanalströmung mit starken viskosen Wandschichten", Mitt. Max-Planck Inst. f. Strömungsforschung, Göttingen No. 48 (1970)

Dynamics of an Unsteady Turbulent Boundary Layer

P. G. Parikh, R. Jayaraman, and W. C. Reynolds

Department of Mechanical Engineering, Stanford University
Stanford, CA 94305, USA

Abstract

The response of a well-defined, steady, flatplate turbulent boundary layer subjected subsequently to two types of oscillations of the freestream velocity in the test section is studied in this investigation. In the first case, the free-stream velocity decreased linearly in the test section in a time-dependent manner such that the magnitude of the streamwise *gradient* of the free-stream velocity varied sinusoidally from zero to a maximum value at frequencies ranging from zero to approximately the bursting frequency. In the second case, the *mean* free-stream velocity decreased in the test section in a nonlinear fashion independent of the oscillation frequency, while the amplitude and phase of the free-stream oscillations varied along the streamwise direction in an arbitrary manner, depending upon the oscillation frequency. Detailed measurements are reported for the first case. It was found that the mean velocity and mean turbulence intensity profiles in the boundary layer were unaffected by the imposed oscillations. The ratio of the amplitude of the periodic velocity component in the boundary layer to that in the free-stream, although as much as 1.7 for quasi-steady oscillations, becomes unity over the outer region of the boundary layer at high frequencies. Also, at high frequencies of imposed oscillations, both the boundary layer thickness and the Reynolds stress distribution across the boundary layer becomes frozen over the oscillation cycle at their mean values. Only the mean velocity profile measurements were carried out in the second case. It was found that, despite the large streamwise variations of the amplitude and phase of the free-stream oscillations, the mean velocity profile in the boundary layer once again remained independent of the imposed oscillation frequency so long as the *mean* free-stream velocity distribution remained unchanged.

Introduction

The objectives of the Stanford Unsteady Turbulent Boundary Layer Program are: to develop a fundamental understanding of such flows, to provide a definitive data base which can be used to guide turbulence model development, and to provide test cases which can be used by computors for comparison with predictions.

Due to space limitations, work of other investigators will not be summarized here, except to note that all the previous experiments are characterized by unsteady flow at the inlet to the unsteady region. For a comparison of the present experimental parameter range with those of other investigations see [1]. The distinctive feature of the present experiments is that the boundary layer at the inlet to the unsteady region is a standard, *steady*, flat-plate turbulent boundary layer. It is then subjected to controlled oscillations of the free stream. This feature is especially important from the point of view of a computor, who needs precise specification of boundary conditions for computation of the flow.

Free-Stream Boundary Condition of the Present Experiment

The free-stream velocity distribution $u_\infty(x,t)$ for the first case in the water tunnel built for this work is shown in Fig. 1. u_∞ remains *steady* and uniform for the first two meters of boundary layer development. It then decreases linearly in the test section; the magnitude of the velocity *gradient* varies sinusoidally from zero to a maximum value during the oscillation cycle. The mean free-stream velocity distribution in the test section is thus linearly decreasing and corresponds to the distribution at the cycle phase angle of 90°, while the amplitude of imposed free-stream oscillations grows linearly in the streamwise direction, starting at zero at the entrance to a maximum value of a_0 at the exit. Hence,

$$u_\infty(x,t) = u_{\infty,0} \quad , \qquad x < x_0$$

$$= u_{\infty,0} - \frac{a_0(x-x_0)}{L}(1 - \cos\omega t) \quad , \qquad x_0 < x < x_0 + L \quad .$$

The important parameters of this problem are the amplitude parameter $\alpha = a_0/u_{\infty,0}$ and the frequency parameter: $\beta_\delta = f\delta_0/u_{\infty,0}$. Here $f = \omega/(2\pi)$ and δ_0 is the thickness of the boundary layer at the inlet to the unsteady region. In the present experiments:

$$u_{\infty,0} = 0.73 \text{ m/s} \quad , \qquad \delta_0 = 0.05 \text{ m} \quad , \qquad 0 < f < 2 \text{ Hz} \quad ,$$

$$0 < \alpha < 0.25 \quad , \qquad 0 < \beta_\delta < 0.14 \quad .$$

It should be mentioned that the value of the frequency parameter β_δ at the so-called "bursting frequency" in turbulent boundary layers is about 0.2 [2]. Thus the imposed oscillation frequencies used in the present experiments cover the range from quasi-steady ($f \cong 0$) to values approaching the bursting frequency. The results reported here are for one non-dimensional amplitude, $\alpha = 0.05$.

In the second case, the *mean* free-stream velocity distribution was nonlinear along x but was independent of the oscillation frequency, as shown in Fig. 2. The amplitude of the imposed free-stream oscillations near the test-section exit was comparable to that in the first case (5%). However, the growth of the amplitude from nearly zero at the test section entrance to the maximum value at the exit was nonlinear and frequency dependent, as shown in Fig. 3. Furthermore, the phase of the free-stream oscillations varied

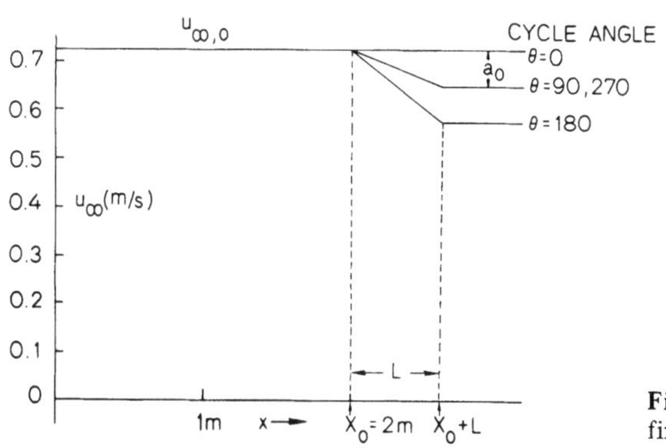

Fig. 1. Free-stream velocity variation first case

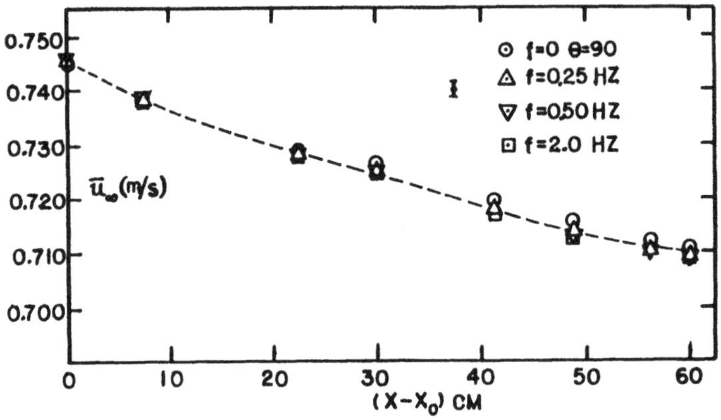

Fig. 2. Free stream velocity variation, second case

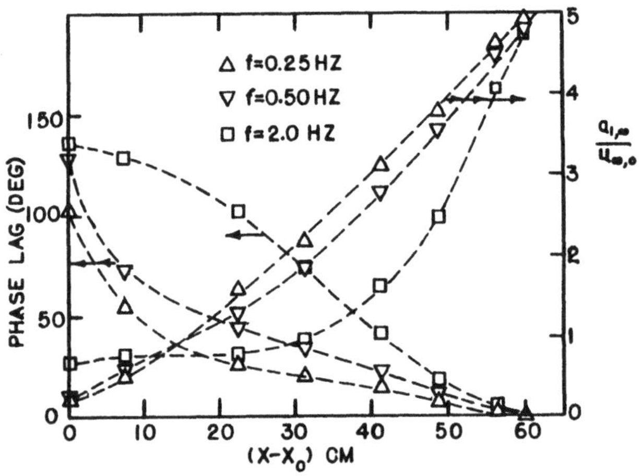

Fig. 3. Streamwise variation of amplitude and phase, second case

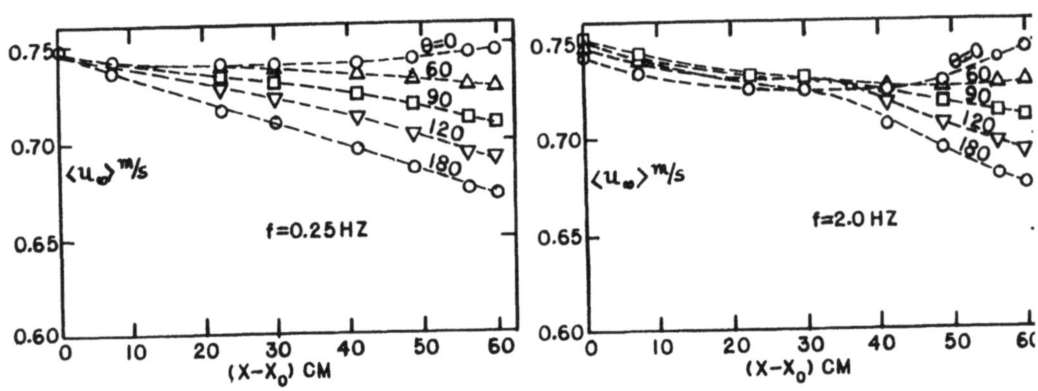

Fig. 4. Phase-averaged free-stream velocity distributions, second case

dramatically along x and was also dependent on the oscillation frequency, as shown in Fig. 3. The net result at high oscillation frequency was a strong departure of the phase-averaged free-stream velocity distribution from a linear one accomplished in the first case (see Fig. 4).

Experimental Facility

Figure 5 is a schematic of the facility. The 16:1 nozzle contraction is followed by a 2 m long development section, where the test boundary layer is grown on the top wall. A constant head and a constant flow resistance provide a constant flow. The free-stream velocity in the development section is maintained uniform along x by bleed from the bottom wall.

The linear decrease in free-stream velocity in the test section, as desired in the first case, is accomplished by uniformly bleeding off some flow through a perforated plate which forms the bottom wall in the test section. The holes in the perforated plate are sized such that the pressure drop across them is large compared to the dynamic pressure in the test section. To make the amplitude and phase of free-stream oscillations dependent on oscillation frequency, the holes were made larger so that the pressure drop across the perforated bottom plate became small compared to the dynamic pressure. The remainder of the flow exits downstream. Each of these two flows exits the tunnel through slots in an oscillating plate. The design ensures that, regardless of the position of the oscillating plate, the total flow area of the slots remains the same. The slots are the controlling resistance of the entire fluid circuit, hence the constant flow. By sinusoidally oscillating the plate, a periodic free-stream distribution is established in the test section, while the upstream flow in the development section remains steady.

Fig. 5. The apparatus

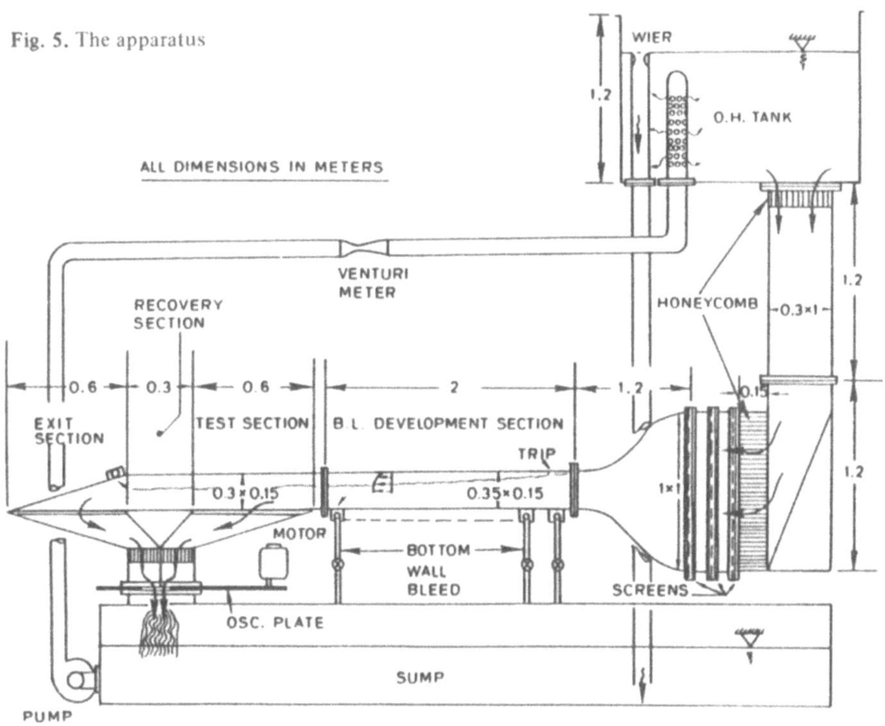

Measurement and Data-Processing Techniques

Pitot tubes are used for mean velocity measurements in steady flow regions. Unsteady velocity measurements use a single-channel, forward-scatter, Bragg-shifted DISA laser anemometer in the tracking mode.

Following *Hussain* and *Reynolds* [3], the instantaneous velocity signal from an unsteady turbulent flow may be decomposed into three parts:

$$u = \bar{u} + \tilde{u} + u' \quad , \tag{1}$$

where $\bar{u}$ is the mean, $\tilde{u}$ is the time-dependent, organized (deterministric) component and u' is the random fluctuation. $\bar{u}$ is determined by long-time averaging of u. Here $\tilde{u}$ is of a periodic nature and may be determined by first phase-averaging the instantaneous velocity signal and then subtracting out the mean. Thus,

$$\tilde{u} = \langle u \rangle - \bar{u} \quad , \tag{2}$$

Here $\langle u \rangle$, the phase average velocity, is determined by averaging over an ensemble of samples taken at a fixed phase in the imposed oscillation. In the present experiments, with harmonic oscillation of the free stream, the response at points within the boundary layer is almost sinusoidal, with higher harmonics contributing less than 5%. Hence, $\tilde{u}$ may also be extracted from the instantaneous signal u by cross-correlation with a sine wave in phase with the oscillation. A digital correlator (HP 3721 A) was used to determine cross-correlations leading to the $\tilde{u}$ data reported here. Currently a DEC MINC-11 laboratory minicomputer system is used for automatic data acquisition and processing, allowing, the determination of phase averages of u and u'^2.

The measurements reported here were taken at a fixed streamwise location near the end of the test section at $x - x_0 = 0.568$ m.

Results and Discussion

Case I. Linear Phase-Averaged Free-Stream Distribution

The mean velocity profiles measured with the oscillating plate in fixed positions $\theta = 0$, 90°, 180° are fit by dashed curves in Fig. 6. These represent phase-averaged profiles at zero frequency, i.e., quasi-steady profiles. At this amplitude ($\alpha = 0.05$), the response of the boundary layer is almost linear, so that the profile corresponding to $\theta = 90°$ lies nearly midway between the $\theta = 0$ and 180° profiles. The 90°profile represents the mean profile for quasi-steady oscillations. The difference between the 0 and 90° profiles at a fixed y-location represents the amplitude of quasi-steady oscillations at that location in the boundary layer. Note that the quasi-steady amplitudes in the boundary layer are larger than the free-stream amplitude.

The mean velocity profiles measured under oscillatory conditions at 0.5 Hz and 2.0 Hz are shown as data points in Fig. 6. Note that the mean velocity profiles at various frequencies are identical with the profile measured under stationary conditions with pulser angle set at $\theta = 90°$. It may be concluded that the mean velocity profile (at a fixed amplitude $\alpha = 0.05$) is independent of the imposed oscillation frequency in the entire range $0 \leqslant f \leqslant 2$ Hz. The same behavior persists all the way up to the wall.

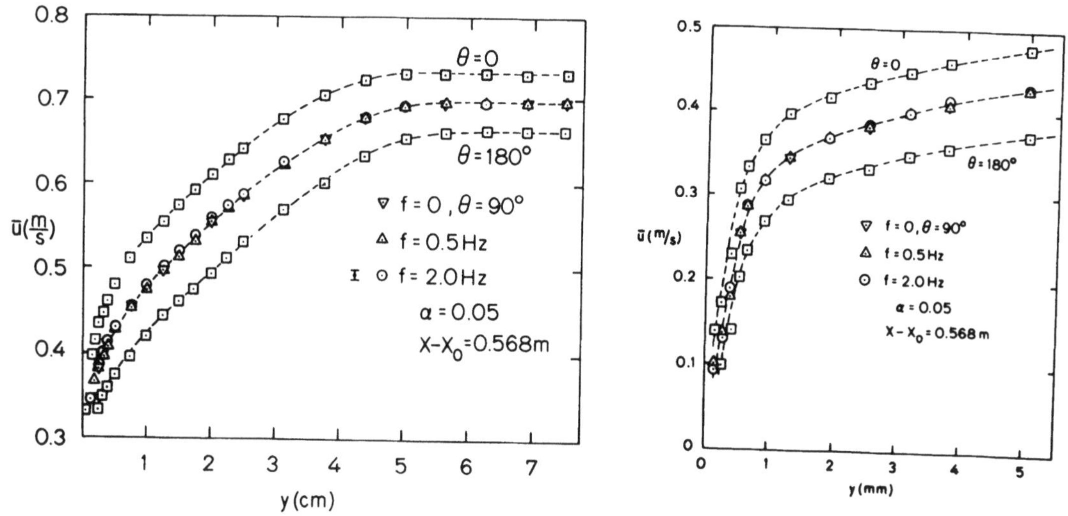

Fig. 6. Mean velocity profiles at three frequencies and profiles at the extremes of the oscillation at zero frequency for $\alpha = 0.05$

This behavior of the mean velocity profile may be explained by an examination of the governing equations. Use of (1) in the momentum equation and time-averaging yields

$$\bar{u}\,\frac{\partial \bar{u}}{\partial x} + \bar{v}\,\frac{\partial \bar{u}}{\partial y} = -\frac{1}{\rho}\frac{\partial \bar{p}}{\partial x} + \nu\,\frac{\partial^2 \bar{u}}{\partial y^2} - \frac{1}{\rho}\frac{\partial}{\partial y}\quad \overline{u'v'} + \overline{\tilde{u}\tilde{v}} \tag{3}$$

Equation (3) may be recognized as the equation governing an ordinary turbulent boundary layer, except for the addition of the term $\overline{\tilde{u}\tilde{v}}$, which represents Reynolds stresses arising from the organized oscillations.

The time-mean pressure gradient $\partial\bar{p}/\partial x$ may be shown to be independent of the imposed oscillation frequency and, except for a higher-order term, it is the same as that obtained for $f = 0$ at $\theta = 90°$. Therefore, the mean velocity field will be frequency-dependent if and only if one or both of the following happen:

- The distribution of Reynolds stress $\overline{u'v'}$ is altered under oscillatory conditions and is dependent on the frequency of imposed oscillations.

- The Reynolds stress $\overline{\tilde{u}\tilde{v}}$ arising from organized fluctuations becomes significant compared with $\overline{u'v'}$.

We shall now argue that neither of the above requirements is met. Figure 7 shows the measured distribution of u'_{rms} under stationary conditions with the pulser at $\theta = 90°$ (the mean position) as well as those measured under oscillatory conditions at frequencies up to 2 Hz. Note that u'_{rms} is independent of the imposed oscillation frequency and, further, that it is the same as that measured at $f = 0$ and $\theta = 90°$. We believe that the same would be true for $\overline{u'v'}$, which at present we cannot measure. Figure 8 gives a comparison between measured values of $\overline{\tilde{u}\tilde{v}}$ at 2 Hz with data on $\overline{u'v'}$ obtained by *Anderson* [4] in a steady adverse pressure gradient boundary layer at comparable conditions. The present data on $\overline{\tilde{u}\tilde{v}}$ were obtained by separate LDA

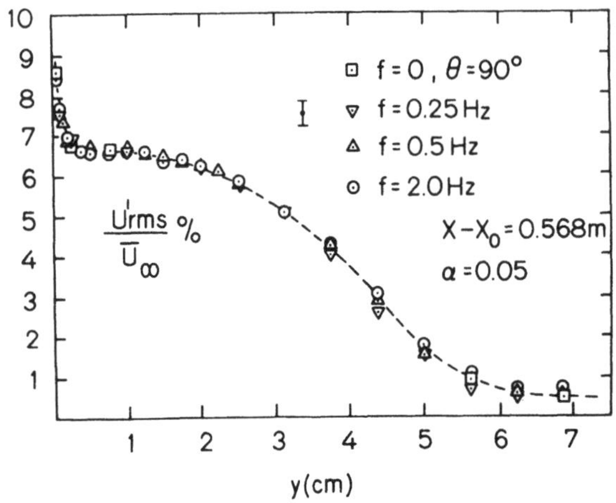

Fig. 7. Mean turbulence profiles at three frequencies for $\alpha = 0.05$

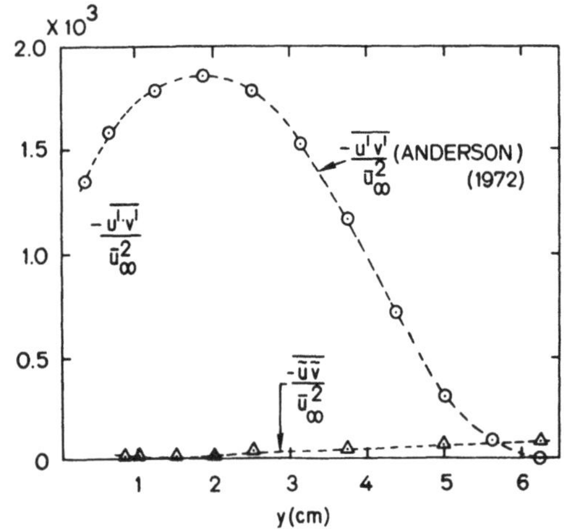

Fig. 8. Turbulent and organized Reynolds stresses for $\alpha = 0.05$

measurements of $\tilde{u}$ and $\tilde{v}$ and their respective phases. It may be seen that the contribution of $\overline{\tilde{u}\tilde{v}}$ to total Reynolds stress is insignificant over almost the entire boundary layer. Hence, $\overline{u'v'}$ is independent of frequency and $\overline{\tilde{u}\tilde{v}}$ is negligible, and so the mean velocity profile is also independent of frequency and is the same as that found at $f = 0$ with $\theta = 90°$.

The behavior of the periodic component $\tilde{u}$ will next be examined. We denote

$$\tilde{u} = a_1(y) \cos[\omega t + \phi(y)] \quad . \tag{4}$$

The profiles of amplitudes a_1 measured in the boundary layer and normalized by the free-stream amplitude $a_{1,\infty}$ are shown in Fig. 6. The profile for quasi-steady ($f \cong 0$) oscillations was determined, as explained earlier, from the mean velocity profiles measured at $f = 0$ with $\theta = 0, 90°$, and $180°$ [see Fig. 6 a,b]. Note that, during quasi-steady oscillations, the amplitude in the boundary layer exceeds the free-stream amplitude by as much as 70%. It may be mentioned that data for $f = 0.1$ Hz, not shown on Fig. 9, do indeed come very close to the quasi-steady behavior.

As the frequency is increased, the amplitude within the boundary layer is attenuated. The amplitude appears to drop as f is increased and then rises again. At high frequencies, the amplitude in most of the boundary layer is the same as in the free-stream; near the wall the amplitude of the periodic component rapidly drops to zero.

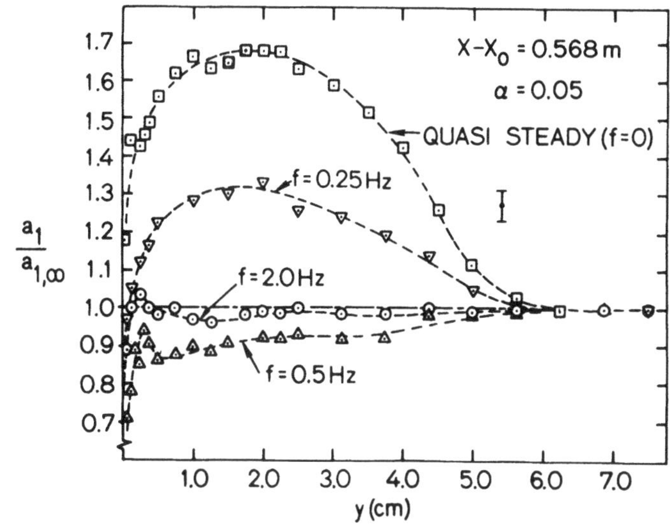

Fig. 9. Amplitude of organized disturbance for $\alpha = 0.05$

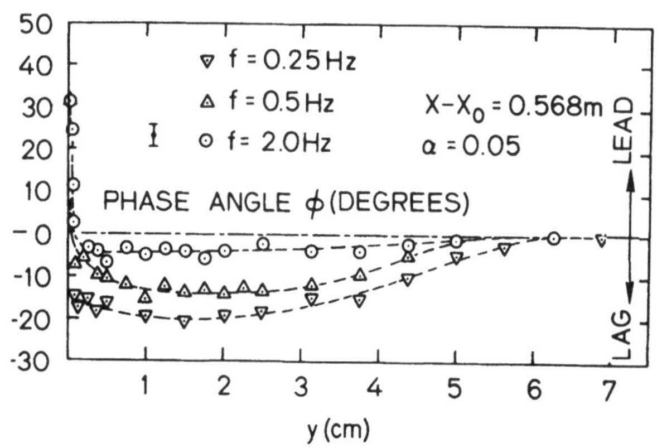

Fig. 10. Phase of organized disturbance for $\alpha = 0.05$

The phase differences between the boundary layer oscillations and free-stream oscillations are shown in Fig. 10. For $f = 0$ there is no phase difference. The largest phase lags in the outer region of the boundary layer were observed at $f = 0.25$ Hz. The effect of increasing the frequency is to reduce the phase lag in the outer region, but to introduce large phase *leads* in the region very close to the wall. Clearly, the asymptotic behavior of the *outer region* for high frequencies is once again a zero phase lag with respect to free-stream oscillations, as in the quasi-steady case.

At high frequencies, the combination of the asymptotic behaviors of $a_1/a_{1,\infty}$ and ϕ in the outer region together with the fact that the mean velocity profile is unaffected by imposed oscillations, has the effect of freezing the boundary layer thickness. This is shown in Fig. 11, where the phase-averaged boundary layer thickness $<\delta_{0.99}>$ is plotted as a function of the cycle phase angle for several frequencies. The quasi-steady behavior of $<\delta_{0.99}>$ is quite obvious: at $\theta = 0$, the boundary layer in the test section continues to develop under a zero pressure gradient and is the thinnest as this point in the entire cycle. As the phase angle is increased, pressure gradients of increasing adversity are imposed on the boundary layer, causing it to thicken. The maximum thickness is attained at $\theta = 180°$ under the maximum adverse pressure gradient. Hence, at $f = 0$, δ oscillates 180° out of phase with u_∞.

Under oscillatory conditions at $f = 0.25$, 0.5, and 2.0 Hz, two things happen: a significant phase lag develops from quasi-steady behavior and the amplitude attenuates with increasing frequency. For the $f = 2.0$ Hz case, the variation over the complete cycle is less than 1% and the boundary layer thickness is practically frozen during the oscillation cycle.

It may be shown by a simple argument based on a mixing length model of boundary layer turbulence that the freezing of the boundary layer thickness at high frequencies is also accompanied by freezing of the Reynolds stress over the oscillation cycle. To prove this, we hypothesize that the phase-averaged Reynolds stress distribution may be related to the phase-averaged velocity profile in the same manner as for a steady boundary layer, i.e.,

$$-\langle u'v'\rangle = \epsilon_m \frac{\partial \langle u\rangle}{\partial y}\,,\quad \epsilon_m = l^2 \frac{\partial \langle u\rangle}{\partial y}\quad . \tag{5}$$

Now , in the *outer region* of the boundary layer, the mixing length l may be modeled as

$$l = \lambda \langle \delta_{0.99}\rangle \tag{6}$$

where λ is nearly a constant. Now

$$\langle u\rangle = \bar{u} + \tilde{u} = \bar{u} + a_1(y)\cos[\omega t + \phi(y)] \tag{7}$$

However, in the high-frequency limit,

$$a_1(y) = a_{1,\infty} = \text{const}\;,\quad \phi(y) = 0\;,\quad \text{and} \tag{8}$$

$$\langle \delta_{0.99}\rangle = \delta_{0.99} = \text{const.}$$

Therefore

$$\frac{\partial \langle u\rangle}{\partial y} = \frac{\partial \bar{u}}{\partial y} \tag{9}$$

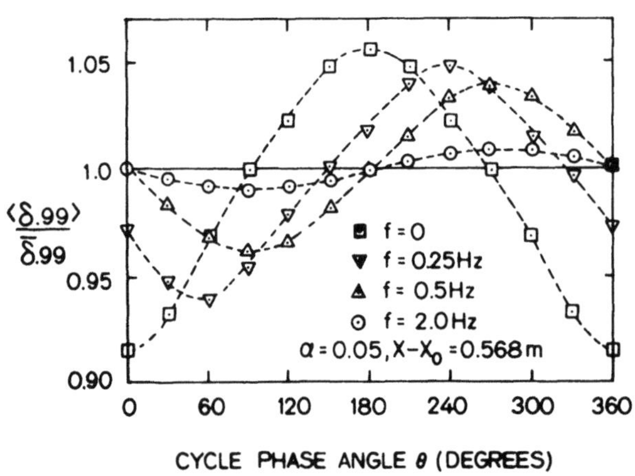

Fig. 11. Phase of boundary layer thickness for $\alpha = 0.05$

Combining that above, one finds

$$-\langle u'v'\rangle = \lambda^2 \bar{\delta}^2_{0.99} \frac{\partial \bar{u}}{\partial y}^2 = \overline{u'v'}$$

i.e, the phase-averaged Reynolds stress in the outer region also becomes frozen at $-\overline{u'v'}$.

Experimental evidence of this stress-freezing behavior was obtained by measurements of phase-averaged normal turbulent stress $\langle u'^2\rangle$. The quasi-steady ($f = 0$) profiles of $\langle u'^2\rangle$ are shown in Fig. 12 for three phase angles $\theta = 0°, 90°$, and $180°$. Note that the distribution for 90° lies nearly midway between those for 0° and 180°. This distribution of $\langle u'^2\rangle$ for 90° is the same as the distribution of $\overline{u'^2}$, as seen earlier. Therefore the difference between the 0° and 90° curves in Fig. 12 represents the amplitude of quasi-steady oscillations of $\langle u'^2\rangle$ at any point in the boundary layer. This amplitude was determined graphically from Fig. 12 and is plotted in Fig. 13 for the case of $f = 0$. Under oscillatory conditions, the amplitude of the normal stress oscillations in the boundary layer attenuates as the frequency of imposed oscillations is increased from $f = 0$. At $f = 2.0$ Hz, the amplitude of stress oscillations across the boundary layer is almost zero over the outer region, as seen in Fig. 13, i.e., the stress is almost frozen over the oscillation cycle.

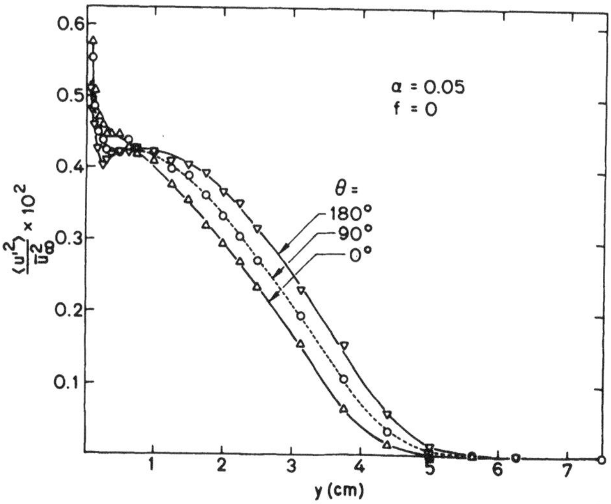

Fig. 12. Phase-average longitudinal fluctuation for $\alpha = 0.05$, f = 0

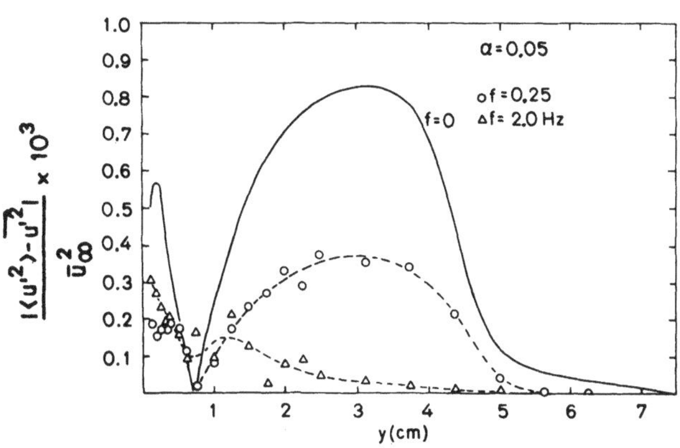

Fig. 13. Reynolds stress oscillations at α=0.05

Case II. Arbitrary Phase-Averaged Free-Stream Distribution

In this case, the phase-averaged free-stream distributions become frequency-dependent, as shown in Fig. 4. However, the *mean* free-stream distribution remained almost the same as that shown in Fig. 2 at all frequencies employed. The mean velocity profiles in the boundary layer were measured at the same axial location as in Case (i). Figure 14 shows that the mean velocity profile was still unaffected by the imposed oscillations at frequencies ranging from zero to 2.0 Hz. This shows that the mean velocity profile is dependent only on the *mean* free-stream velocity distribution ahead of the measurement station and is independent of the amplitude and phase distributions of the imposed oscillatory component in the free stream.

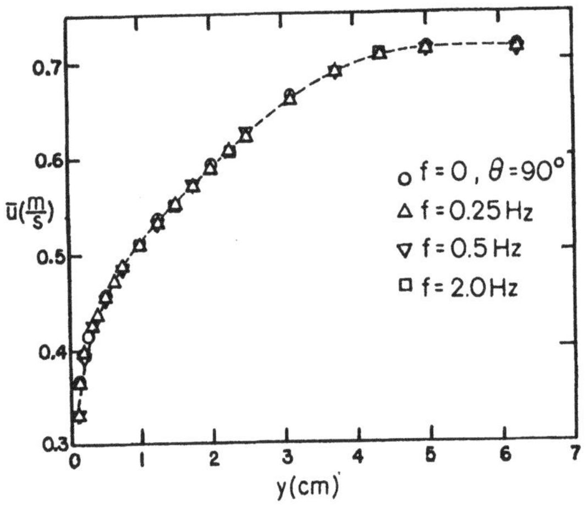

Fig. 14. Mean velocity profiles at different frequencies, second case.

Conclusions

The conclusions from our experiments to date may be summarized as follows:

1. The mean velocity profile in the boundary layer is unaffected by imposed free-stream oscillations in the range of frequencies employed, and it is the same as the one measured with a free-stream velocity distribution held steady at its mean value. This is true even if the amplitude and phase of the imposed oscillatory component in the free stream vary arbitrarily along x as functions of imposed oscillation frequency.

2. This behavior of the mean velocity field is a consequence of two observations: (a) the time-averaged Reynolds stress distribution across the boundary layer is unaffected by the imposed oscillations and is indeed the same as the one measured with the free-stream velocity distribution held steady at the mean value; and (b) the Reynolds stresses arising from the organized velocity fluctuations under imposed oscillatory conditions are negligible compared to the Reynolds stresses due to the random fluctuations.

3. The amplitude of the periodic component in the boundary layer under quasi-steady oscillations ($f \to 0$) is as much as 70% larger than the imposed free-stream amplitude. However, at higher frequencies the peak amplitude in the boundary layer is rapidly attenuated toward an asymptotic behavior where amplitudes in the outer region of the boundary layer become the same as the free-stream amplitude, dropping off to zero in the near-wall region.

4. Quasi-steady boundary layer velocity response is in phase with the imposed free-stream oscillations. As the frequency is increased, phase lags begin to develop in the outer region of the boundary layer. The magnitude of this phase lag reaches a maximum and then decreases with increasing frequency until an asymptotic limit is reached where the outer region once again responds in phase with the free stream. Near the wall, however, large lead angles are present at higher oscillation frequencies.

5. A consequence of (3) and (4) above is that the boundary layer thickness becomes nearly frozen over the oscillation cycle at higher frequencies.

6. A consequence of (3), (4), and (5) above is that the Reynolds stress distribution in the outer region of the boundary layer also becomes frozen over the oscillation cycle at higher frequencies.

Acknowledgments. This research is carried out at Stanford in cooperation with and under the sponsorship of the Army Aeromechanics Laboratory, the NASA-Ames Research Center, and the Army Research Office. The authors wish to express their gratitude to Drs. Larry Carr and James McCroskey (AML), Mr. Leroy Presley (NASA-Ames), and Dr. Robert Singleton (ARO) for their continued assistance.

References

1. Parikh, P.G., et al.: On the behavior of an unsteady turbulent boundary layer. AIAA. (1981)
2. Rao, K. N., et al.: The bursting phenomenon in a turbulent boundary layer. J. Fluid Mech. **48**, 339-352 (1971)
3. Hussain, A.K.M.F., Reynolds, W.C.: The mechanics of an organized wave in turbulent shear flow. J. Fluid Mech., **41**, 241-258 (1970); **54**, 241-288 (1972)
4. Andersen, P.S., Kays, W.M., Moffat, R.J.: 'The Turbulent Boundary Layer on a Porous Plate: An Experimental Study of the Fluid Mechanics for Adverse Free-Stream Pressure Gradients', Report No. HMT-15, Dept. of Mech. Eng., Stanford University (1972)

Influence of Strouhal Number on the Structure of Flat Plate Turbulent Boundary Layer

J. Cousteix, J. Javelle, and R. Houdeville

Aerothermodynamics Department ONERA/CERT, 2 av. E. Belin
F-Toulouse Cedex, France

Abstract

The development of an oscillating turbulent boundary layer is studied along a flat plate. The Strouhal number, based on a fictitious origin of the boundary layer, varies from 1.5 to 18. Systematic measurements of the velocity profiles are carried out and are presented in the form of the harmonic analysis of the velocity, as well as of the displacement thickness. It is shown that any quantity, for example the phase and the amplitude of the displacement thickness, oscillates as the Strouhal number increases. The period of this oscillation can be correctly estimated from a small perturbation development of the Von Kàrmàn momentum integral equation.

The evolution of the phase of velocity profiles near the wall is carefully examined. An estimation of the wall shear stress is deduced from the law of the wall, and is compared to direct measurements obtained with a skin friction gauge.

The evolution in time of the u-component of the kinetic energy of turbulence and of the correlation coefficient, are presented in a typical case.

Introduction

The study of unsteady turbulent flows has developed greatly during the last few years, due to the improvement in prediction methods and closure models. Although the number of experimental investigations is also increasing, there exists a lack of experimental data to validate the theoretical work and also, to obtain a better understanding of the physics of the phenomenon. For these reasons, it was necessary to undertake a study of the response of a turbulent boundary layer over a flat plate to an oscillating external velocity field. The simplicity of the boundary conditions, fixed walls and periodic external velocity, allows an extensive and systematic investigation of the boundary layer. This simplicity is also helpful for comparison with other experiments and with predictions.

In the present experiment, the range of the Strouhal number is quite large, between 1.5 and 18, and the velocity profiles are measured at close intervals of the Strouhal number, to obtain a detailed description of the response of the boundary layer to the perturbation induced by the oscillation of the external velocity.

This work was sponsored by the Direction de la Recherche et Etudes Techniques under Grant DRET n° 80.432 (lot n° 6)

Nomenclature

f, T	frequency and period	τ	shear stress
U	longitudinal velocity	μ,ν	dynamic and kinematic viscosity
X, Y	longitudinal and normal coordinates	ρ	density
X_0	distance from the fictitious origin of the turbulent boundary layer	ΔF	amplitude of F
S	Strouhal number $S = \omega X_0/\bar{U}_e = 2\pi f X_0/\bar{U}_e$	φ_F	phase of F
H	shape parameter	$(\bar{\ })$	time mean value
R_x	Reynolds number $R_x = U_e X/\nu$	$<\ >$	ensemble average
C_f	skin friction coeffient $C_f = \tau_w / \frac{1}{2} \rho_e U_e^2$	$(')$	random fluctuation
$U_\tau = \sqrt{\tau_w/\rho}$			
$u^+ = U/U_\tau$			
$Y^+ = YU_\tau/\nu$			
δ_1	displacement thickness		
θ	momentum thickness		

Subscripts

0 time mean value
1 complex amplitude
e external value
w value at the wall

Experimental Set-Up and Flow Parameters

The boundary layer under consideration develops on a 0.8 m long, 0.22 m wide and 0.15 mm thick flat plate. The aspect ratio of the elliptic leading edge is 6. A transverse wire, 0.8 mm in diameter and located 70 mm from the leading edge of the plate, trips the transition of the boundary layer.

The pulsation of the flow is induced by a simple rotating butterfly valve, located 1.3 m downstream the trailing edge of the flat plate. This valve generates a periodic pressure loss in the suction facility. By choosing the r.p.m. of the butterfly corresponding to the resonance frequencies of the wind tunnel, sinusoidal variation of the external velocity is easily obtained, with a small harmonic distortion within a few per cent of the fundamental.

The variation of the Strouhal number $\omega X/\bar{U}_e$ cannot be obtained only by changing X, which corresponds to a displacement along the plate. Practically, a variation of S from 1.5 to 18 is achieved by choosing two combinations of ω and U_e. Table 1 gives the parameters of the two experiments, as well as a description of the external flow fields in terms of their harmonic analysis. In both experiments, the mean velocity remains constant in the X direction, but the relative amplitude of the fluctuation decreases linearly by a factor 1.5 at $f = 38$ Hz and 3 at $f = 62$ Hz, over a distance of 700 mm. At 38 Hz, the phase difference between the two extreme measurement stations is very small, but it reaches 45° at 62 Hz, which corresponds to a travelling wave with an apparent velocity equal to 450 m/s.

For the calculation of the Strouhal number, a difficulty arises due to the definition of the origin of the turbulent boundary layer. This origin is calculated from the mean momentum thickness, obtained at the first measurement station by using the steady flow flat plate approximation:

$$\frac{\theta}{X_0} = \frac{0.0221}{R_{X_0}^{1/6}} \tag{1}$$

The quantity X_0 in the figures is the distance from this fictitious origin.

Table 1. External velocity and main parameters of the flows

	$u_e = u_e + \Delta u_e \sin(\omega t + \varphi_{u_e})$			
F[Hz]	38		62	
$\bar{u}_e$[m/s]	21.9		16.8	
$\Delta u_e/\bar{u}_e$	0.152 - 0.0743 x $(X_0$ - 0.047)		0.118 - 0.114 x $(X_0$ - 0.047)	
$\varphi°u_e$	7.6 x $(X_0$ - 0.047)		88.9 x $(X_0 - 0.047)^2$ + 6.66 x $(X_0$ x 0.047)	
X_0 [m]	0.14	to 0.77	0.137	to 0.762
$\omega X_0/\bar{u}_e$	1.52	to 8.35	3.2	to 17.8
R_{x0}	0.2 x 10^6	to 1.1 x 10^6	0.15 x 10^6	to 0.87 x 10^6
$u'_{max}/\delta[s^{-1}]$	350	to 120	270	to 90

Measurements and Data Reduction

Measurements of the longitudinal component of the velocity are done by constant temperature hot wire anemometry. The average velocity is defined using a conditional sampling technique. U being the instantaneous velocity, φ the phase angle in the cycle and n the number of the cycle, we have :

$$< U> = \sum_{n=1}^{N} U(nT + \varphi) \quad \text{with} \quad U = < U> + u' \quad \text{and} \quad < u'> = 0$$

The u-component of the turbulence energy is $<u'^2>$ by definition.

In practice, the rotation of the butterfly valve which induces the pulsation of the flow gives the phase information. $< U>$ and $<u'^2>$ are obtained at 24 instants per cycle by averaging over 600 cycles [1].

Average Flow Field

Mean Flow Field

For both experiments, corresponding to f = 38 and 62 Hz, the average flow field has been compared with the steady configuration at the same Reynolds number. Figure 1 presents only the results obtained at f = 38 Hz. No influence of the unsteadiness of the flow can be detected at the considered Strouhal numbers, namely between 1.5 and 8. Results corresponding to the case f = 62 Hz extend this conclusion up to S = 18. The skin friction presented in the figure has been obtained using the tecnique of 'Clauser plot', which consists in fitting the experimental velocity profiles by the law of the wall. The results are in good agreement with predictions using an integral method. Moreover, the slope of $\bar{\theta}(X)$ agrees well with the estimated values of the skin friction coefficient, which indicates a good two-dimensionality of the flow.

Harmonic Analysis of the Velocity

Profiles of the relative amplitude of fluctuation and of phase shift with respect to the external velocity, are presented in Figs. 2 and 3 for several stations along the plate.

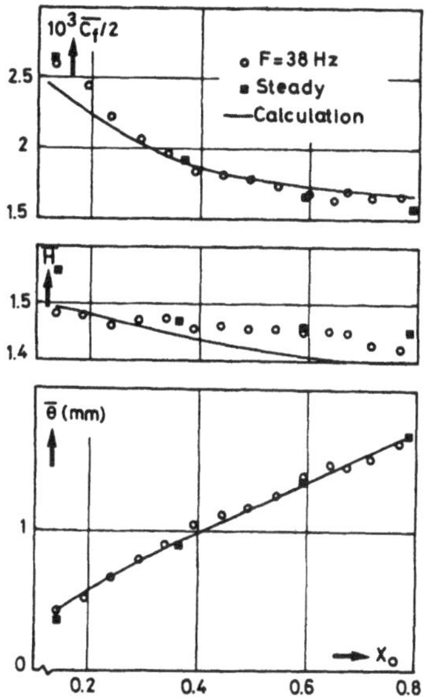

Fig. 1. Mean flow field. Evolution of momentum thickness, shape parameter and skin friction along the plate $- f = 38$ Hz

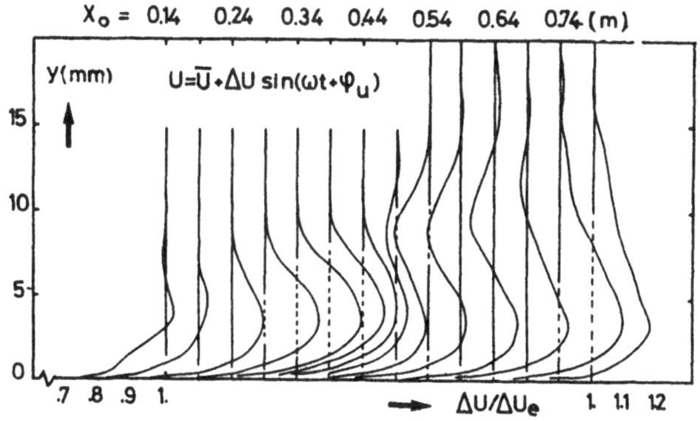

Fig. 2. Harmonic analysis of the velocity – Relative amplitude of fluctuation $- f = 38$ Hz

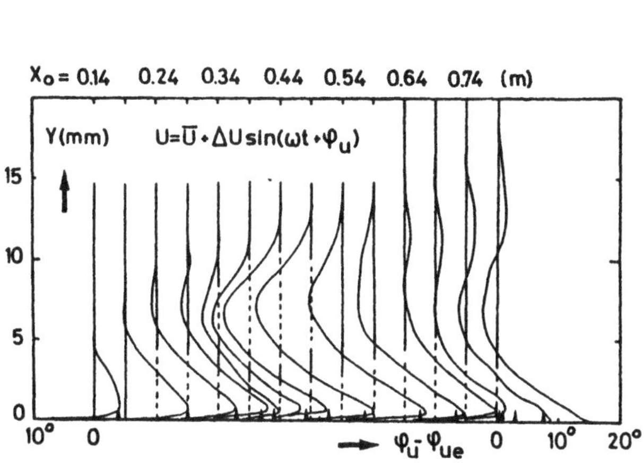

Fig. 3. Harmonic analysis of the velocity – Phase shift – ▲: φ_{τ_w} calculated from the law of the wall $- f = 38$ Hz

A slight overshoot of $\Delta U/\Delta U_e$ exists at the first station. It increases to reach 20% at $X_0 = 0.34$ m. Further downstream, it decreases and disappears completely at $X_0 = 0.54$ m. Then, this evolution repeats. If the boundary layer is considered as a resonant mechanical system, the observed periodicity along X can be interpreted as the response of the boundary layer to the perturbation induced by the pulsation of the external flow.

By assuming the existence of a travelling wave which affects the successive profiles, the convective velocity of this wave is of the order of $0.75\ \overline{U}_e$. The results obtained at 62 Hz lead to the same interpretation.

The evolution with Strouhal number of the profiles of the phase shift of the velocity, also displays a periodicity, similar to that of the profiles of $\Delta U/\Delta U_e$. An interesting feature of the profiles $\varphi_U - \varphi_{U_e}$ is the maximum phase lead observed around $Y^+ = 50$, corresponding to the logarithmic region of the velocity profile. *Simpson* [2] has also observed an extremum in the logarithmic region. This property will be interpreted later.

Integral Boundary Layer Thickness

The relative amplitude of fluctuation of the displacement thickness δ_1, divided by the relative amplitude of the external velocity, and the phase difference with respect to U_e, are presented in Fig. 4. All the experimental results obtained at ONERA are plotted with those of *Karlsson* [3]. For the sake of uniformity, the Strouhal number is calculated using a fictitious origin of the turbulent boundary layer. Despite large differences in the initial conditions and the transition process, and although the Strouhal number is not the unique parameter to take into account, an overall good agreement is observed between the various experiments, especially the phase difference with respect to the external velocity. In experiments, this last quantity is measurable with a better accuracy than the relative amplitude of oscillation.

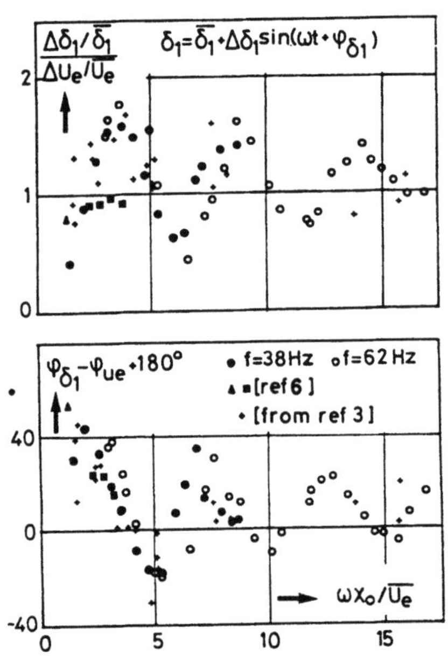

Fig. 4. Harmonic analysis of the displacement thickness versus Strouhal number

The striking result is the periodic oscillation of the quantities with a small damping as the Strouhal number increases. This oscillation has been previously observed for the shape of the profiles of relative amplitude and phase shift of the velocity. Its period is close to 5. This means that it can be detected only if the boundary layer is measured at close intervals of Strouhal number, which was not the case in the pioneering experiments of *Karlsson*.

The general feature of the response of the boundary layer to the oscillation of the external velocity, can be easily illustrated from a small perturbation expansion of the von Kàrmàn's equation:

$$\frac{C_f}{2} = \frac{\partial \theta}{\partial X} + \theta \; \frac{H + 2 \partial U_e}{U_e \; \partial X} + \frac{1 \partial}{U_e^2 \partial t} (U_e \delta_1) \tag{2}$$

By assuming that all the quantities appearing in Eq. (2) can be written in the form:

$$f = f_0 + f_1 \exp(\mathrm{i}\omega t) \quad ,$$

and the modulus of f_1, the complex amplitude, being supposed small compared to f_0, the Kàrmàn's equation reads:

$$\frac{d\theta_1}{dx} = \frac{C_{f1}}{2} - \frac{\mathrm{i}\omega}{U_{e0}} \left(\delta_{11} + \delta_{10} \frac{U_{e1}}{U_{e0}} \right) \tag{3}$$

To close Eq. (3), it is assumed that the shape parameter remains constant and that the skin friction is a function only of R_θ and H, or equivalently in the present case of θ and U_e, but is not explicitly a function of time. Then, Eq. (3) becomes:

$$\frac{d\theta_1}{dX} + \theta_1 \left[\mathrm{i}\omega \frac{H_0}{U_{e1}} - \frac{1}{2} \left(\frac{\partial C_f}{\partial \theta} \right)_0 \right] = \frac{U_{e1}}{2} \left(\frac{\partial C_f}{\partial U_e} \right)_0 - i\omega H_0 \theta_0 \frac{U_{e1}}{U_{e0}} \tag{4}$$

Using the Strouhal number $S = \omega X / U_{e0}$ instead of X, Eq. (4) has a solution of the form:

$$\theta_1 = \Theta \exp \left\{ \left[-\mathrm{i} H_0 + \frac{U_{e0}}{2\omega} \left(\frac{\partial C_f}{\partial \theta} \right)_0 \right] S \right\} \tag{5}$$

The period of θ_1 is equal to $2\pi / H_0 \cong 4.5$, which is close to the experimentally observed value of 5. Moreover, the above calculation shows that the damping factor is strongly dependent on the chosen skin friction law.

Wall Shear Stress – The Logarithmic Region

If the well known results in steady flows are extrapolated to the unsteady case, it can be thought that the flow behaviour near the wall is controlled by the existence of a logarithmic law. At least, it is expected that such a law is valid for small Strouhal numbers. Then, at each instant in the cycle, the following law would hold:

$$U^+ = \frac{1}{x}\ln Y^+ + C \quad \text{with} \quad \chi = 0.41 \quad ; \quad C = 5.25 \quad . \tag{6}$$

To check the validity of this law, the scaling parameter U_T (the so-called skin friction velocity) must be known. To determine U_T , two ways have been investigated : I) the use of the 'Clauser plot' technique which is based on the existence of Eq. (6) II) direct measurement of the skin friction. First, let us discuss the results obtained from the 'Clauser plot' technique. In Fig. 5, profiles measured at a given location but at different instants during a cycle, are plotted in semi-logarithmic coordinates. At each instant, as it is the principle of the Clauser plot technique, the value of U_T which gives the best fit with Eq. (6) is used to calculate Y^+ and U^+. A good agreement is observed between the experimental velocity profiles and formula (6) over a rather wide range of Y^+ as in steady flow. Let us note, however, that the quality of agreement is sometimes a matter of appreciation: for example, at instant $t/T = 0.67$ in Fig. 5, it is not so obvious that the slope of the experimental profile is exactly 1/0.41. With these restrictions, the results obtained for all the Strouhal numbers investigated seem to indicate that a logarithmic law remains valid in unsteady flow.

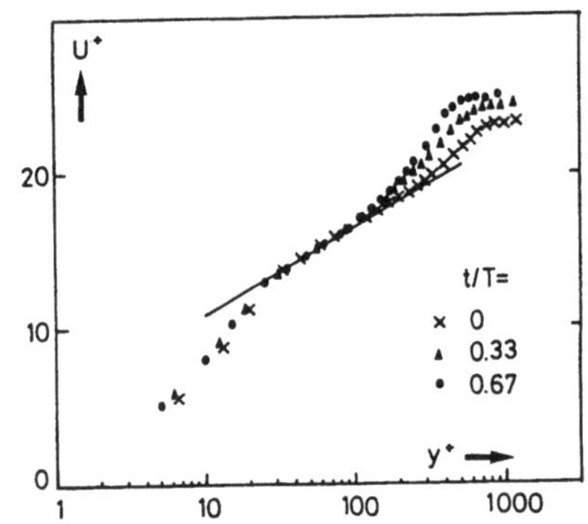

Fig. 5. Velocity profiles at different instants. $S = 5.9$, $F = 38$ Hz; –: Law of the wall (Eq. 6)

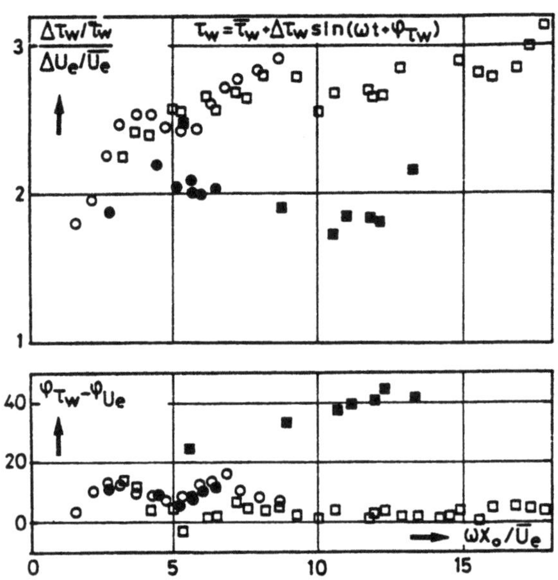

Fig. 6. Harmonic analysis of the wall shear stress; ○ □ : from the law of the wall $F = 38 - 62$ Hz; ● ■: from hot wire at the wall $F = 38$ Hz

As far as it is legitimate to assume that the scaling parameter U_τ is the friction velocity, the Clauser plot technique has been used to estimate the skin friction τ_w and its harmonic analysis which is presented in Fig. 6. The phase shift of τ_w with respect to U_e remains small, less than 20° and decreases as the Strouhal number increases. As regards the relative amplitude of fluctuation of τ_w, the general trend is an increase, to which an oscillation with a period equal to 5.5 is superimposed.

Unfortunately, the trend of the results as the Strouhal number increases is not coherent with those obtained by *Binder* and *Kueny* at higher Strouhal numbers [7]. These authors have shown that, at high frequency, the velocity profiles behave like in a laminar Stokes layer. In particular, the wall velocity phase angle is 45° and not 0°, as it could be thought from the above described results (Fig. 6).

Law of the Wall

To explain this problem, a further analysis is needed. As shown by *Simpson* [2], the existence of a logarithmic region of overlap, between the law of the wall and the law of the wake, implies that the velocity phase shift remains constant in such a region. In a more general manner, let us assume that the near wall flow is governed by a universal law, independent of time:

$$U^+ = f(Y^+) \tag{7}$$

which could be valid not only in the logarithmic region, but also down to the wall.

A small perturbation expansion of Eq. (7) leads to the following equation for the time mean velocity and the complex amplitude:

$$\frac{U_0}{U_{\tau 0}} = \frac{f(YU_{\tau 0})}{\nu} \tag{8}$$

$$U_1 = U_{\tau 1} \left[\frac{f(YU_{\tau 0})}{\nu} + \frac{YU_{\tau 0}}{\nu} \frac{df}{dY^+} \right] \tag{9}$$

Equation (8) indicates that the time-mean velocity profile obeys the steady universal law, which is not surprising. Equation (9) is more interesting: it shows that the velocity phase shift is constant and equal to the phase of the friction velocity, or equivalently to the phase of the wall shear stress.

Experimentally, the velocity phase shift is nearly constant in the region where a logarithmic behaviour exists. Actually, a local maximum of the velocity phase lead occurs around $Y^+ = 50$. In addition, the value of the phase shift in this region is in good agreement with the phase shift of the skin friction determined from the Clauser plot technique: these latter values are indicated by triangles in Fig. 3.

As discussed just above, the velocity phase shift is nearly constant in the logarithmic region but experiment shows that the phase is not constant nearer the wall. This invalidates the assumption of a universal law such as Eq. (7) between $Y^+ = 0$ and $Y^+ = 40$.

At the very most, the existence of a universal logarithmic law is possible. It has been shown that from this hypothesis, the velocity phase angle in the logarithmic region should be equal to the skin friction phase angle. Then to be coherent with experimental results shown in Fig. 3, the evolution of the velocity phase angle near the wall should be imagined as drawn in Fig. 7a.

Sublayer Region

Indeed, a very rapid decrease of the velocity phase angle in the sublayer is quite compatible with the boundary layer equations. Let us assume that the sublayer behaviour is governed by the simplified following equation:

$$-\frac{1}{\rho}\frac{\partial P}{\partial X}+\frac{\partial}{\partial Y}\frac{(\mu\partial U)}{\partial Y}=0\ . \tag{10}$$

By setting

$$-\frac{1}{\rho}\frac{\partial P}{\partial X}=\frac{\partial U_e}{\partial t}$$

and using a small perturbation expansion, the phase of the velocity is given by:

$$\tan(\varphi_U-\varphi_{U_e})=\tan(\varphi_{\tau_w}-\varphi_{U_e})-\frac{\omega\nu}{2U_{e0}^2}\frac{(C_f)^{-3/2}}{2}\frac{|U_{e1}|/U_{e0}}{|\tau_{w1}|/\tau_{w0}}\frac{yU_\tau}{\nu}\frac{1}{\cos(\varphi_{\tau_w}-\varphi_{U_e})} \tag{11}$$

This equation shows that the velocity phase shift is not constant near the wall, but decreases as Y increases. Substitution of typical experimental values indicates that for $f = 38$ Hz, the velocity phase shift decreases by about 6° between $Y^+ = 0$ and $Y^+ = 5$. This decrease is about 13° for $f = 62$ Hz. Therefore, the scheme of Fig. 7a is quite coherent with experimental results of Fig. 3.

However, this model is no longer valid as the frequency increases. Measurements by *Binder* and *Kueny* at high frequency [7] show that the behaviour of the velocity phase angle near the wall is nearly monotonic as indicated in Fig. 7c. In these experiments, the variations of velocity phase angle and of amplitude are mainly confined in the sublayer and are well described by the Stokes solution. In this case, a relevant parameter (*Binder* and *Kueny*) is the ratio of the thickness of the Stokes layer to the thickness of the sublayer or in other words, this parameter is the thickness of the Stokes layer expressed in terms of wall variables:

$$l_s^+=\left(\frac{2\nu}{\omega}\right)^{1/2}U_{\tau 0}/\nu$$

When l_s^+ is small (say of order 5), the external region of the boundary layer is frozen.

Undoubtedly, the near wall behaviour depends on the value of l_s^+. Let us note that this parameter can be related to the Reynolds number and the Strouhal number. If the mean value of the skin friction is expressed by means of a classic law.

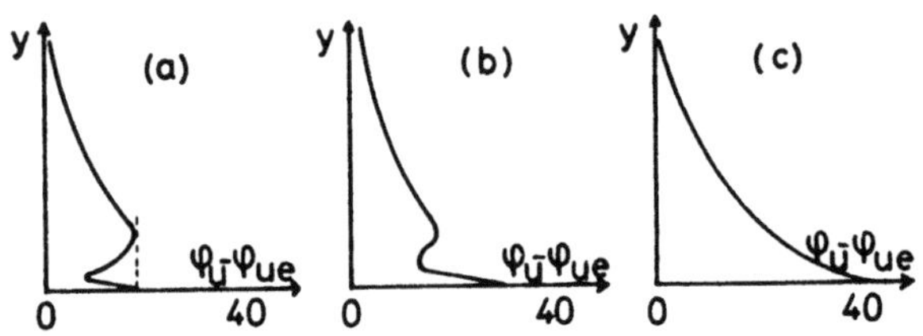

Fig. 7. Profiles of the velocity phase-shift near the wall: possible behaviours

$$\frac{C_f}{2} = \frac{0.0184}{R_x^{1/6}} \quad ,$$

the parameter l_s^+ is

$$l_s^+ = 0.192 \frac{R_x^{5/12}}{(\omega X/U_e)^{1/2}}$$

Table 2. Values of l_s^+ in the present experiment

f [Hz]	38		62	
X_0[m]	0.14	0.77	0.137	0.762
$\omega X_0/U_e$	1.52	8.35	3.2	17.8
R_{x0}	0.2×10^6	1.1×10^6	0.15×10^6	0.87×10^6
l_s^+	25.2	21.9	15.4	13.8

Between the high frequency regime (Fig. 7c) and the low frequency regime (Fig. 7a), an intermediate behaviour of the velocity phase angle can be imagined (Fig. 7b). In this case, a logarithmic law can exist because the phase angle is nearly constant in the region of its maximum, but the scaling parameter is not the friction velocity since the value of the maximum is not equal to the wall value. In other words, this means that as the frequency increases, the boundary layer has an unsteady structure near the wall which does not overlap with the external region by means of the classic logarithmic properties.

Direct Measurements of Skin-Friction

To confirm the various hypotheses made above on the evolution of the skin-friction, and also to try to obtain coherence between our results and those of *Binder* and *Kueny* [7] at higher frequencies, direct measurements of the skin friction have been performed.

For this, hot wire skin friction gauges similar to those developed by other authors have been used [8, 9]. In steady flow, the main difficulty is the sensitivity of the gauges to the temperature. If the temperature is carefully controlled, rather accurate measurements of skin friction are possible. Unfortunately, the dynamic response of these gauges is very poor.

To improve the response, a newly designed gauge has been constructed. Before soldering the wire, a small cavity is made in the substrate between the prongs. Then the wire is mounted at the level of the surrounding surface. This gauge has been tested in an oscillating laminar boundary layer. Away from the wall, the velocity has been measured by hot wire anemometry. In laminar flow, the theory shows that the properties at the wall can be deduced by extrapolation of the results in the boundary layer. It this way, it has been shown that the skin friction measurements are quite coherent with the boundary layer data.

This has been checked for several values of the Strouhal number and at frequencies equal to 17, 38 and 62 Hz. At large Strouhal numbers, the measured phase lead of the wall shear stress with respect to the external velocity is equal to 45°, in good agreement both with the experimental profile of phase and with the theoretical results. As concerns the amplitude of fluctuation of τ_w, the accuracy seems also to be good, although this quantity is more difficult to extrapolate from velocity measurements. Moreover, the correct response of the wall stress gauge has been tested at frequenciesof the order 1300 Hz, corresponding to Tollmien-Schlichting instability waves. Once again, the fluctuations of τ_w have a 45° phase-lead with respect to the fluctuations of the velocity, as predicted by the linear stability theory, in the conditions of the experiment.

Preliminary results on the phase angle and the amplitude of the wall shear stress are given in Fig. 6. In a general manner, it appears that the direct measurements give a lower level of the amplitudes of wall shear stress than those determined from the Clauser plot technique. However, at low Strouhal numbers ($\omega X_0/\overline{U}_e < 5$ or 7), the two evaluations of phase angle and amplitude are in not too bad agreement. At higher Strouhal numbers, it is interesting to observe that the evolutions of the phase angle are very different: from the Clauser plot technique, the phase angle tends to zero whereas the direct measurements indicate an increase in the wall shear stress phase angle, as the Strouhal number increases. At a Strouhal number of 12, the measured phase angle is 42°. Qualitatively, this trend is in agreement with the measurements by *Binder* and *Kueny* [7].

Skin Friction Law

The validity of a quasi-steady skin friction law has been tested by using the same law as in a steady flow [4].

$$\left(\frac{2}{C_f}\right)^{1/2} = \frac{1}{\chi} \ln \frac{U_e \delta_1}{\nu} + D^* \qquad (12)$$

with $D^* = 2G - 4.25 G^{1/2} + 2.12$ and $G = (H-1)/[H\sqrt{C_f/2}\,]$

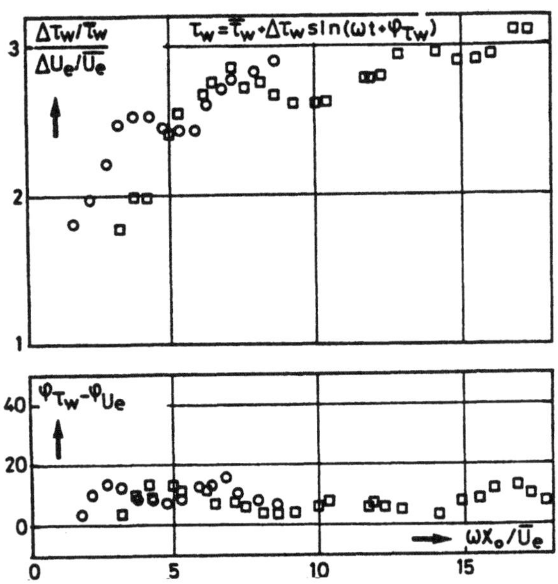

Fig. 8. Harmonic analysis of the wall shear stress from the law of the skin friction (Eq. 12)
○: f = 38 Hz □: f = 62 Hz

It is assumed that this law is valid at each instant.

By carrying out a small perturbation expansion of Eq. (12), the phase and the amplitude of τ_w can be expressed as function of the harmonic analysis of the displacement and momentum thicknesses and of the external velocity [5]. The results obtained from this calculation compare well with those deduced from the Clauser plot technique (Fig. 8). They also compare well with the velocity phase shift in the logarithmic region. When compared with the direct measurements of skin friction, substantial deviations are observed as the Strouhal number increases ($\omega X_0/\bar{U}_e > 5$ or 7).

It is concluded that a quasi-steady skin friction law can be used with confidence up to Strouhal number of about 5 or 7, which in fact covers a very interesting range of practical situations.

Turbulence Quantities

Figure 9 shows the cycles of the evolution in time of the u-component of the kinetic energy of turbulence, divided by its mean value. The distance to the wall corresponds to a constant value of y/δ equal to 0.8, and the distance between two successive stations is equal to 5 cm. The stars indicate the instants when $\langle u'^2 \rangle$ is maximum. The fact that they are along the dotted straight lines, the slope of which corresponds to 0.85 U_e, leads to a result similar to that obtained in steady flow from space-time correlation measurements.

Another interpretation of the results of Fig. 9 is that the free edge of the boundary layer oscillates at the frequency of the external flow, and can be represented by a travelling wave, the velocity of which being equal to 0.85 U_e.

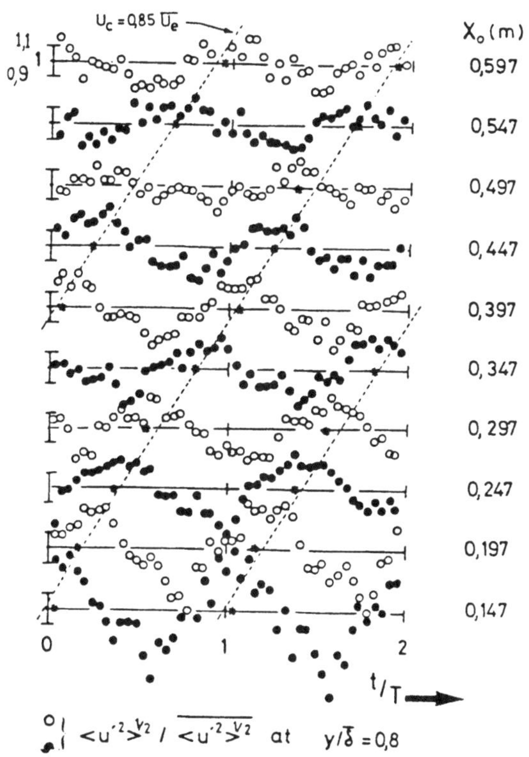

Fig. 9. Evolution in time of the turbulence intensity at $y/\delta = 0.8$ for different stations along x $F = 62$ Hz; *: instants of maximum of turbulence

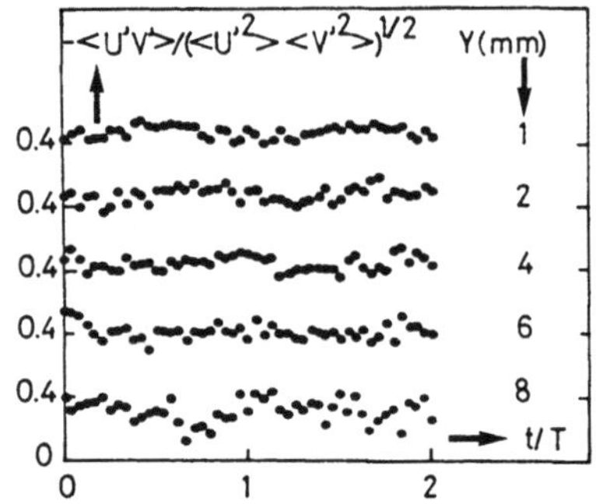

Fig. 10. Evolution in time of the correlation coefficient at different distances from the wall. $S = 5$

An important turbulence parameter is the correlation coefficient. In the closure models for calculation methods, it is more or less explicitly assumed constant. The evolution in time of this coefficient is plotted in Fig. 10 for two cycles at increasing distances from the wall. Except in the external region where $\langle u' v' \rangle$ vanishes, the correlation coefficient remains constant and equal to the value commonly found in steady turbulent boundary layers, which is 0.45. This result tends to prove that, up to Strouhal number of 5, the development of the turbulence is not strongly affected by the unsteadiness of the flow.

Conclusions

The experimental study of the response of a turbulent boundary layer over a flat plate to an oscillating external velocity field has confirmed that the mean flow field is not affected by the unsteady effects up to Strouhal number equal to 18. A possible interpretation of the measurements is that the boundary layer responds to the perturbation induced by the pulsation of the external flow as a mechanical system with a small damping.

Direct measurements of the wall shear stress have been done using a modified hot wire gauge. The comparison of the obtained results with the values of the wall shear stress deduced from the law of the wall, shows that at reduced frequencies larger than 5 to 7, the law of the wall is no longer scaled on the skin friction velocity. Moreover, the parameter l_s^+ has to be taken into account to describe the wall region. It has been shown that the evolution of the phase of the velocity near the wall ($y^+ < 10$) can change very rapidly in a complicated manner.

References

1. Houdeville, R., Cousteix, J.: "Couches Limites Turbulentes en Ecoulement Pulsé avec Gradient de Pression Moyen Défavorable", *La Recherche Aérospatiale,* 1979-1, pp. 33–48 NASA Tr. TM 75799-N 8017400
2. Simpson, R.L.: "Features of Unsteady Turbulent Boundary Layers as Revealed From Experiments", *Proc. Conf. on Unsteady Aerodynamics,* Ottawa, Sept. 1977, AGARD-CP-227
3. Karlsson, S.K.F.: An unsteady turbulent boundary layer. J. Fluid Mech. **5**, 622–636 (1959)
4. Cousteix, J., Houdeville, R.: "Turbulent Boundary Layer Calculations in Unsteady Flow", in *Numerical Methods in Applied Fluid Dynamics,* (Academic, New York 1980) pp. 615–644
5. Cousteix, J., Houdeville, R., Javelle, J.: "Response of a Turbulent Boundary Layer to a Pulsation of the External Flow With and Without Adverse Pressure Gradient", *Proceedings of the IUTAM Symposium on Unsteady Turbulent Shear Flows,* Toulouse, May 1981, (Springer, Berlin, Heidelberg, New York 1981)

6. Cousteix, J., Desopper, A., Houdeville, R.: "Structure and Development of a Turbulent Boundary Layer in an Oscillatory External Flow", in *Turbulent Shear Flows I*, ed. by F. Durst, B.E. Launder, F.W. Schmidt, J.H. Whitelaw (Springer, Berlin, Heidelberg, New York 1979) pp. 154–171
7. Binder, G., Kueny, J.L.: "Measurements of the Periodic Velocity Oscillations near the Wall in Unsteady Turbulent Channel Flow", *IUTAM Symposium on Unsteady Turbulent Shear Flows*, Toulouse May 1981 (Springer, Berlin, Heidelberg, New York 1981) pp. 100–108
8. Rubesin, M.W., Okuno, A.F., Mateer, C.C., Brosh, A.: "A Hot-Wire Surface Gauge for Skin Friction and Separation Detection Measurements", NASA TM X62465, 1975
9. Cousteix, J., Juillen, J.C.: Jauges à fil chaud pour la mesure du frottement pariétal (réalisation, étalonnage, application). Rech. Aérosp. **3**, 207–216 (1982)

A Theoretical Model of the Coherent Structure of the Turbulent Boundary Layer in Zero Pressure Gradient

Z. Zhang* and G.M. Lilley
Aeronautics and Astronautics Department, University of Southampton,
Southampton, S09 5NH, England

Abstract

A self-generating deterministic coherent structure is shown to arise in calculations on a model of a turbulent boundary layer. The solution commences with the evaluation of the linear damped periodic eigen mode for the vertical pertubation which leads to an initial growth in the vorticity perturbations. These lead to changes in the Reynolds stresses and hence to a distortion of the time dependent mean velocity distribution. Our analysis in this respect is somewhat similar to the Benney-Lin investigation of the non-linear interaction of two symmetric oblique waves in unstable laminar boundary layers. The distorted mean velocity profile leads to a strong growth in the Reynolds shear stress, provided the initial disturbance exceeds a certain amplitude of the order of 10% of the freestream velocity. The numerical results display a spanwise periodic structure with a spacing of the order of 100 wall units and have the form of side by side ejections and sweeps. The maximum distortion in the mean velocity occurs near $y^+ = 20$ over a wide range of Reynolds numbers. It is concluded that the distorting mean velocity profile in the inner region of the boundary layer is grossly unstable and will lead to a catastrophic breakdown of the flow into smaller scales,which therefore represents the high production of turbulent energy and its dissipation. This process is self-generating and occurs randomly throughout the turbulent boundary layer.

Introduction

During the past 25 years experimental studies of the incompressible turbulent boundary layer have shown that over a wide range of Reynolds numbers the seemingly random structure of convecting, distorting eddies is nevertheless controlled by repeating, coherent large structures which occur naturally at a random.

Many theoretical attempts have been made, based on quasi-laminar linear perturbation theory, to uncover the production and subsequent development of these coherent structures. These have met with only limited success. Some non-linear approaches have been attempted such as in *Landahl* (1973), *Walker* (1978) and *Hanratty* (1979). A detailed commentary on these methods is described elsewhere, *Zhang* (1981).

In this work we describe a new approach which is based on the full equations of fluid flow in which the small scale turbulence is modelled using an eddy viscosity. Although it is shown that the first order mode is damped, nevertheless there is an initial linear growth in part of the flow structure leading to a strong non-linear interaction and consequent

* Present address: Department of Engineering Mechanics, Tsinghua University, Peking, China

distortion to the mean flow. The distorted mean profile suffers a secondary instability and subsequent breakdown of the flow within a streamwise length scale of the order 3 to 5 boundary layer thickness and which is a weak function of Reynolds number. Our work also describes the formation of streamwise vortices which suffer a secondary instability and breakdown referred to above.

The model is incomplete in that the breakdown of the flow arising from the secondary instability is only described qualitatively. In addition, no attempt is made to model the mean growth of the boundary layer and the necessary entrainment of irrotational fluid from the freestream. Since our model describes the self-generation of unstable and repeatable flow structures as a result of instabilities that arise within the inner region of the boundary layer, the neglect of such features as boundary layer growth and entrainment are, we believe, not of prime importance in the determination of turbulent boundary layer flow structure. However, these features and those relating to the break-up of the secondary instability are subjects for further work.

The preliminary work leading to this model was completed, when one of us (GML) was a Visiting Professor in the Department of Aeronautics and Astronautics at Stanford University (1977–1978).

Description of Model

In order to model the self-generated cell-like structure, which we refer to as a coherent structure, in a turbulent boundary layer, as described in Fig.1, we have to make several assumptions. These are as follows: (a) the large scale features of the coherent structures are treated as a conditional ensembled average of perturbations to the turbulent flow, (b) three-dimensional linear perturbations are introduced which are composed of a series of eigen functions of a modified Orr-Sommerfeld equation, (c) distortions to the mean flow arise from non-linear interactions between the linear perturbations, (d) the distorted mean profile, which is time dependent, suffers a secondary instability resulting in a flow breakdown, (e) non-linear distortion and energy considerations will lead to a finite maximum amplitude for the secondary instability. All the characteristics of our model, except (e), have been treated in the following sections. An important aspect of our model is the use of an eddy viscosity to model the small scale random disturbances in the flow as well as the non-linear interactions. Several forms of eddy viscosity have been used in predictions of unsteady turbulent flow (*Reynolds*, 1972; *Smith* and *Cebeci*, 1974). We have used the Smith and Cebeci model for both the steady and unsteady mean flow. Although

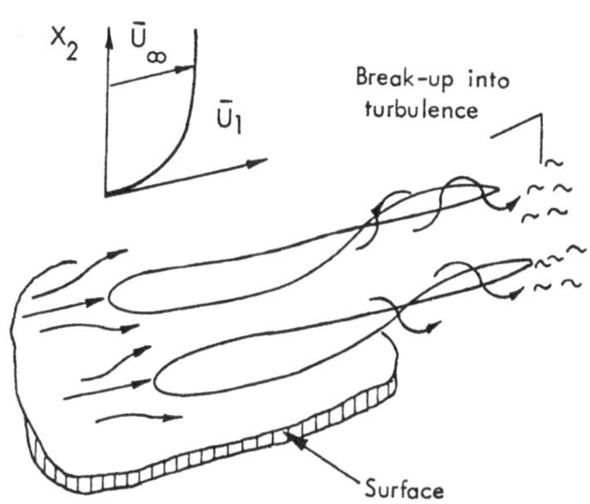

Fig. 1. Development of streamwise vortices and the lift-up of wall streaks and their subsequent break-up. (After *Kline* and *Runstadler*, 1959)

other models may have certain advantages, including that of a time dependent eddy viscosity, we believe that the model we have used is plausible and reflects most of the observed features of turbulent flows.

Finally, we note that the non-linear interaction arising from the linear growing modes and the subsequent development of the secondary perturbations and mean flow distortion have a related form to the Benney-Lin analysis (*Benney* and *Lin*, 1960; *Benney*, 1964) of non-linear interactions of a pair of symmetrical oblique linear waves in unstable laminar shear flows.

Analysis

Mean Flow

The turbulent flow in the boundary layer is assumed to be incompressible (density equal to unity) and Newtonian so that at all (***x***, *t*) the basic equations for the unsteady flow are:

$$\left(\frac{\partial}{\partial t} - \nu\nabla^2\right) v_i = \frac{\partial}{\partial x_j}(p\,\delta_{ij} + v_i v_j) = 0 \quad ; \quad \frac{\partial v_j}{\partial x_j} = 0 \; . \tag{1}$$

The ensemble average taken over an infinite number of realisations of the flow reduces these equations to:

$$\frac{\partial}{\partial x_j}(\bar{P}\,\delta_{ij} + \bar{U}_i\,\bar{U}_j + \overline{u_i' u_j'}) = \nu\nabla^2\,\bar{U}_i \quad ; \quad \frac{\partial \bar{U}_i}{\partial x_j} = 0 \tag{2}$$

where $\langle v_i \rangle = \bar{U}_i$; $\langle p \rangle = \bar{P}$ and with $v_i = \bar{U}_i + u'_i$, $\langle v_i v_j \rangle = \bar{U}_i \bar{U}_j + \overline{u_i' u_j'}$.
We assume the mean flow velocity $\bar{U}_i$ is govered by Prandtl's boundary layer approximations for a two-dimensional mean flow so that $\bar{U} \equiv (\bar{U}_1, \bar{U}_2, 0)$ with $\bar{U}_1 \gg \bar{U}_2$ and $\partial/\partial x_2 \gg \partial/\partial x_1$. The flow outside the boundary layer is constant and equal to $\bar{U}_\infty$.

For the purpose of evaluating U_i we introduce an empirically derived eddy viscosity ν_T for which *Cebeci* and *Smith* (1974) suggested the following formula:

$$\frac{\nu_T}{\nu} = K^2 y^{+2} \frac{du^+}{dy^+}\left[1 - \exp\left(-\frac{y^+}{26}\right)\right]^2 \qquad 0 \leqslant y \leqslant y_c \tag{3}$$

$$= K_1\, R_\theta\, H \qquad\qquad y_c \leqslant y \leqslant \delta$$

where K is von Karman's constant,

$$K_1 = 0.016\,[1 - 0.355\,\exp(-\,0.243\,z^{1/2} - 0.298\,z)]^{-1}$$

with $z = R_\theta/425 - 1$

and R_θ, H are the Reynolds number $[\bar{U}_\infty \theta/\nu]$ based on the momentum thickness, θ, and the form factor respectively. In Eq. (3) $u^+ = \bar{U}_1/u_T$ and $y^+ = x_2 u_T/\nu$ where $u_T = \sqrt{\tau_w/\rho}$ is the shear velocity. The same formula for ν_T will be used below for the determination of the Reynolds stress when the mean flow suffers a distortion. In the present case:

$$-\overline{u_i'u_j'} = \nu_T \left(\frac{\partial \bar{U}_i}{\partial x_j} + \frac{\partial \bar{U}_j}{\partial x_i}\right) \quad \text{for } i \neq j \quad . \tag{4}$$

The Unsteady Flow Model

We next assume that the turbulent boundary layer develops at each position a repeatable coherent structure which is self-generating. These structures occur randomly in space and time. The coherent structure is of finite amplitude and during its development a measurable distortion of the mean flow occurs. At any given station within the boundary layer we therefore take a conditionally sampled ensemble average of the flow field such that its trigger is associated with the commencement of a new coherent or cell-like structure. At the commencement of the coherent structure the flow is such that its ensemble average velocity is $\bar{U}_i$ and its Reynolds stress is $\overline{u_i'u_j'}$. During the development of the coherent structure the conditionally sampled ensemble average (written $\langle\!\langle\ \rangle\!\rangle$) shows a slow variation in velocity in space and time given by $\langle\!\langle v_i \rangle\!\rangle = U_i(\boldsymbol{x},t)$ and a fast fluctuation $u_i'(\boldsymbol{x},t)$ where $v_i = U_i(\boldsymbol{x},t) + u_i(\boldsymbol{x},t)$ and $\langle\!\langle v_i v_j \rangle\!\rangle$ $U_iU_j + \widetilde{u_i'u_j'}(\boldsymbol{x},t)$. We assume that Reynolds stress $\widetilde{u_i'u_j'}$ can be modelled by the use of the same eddy viscosity ν_T, as was used for the mean flow analysis leading to the evaluation of $\bar{U}_i$. Thus we write

$$-\widetilde{u_i'u_j'} = \nu_T S_{ij} \quad \text{for } i \neq j \tag{5}$$

where

$$S_{ij} = \frac{\partial U_i}{\partial x_j} + \frac{\partial U_j}{\partial x_i}$$

is the conditionally sampled ensemble averaged rate of strain tensor. This assumption that ν_T is governed by the mean flow is not critical to the remainder of the analysis and although other approximations may be used, that above is the simplest and most plausible. Hence in terms of these conditionally sampled ensemble averages, Eq. (1) becomes

$$\frac{\partial U_i}{\partial t} + \frac{\partial}{\partial x_j}(P\delta_{ij} + U_iU_j) = \frac{\partial}{\partial x_j}(\nu + \nu_T)S_{ij} \qquad \text{with} \tag{6}$$

$$\frac{U_iU_j}{\partial x_j} = 0 \quad . \tag{7}$$

The mean flow distortion $\Delta U_i = U_i - \bar{U}_i$ is governed both by large scale disturbances of order of the cell size and the more rapid fluctuations which generate the Reynolds stress $\widetilde{u_i'u_j'}$ which itself is governed by the large scale distortion.

So far our model is completely general but being non-linear is incapable of exact solution. We therefore simplify our model by the introduction of a perturbation scheme taken up to second order. A higher order analysis may have certain advantages over that adopted here. Nevertheless, we have no evidence that a third order analysis will lead to improved results. Thus, the conditionally sampled ensemble averaged velocity and pressure are written:

$$U_i = \bar{U}_i(\boldsymbol{x}) + \epsilon\, U_i^{(1)}(\boldsymbol{x},t) + \epsilon^2\, U_i^{(2)}(\boldsymbol{x},t) \tag{8}$$

$$P = \bar{P}(\boldsymbol{x}) + \frac{1}{3}\,\overline{u_i' u_i'} + \epsilon P^{(1)}(\boldsymbol{x},t) + \epsilon^2 P^{(2)}(\boldsymbol{x},t) \tag{9}$$

where ϵ is a small parameter. In our flow model $\bar{P} = 0$ and $\bar{U}_i = \bar{U}_1(x_2)\delta_{i1}$. When substituted into Eq. (6) we find the zeroth, first and second order approximations give respectively the mean flow, the linear perturbation and the secondary perturbation to the mean flow.

Linear Perturbation

The equation for $U_i^{(1)}$ becomes

$$\left(\frac{\partial}{\partial t} + \bar{U}_1(x_2)\frac{\partial}{\partial x_1}\right) U_i^{(1)} + \delta_{i1} U_2^{(1)} \frac{d\bar{U}_1}{dx_2} + \frac{\partial P^{(1)}}{\partial x_i} = \frac{\partial}{\partial x_j}(\nu + \nu_T) S_{ij}^{(1)} \tag{10}$$

with

$$\frac{\partial U_j^{(1)}}{\partial x_j} = 0 \quad . \tag{11}$$

We next find the Fourier Transformation of these equations with respect to the spatial coordinates x_1, x_3 where

$$U_i^{(1)} = \int\!\!\int_{-\infty}^{\infty} \hat{U}_i(\alpha,\beta;x_2,t)\, \exp[\mathrm{i}(\alpha x_1 + \beta x_3)]\, d\alpha\, d\beta \tag{12}$$

and a similar relation for $P^{(1)}$.

The equations for $\hat{U}_i$ are given by:

$$\left(\frac{\partial}{\partial t} + \mathrm{i}\alpha\bar{U}_1\right)\begin{Bmatrix}\hat{U}_1\\ \hat{U}_2\\ \hat{U}_3\end{Bmatrix} + \begin{Bmatrix}\bar{U}_1'\hat{U}_2\\ 0\\ 0\end{Bmatrix} + \begin{Bmatrix}\mathrm{i}\alpha\hat{P}\\ \hat{P}'\\ \mathrm{i}\beta\hat{P}\end{Bmatrix}$$

$$= (\nu + \nu_T)\left(\frac{\partial^2}{\partial x_2^2} - k^2\right)\begin{Bmatrix}\hat{U}_1\\ \hat{U}_2\\ \hat{U}_3\end{Bmatrix} + \nu_T'\begin{Bmatrix}\hat{U}_1' + \mathrm{i}\alpha\hat{U}_2\\ 2\hat{U}_2'\\ \hat{U}_3' + \mathrm{i}\beta\hat{U}_2\end{Bmatrix} \tag{13}$$

with

$$\mathrm{i}(\alpha\hat{U}_1 + \beta\hat{U}_3) + \frac{\partial\hat{U}_2}{\partial x_2} = 0 \tag{14}$$

where $k^2 = \alpha^2 + \beta^2$ and primes denote differentiation with respect to x_2.

If we introduce the vertical component of the vorticity $\eta^{(1)}$ where

$$\eta^{(1)} = \frac{\partial U_1^{(1)}}{\partial x_3} - \frac{\partial U_3^{(1)}}{\partial x_1} \qquad \text{and} \tag{15}$$

$$\eta^{(1)} = \int\!\!\int_{-\infty}^{\infty} \hat{\eta}\,(\alpha,\beta;x_2,t)\exp[\mathrm{i}(\alpha x_1 + \beta x_3)]\,d\alpha\,d\beta \tag{16}$$

it can be shown that since we are dealing with oblique waves, our set of equations for $\hat{U}_1$, $\hat{U}_2$, $\hat{U}_3$ and $\hat{P}$ reduce to two equations in $\hat{U}_2$ and $\hat{\eta}$ only. These equations are

$$\left(\frac{\partial}{\partial t} + \mathrm{i}\alpha\bar{U}_1\right)\left(\frac{\partial^2}{\partial x_2^2} - k^2\right)\hat{U}_2 - \mathrm{i}\alpha\bar{U}_1''\hat{U}_2 = (\nu + \nu_T)\left(\frac{\partial^2}{\partial x_2^2} - k^2\right)^2\hat{U}_2$$
$$+ 2\nu_T'\frac{\partial}{\partial x_2}\left(\frac{\partial^2}{\partial x_2^2} - k^2\right)\hat{U}_2 + \nu_T''\left(\frac{\partial^2}{\partial x_2^2} + k^2\right)\hat{U}_2 \tag{17}$$

and

$$\left(\frac{\partial}{\partial t} + \mathrm{i}\alpha\bar{U}_1\right)\hat{\eta} - (\nu + \nu_T)\left(\frac{\partial^2}{\partial x_2^2} - k^2\right)\hat{\eta} + \nu_T'\frac{\partial\hat{\eta}}{\partial x_2} = -\mathrm{i}\beta\bar{U}_1'\hat{U}_2 \tag{18}$$

In addition

$$\hat{U}_1 = \left(\mathrm{i}\alpha\frac{\partial}{\partial x_2}\hat{U}_2 - \mathrm{i}\beta\hat{\eta}\right)\Big/k^2 \tag{19}$$

$$\hat{U}_3 = \left(\mathrm{i}\beta\frac{\partial\hat{U}_2}{\partial x_2} + \mathrm{i}\alpha\hat{\eta}\right)\Big/k^2 \; . \tag{20}$$

Our method of solving these equations is as follows. We assume that at time $t = 0$ an initial disturbance $[U_i^{(1)}(x_2)]$ is prescribed. The stability and ultimate growth of that disturbance is then investigated. Now the general properties of Eq. (13) with constant viscosity and with appropriate boundary conditions have been investigated by several authors. It has been shown by *Murdock* and *Stewartson* (1977) and *Mack* (1976) that the linear operator of these equations has a finite discrete set of eigen modes as well as a continuous spectrum and moreover the continuous spectrum has higher damping coefficients. We have found similar properties for our equations, with the terms in the eddy viscosity included by direct numerical calculation. Hence we deduce that for any small finite initial disturbance, it is only the least damping mode that can dominate the later development of the disturbance. Thus we choose

$$\hat{U}_2 = \phi(\alpha,\beta;x_2)\exp(-\mathrm{i}\omega t) \tag{21}$$

where $\phi(\alpha,\beta;x_2)$ is the least damped eigen function of the modified Orr-Sommerfeld equation derived from Eq. (17) above. The eigenvalue is ω and we write

$$c = \omega/\alpha \tag{22}$$

for real α. The real part of ω is the frequency and $c_i < 0$ denotes a damped mode. The least damped mode is denoted by $c_i^*(\alpha,\beta;R)$ where $R = \bar{U}_\infty\delta/\nu$.

Unlike the Eq. (17) above for $\hat{U}_2$ the vorticity Eq. (18) is inhomogeneous with the forcing term $-\mathrm{i}\beta\bar{U}_1'\;\hat{U}_2(\alpha,\beta;x_2 t)$. It has also been shown by *Murdock* et al. that the spectrum of the linear operator of Eq. (18) has a similar spectrum to that of the Orr-Sommerfeld equation and thus there are a finite number of discrete eigen modes plus a continuous spectrum. Furthermore, we have found numerically that the least damping mode of the vorticity equation usually has a larger damping coefficient than that of the Orr-Sommerfeld equation. Hence, only the particular solution from the forcing term can play an important role in the later development of the disturbance which we will show involves a non-linear interaction. The particular solution of Eq. (18) is in the form:

$$\hat{\eta} = H(x_2,t)\exp(-\mathrm{i}\omega t) \tag{23}$$

where the amplitude function H satisfies the equation

$$\frac{\partial H}{\partial t} + \mathrm{i}\alpha(\bar{U}_1 - c)H - (\nu + \nu_T)\left(\frac{\partial^2}{\partial x_2^2} - k^2\right)H - \nu_T' \frac{\partial H}{\partial x_2} = -\mathrm{i}\beta\bar{U}_1'\phi(x_2) \tag{24}$$

with the initial and boundary conditions

$$H(0,t) = H(\infty,t) = 0 \quad ; \quad H(x_2,0) = 0 \quad .$$

The inviscid approximation to Eq. (24) shows that at the critical layer where $\bar{U}_1 = c_r$ we find

$$H(x_2,t) \sim \frac{\mathrm{i}\beta\bar{U}_1'\phi}{\alpha c_i^*}\left[\exp(-\alpha c_i^* t) - 1\right] \tag{25}$$

showing that for small time $\hat{\eta} \sim t$.

This linear growth in the amplitude of $\hat{\eta}$ has also been found in a related problem investigated by *Gustavsson* (1980). We find moreover from the numerical solution of Eq. (24), when viscous terms are included, a similar growth in the vertical component of vorticity during the initial period and that such linear growth makes an important contribution to the developing non-linear interaction. Our solution for the three components of velocity can therefore be written:

$$\begin{aligned}\hat{U}_1 &= (\mathrm{i}\alpha\phi - \mathrm{i}\beta H)\exp(-\mathrm{i}\omega t)\\ \hat{U}_2 &= \phi\exp(-\mathrm{i}\omega t)\\ \hat{U}_3 &= (\mathrm{i}\beta\phi + \mathrm{i}\alpha H)\exp(-\mathrm{i}\omega t) \quad .\end{aligned} \tag{26}$$

It is easy to show from the Orr-Sommerfeld equation that

$$\phi(x_2,t;\alpha,-\beta) = \phi(x_2,t;\alpha,\beta)$$

and from Eq. (24)

$$H(x_2,t;\alpha,-\beta) = -H(x_2,t;\alpha,\beta) \quad .$$

Thus

$$\hat{U}_1(\alpha,-\beta) = \hat{U}_1(\alpha,\beta) \qquad \hat{U}_2(\alpha,-\beta) = \hat{U}_2(\alpha,\beta) \qquad \hat{U}_3(\alpha,-\beta) = -\hat{U}_3(\alpha,\beta)$$

showing that $\hat{U}_1$, $\hat{U}_2$ are even functions of β whereas $\hat{U}_3$ is an odd function of β. These properties of $\hat{U}_1$, $\hat{U}_2$, $\hat{U}_3$ will be used in our consideration of the non-linear interaction which is treated in the next section.

Second Order Perturbation and Mean Flow Distortion

The equation for the secondary perturbations is derived from Eq. (6) above, and is

$$\left(\frac{\partial}{\partial t} + \bar{U}_1 \frac{\partial}{\partial x_1}\right) U_i^{(2)} + \delta_{i1} \bar{U}_1' U_2^{(2)} + \frac{\partial P^{(2)}}{\partial x_i} = -\frac{\partial}{\partial x_j} U_i^{(1)} U_j^{(1)} + \frac{\partial}{\partial x_j} (\nu + \nu_T) S_{ij}^{(2)} \tag{27}$$

with

$$\frac{\partial U_j^{(2)}}{\partial x_j} = 0 \tag{28}$$

where $U_i^{(1)}$ has been derived from Eqs. (12, 26) for the least damped eigen mode. When we make use of the symmetrical properties found for $U_1^{(1)}$ and $U_2^{(1)}$ and the antisymmetrical properties for $U_3^{(1)}$, with respect to β, we find the following average properties for the Reynolds stress $U_i^{(1)} U_j^{(1)}$ produced by symmetrical oblique waves given spanwise wave number $\pm\beta$ and streamwise wave number α

$$\begin{aligned} \sigma_{ij} &= U_i^{(1)} U_j^{(1)} = A(\alpha,\beta)(\hat{U}_i \hat{U}_j^* + \hat{U}_i^* \hat{U}_j) \cos^2 \beta x_3; \quad i,j = 1 \text{ or } 2 \\ \sigma_{33} &= U_3^{(1)} U_3^{(1)} = A(\alpha,\beta) 2\hat{U}_3 \hat{U}_3^* \sin^2 \beta x_3 \\ \sigma_{k3} &= U_k^{(1)} U_3^{(1)} = A(\alpha,\beta)(\hat{U}_k^* \hat{U}_3^*) \sin\beta x_3 \cos\beta x_3; \quad k = 1 \text{ or } 2 \end{aligned} \tag{29}$$

where $A(\alpha,\beta)$ is related to the amplitude of the initial disturbance. These values of the first order Reynolds stresses are all independent of x_1. With these Reynolds stress properties to solution of Eq. (27) has the form

$$\begin{aligned} U_1^{(2)} &= A(\alpha,\beta)(u_{10} + u_{11} \cos 2\beta x_3) \\ U_2^{(2)} &= A(\alpha,\beta)(u_{21} \cos 2\beta x_3) \\ U_3^{(2)} &= A(\alpha,\beta)(u_{31} \sin 2\beta x_3) \\ P^{(2)} &= A(\alpha,\beta)(p_0 + p_1 \cos 2\beta x_3) \end{aligned}$$

When we substitute these values into Eq. (27) above we find writing $\tilde{\sigma}_{ij} = \hat{U}_i \hat{U}_j^* + \hat{U}_i^* \hat{U}_j$ that

$$\frac{\partial}{\partial t}\begin{Bmatrix} u_{10} \\ u_{11} \\ u_{21} \\ u_{31} \end{Bmatrix} = \begin{Bmatrix} 0 \\ -u_{21}\bar{U}_1' \\ -p' \\ 2\beta p_1 \end{Bmatrix} + (\nu + \nu_T)\begin{Bmatrix} u_{10}'' \\ u_1'' - 4\beta^2 u_1 \\ 2u_{21}'' - 4\beta^2 u_{21} + 2\beta u_{31}' \\ u_{31}'' - 8\beta^2 u_{31} - 2\beta u_{21}' \end{Bmatrix}$$

$$+ \nu_T' \begin{Bmatrix} u_{10}' \\ u_{11}' \\ 2u_{21}' \\ u_{31}' - 2\beta u_{21} \end{Bmatrix} - \begin{Bmatrix} \tilde{\sigma}_{12}'/2 \\ \tilde{\sigma}_{12}'/2 + \beta\tilde{\sigma}_{13} \\ \tilde{\sigma}_{22}'/2 + \beta\tilde{\sigma}_{23} \\ \tilde{\sigma}_{32}'/2 + \beta\tilde{\sigma}_{33} \end{Bmatrix} \quad \text{with} \tag{30}$$

$$p_0 \doteq \tilde{\sigma}_{22}/2 \quad \text{and} \tag{31}$$

$$u'_{21} + 2\beta u_{31} = 0 \tag{32}$$

where primes denote differentiation with respect to x_2. The initial and boundary conditions are

$$U_i^{(2)}(x_2,0) = 0 \; ; \; U_i^{(2)}(0,t) = U_i^{(2)}(\infty,t) = 0 \quad .$$

Of particular interest is the equation for the growth of the streamwise component of vorticity $\xi^{(2)} = \partial U_3^{(2)}/\partial x_2 - \partial U_2^{(2)}/\partial x_3$ which can be found from Eq. (30) above, and the time dependent distortion of the streamwise component of velocity given by $\epsilon^2 U_1^{(2)}$.

Secondary Instability Associated with the Mean Flow Distortion

The stability of the 'slowly' varying time dependent conditionally sampled ensemble averaged mean velocity distribution $\overline{U}_1 \delta_{i1} + \epsilon^2 U_i^{(2)}(x_2,t)$ can be determined for different values of the initial disturbance amplitude. The equation we use is the modified Orr-Sommerfeld Eq. (17), including the terms in the eddy viscosity ν_T, as used above in the determination of the eigen modes for the linear parturbation. The only difference is that $\overline{U}_1$ and $\overline{U}_1''$ are to be replaced by $\overline{U}_1 + \epsilon^2 U_1^{(2)}$ and its second derivative with respect to x_2.

Numerical Method

All the numerical calculations were performed on the ICL 2970 at Southampton University.

We use an orthogonalized Runge-Kutta integration scheme to evaluate the eigen values and eigen functions of the modified Orr-Sommerfeld equation for the least damped mode of the mean flow and for the most unstable mode of the distorted flow velocity profile.

An explicit finite difference scheme was used to solve the parabolic partial differential equation for the vertical vorticity component. The same integration scheme was used to solve for the velocity distribution.

Results

Stability Characteristics of the Mean Velocity Profile

It was found that the mean velocity profile (a typical profile is shown in Fig. 2 for $R = 40{,}000$) was stable, in agreement with the results of several authors, over a wide range of Reynolds number, $R = \overline{U}_\infty \delta/\nu$, from 6000 to 400,000. However, our results are, we believe, new in that the terms involving a variable eddy viscosity are included in the modified Orr-Sommerfeld equation. We found that the least damping modes have phase (wave) speeds ranging from 0.66 to 0.75 $\overline{U}_\infty$ for wave numbers $\alpha\delta^*$ up to 2 and $\beta\delta^*$ up to 3.5. These results, as shown in Fig. 3, are in fair agreement with the corresponding wave speeds derived from the experiments of *Willmarth* and *Wooldridge* (1962) and others.

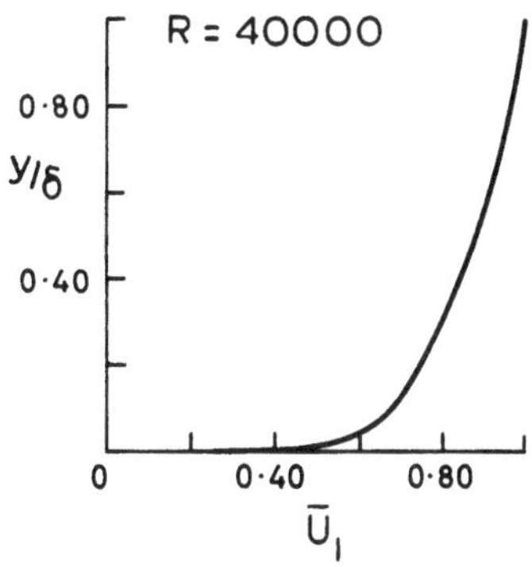

Fig. 2

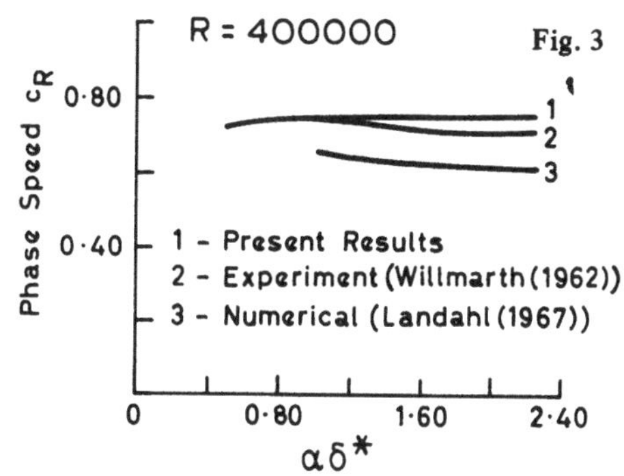

Fig. 3

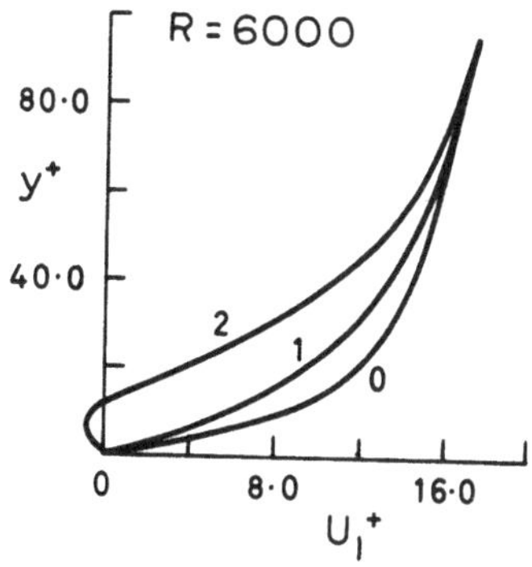

Fig. 4

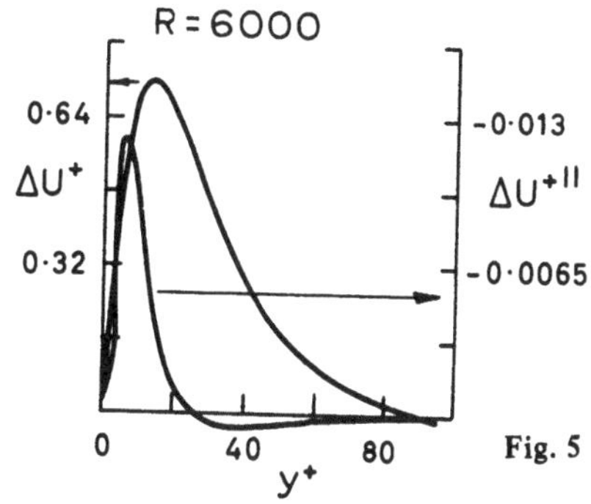

Fig. 5

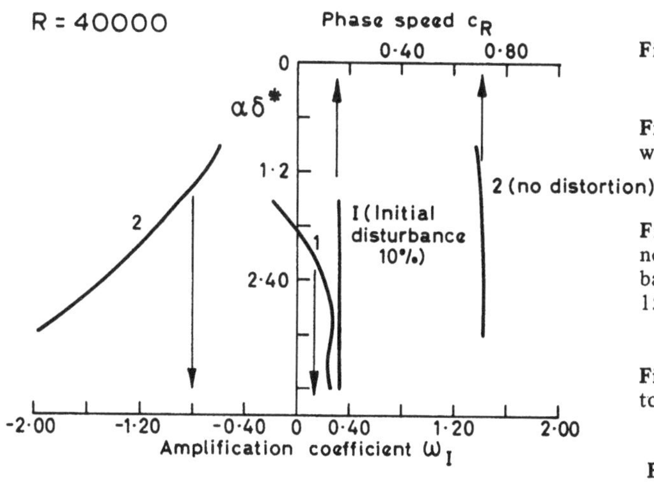

Fig. 2. Mean velocity profile

Fig. 3. Variation of phase with wave number

Fig. 4. Mean flow distortion (0) no distortion; (1) initial disturbance 10%; (2) initial disturbance 15%

Fig. 5. Maximum mean flow distortion (initial disturbance 5%)

Fig. 6. Secondary instability

Mean Flow Distortion

The distorted mean velocity profile during the initial development of a coherent structure is shown in Fig. 4 for two initial disturbance amplitudes whilst in Fig. 5 the distortion and its second derivative are shown for an initial disturbance of 5% of the free-stream velocity. It has been shown that these distortions are qualitatively in agreement with the experimental results of *Blackwelder* et al. (1972) and others where results of conditionally sampled flow velocity profiles have been obtained throughout the complete development and decay cycle of coherent structures. It is shown that the distortions are quite marked for initial disturbance amplitudes as low as 10% of $\bar{U}_\infty$. When the initial disturbance amplitude exceeds 10% of $\bar{U}_\infty$ the calculations show a tendency to flow reversal in the region close to the wall within the complete cycle of the coherent structure formation. Since experimental results do not show flow reversal we can only assume that our results for 10% initial disturbance or less are of significance in forming a composite picture of the development of the coherent structure.

Figure 5 shows that the maximum distortion occurs near the wall and for the case shown is at $y^+ = 16$. This result is typical for a large range of Reynolds numbers.

Secondary Instability Arising from Mean Flow Distortion

In Fig. 6 we show the results of phase speed and damping coefficient for $R = 40{,}000$ and for various values of $\alpha\delta^*$. Our numerical results show that when the initial disturbance amplitude is of the order of 5% or less, the distorted profile is always stable throughthout the cycle of a coherent structure. However, when the initial disturbance amplitude is of order 10% when $R = 6000$, changing to 5% when R is greater than 40,000 the distorted profile becomes unstable within the time to reach a maximum distortion. This time is shorter as the Reynolds number is increased.

In Fig. 7 we show typical results for the initially linear growth of H when the least stable (linear) mode has an initial disturbance of 10% of $\bar{U}_\infty$. In Fig. 7b we show typical values of the mean flow distortion as a function of $x_1 u_\tau / \nu$ on the

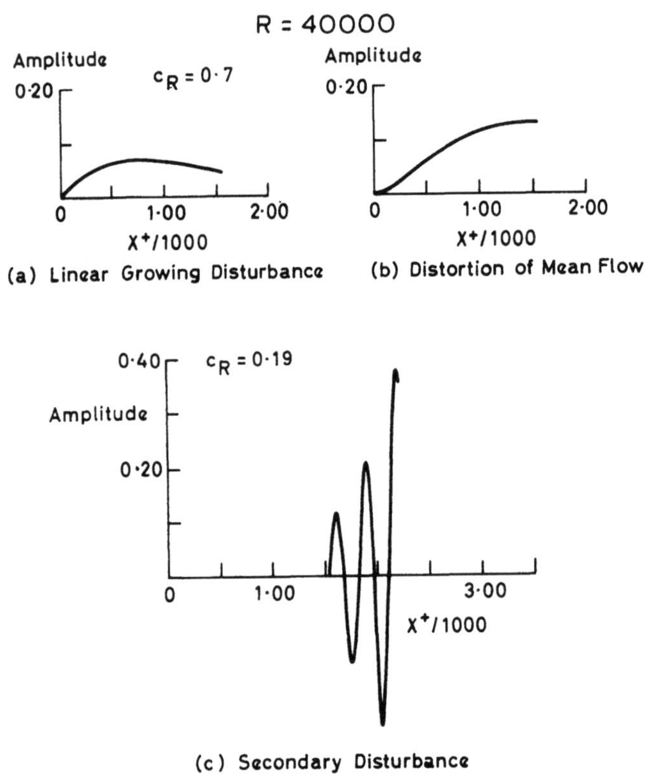

Fig. 7 a–c. Development of flow distortion

assumption that the spatial growth can be determined approximately from the caculated temporal growth and assuming the convection speed is equal to the group velocity, which in our case is nearly equal to the phase velocity. In Fig. 7c the secondary instability is shown to lead rapidly to a breakdown of the flow and occurs within a distance of about $x_1 u_\tau/\nu = 2000$ following the commencement of the coherent structure.

The Development of Strong Streamwise Vorticity

For a given initial disturbance amplitude we found that a strong streamwise vorticity is generated as a second order effect. A typical result is shown in Fig. 8 in which the calculated streamlines are shown in the plane perpendicular to the freestream direction. The non-dimensional spanwise wave length, $\lambda u_\tau/\nu$, of a cell is shown to be of order 100 over a wide range of Reynolds numbers. We note that the flow in transverse planes superimposed on the streamwise component of the distorted mean flow has the appearance of side-by-side ejection events ($\Delta U_1 < 0, \Delta U_2 > 0$) with sweep events ($\Delta U_1 > 0, \Delta U_2 < 0$) and are similar to the 'finger' structure observed by *Brodkey* et al. (1978).

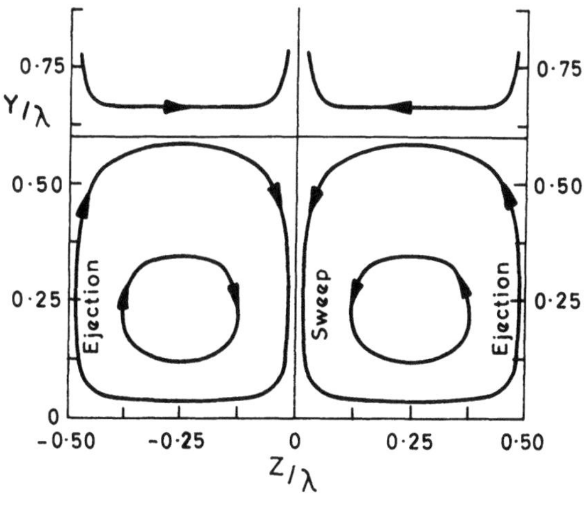

Fig. 8. Streamlines in transverse plane ($\lambda^+ \cong 100$)

Conclusions

This preliminary account of a model of the coherent eddy structure of a turbulent boundary layer has demonstrated the success of this model in describing a plausible sequential set of events, which together describe the self-generating mechanism of turbulence in a boundary layer.

References

Benney, D.J.: Phys. Fluids **7**, 319 (1964)
Benney, D.J., Lin, C.C.: Phys. Fluids **3**, 656 (1960)
Blackwelder, R.F., Kaplan, R.E.: 12th Conf. Int. Union, Theor. Appl. Mech. 1972
Cebeci, T., Smith, A.M.O.: *Analysis of Turbulent Boundary Layers* (Academic Press, New York 1974)

Gustavsson, L.H., Hultgren, L.: J. Fluid Mech. **98**, 149 (1980)
Hatziavramidis, D.T., Hanratty, T.J.: J. Fluid Mech. **95**, 655 (1979)
Kline, S.J., Rundstadler, P.W.: J. Appl. Mech. **26**, 166 (1959)
Landahl, M.T.: Symp. on Mod. Dev. in Fluid Dyn., 1973
Landahl, M.T.: J. Fluid Mech. **29**, 441 (1967)
Mack, L.M.: J. Fluid Mech. **73**, 497 (1976)
Murdock, J.W., Stewartson, K.: Phys. Fluids **20**, 1404 (1977)
Praturi, A.K., Brodkey, R.S.: J. Fluid Mech. **89**, 251 (1978)
Reynolds, W.C., Hussain, A.: J. Fluid Mech. **54**, 263 (1972)
Walker, J.D.A.: Proc. R. Soc. London **A359**, 167 (1978)
Willmarth, S., Wooldridge, C.F.: J. Fluid Mech. **14**, 187 (1962)
Zhang, Z.: Ph.D. Thesis, Southampton University (1981)

The Mechanism of Turbulent Mass Transfer at a Boundary

J. A. Campbell and T. J. Hanratty

School of Chemical Sciences, University of Illinois,
Urbana IL 61801, USA

Abstract

The fluctuating concentration field is calculated from the mass balance equation using measured values of the fluctuating velocity field. It is found that the concentration boundary-layer acts as a filter in that only velocity fluctuations of much lower frequency than the most energetic velocity fluctuations are effective in transporting mass at large Schmidt numbers. Two parameters characterizing the velocity field emerge as being quite important. These are the limiting behavior of the spectral density function of the normal velocity fluctuations for frequency approaching zero and a scale characterizing the spanwise mixing.

Nomenclature

- C Concentration of transported species [mole/m^3]
- C_B Bulk concentration of transported species [mole/m^3]
- D Molecular Diffusion Coefficient [m^2/s]
- k Fluctuation part of mass transfer coefficient [m/s]
- K Mass transfer coefficient [m/s]
- N Mass transfer rate [mole/m^{2}s]
- n Frequency [1/s]
- Re Reynolds number $= \frac{d < U >}{\nu}$
- S Schmidt number $= \frac{\nu}{D}$
- T_B Period of coherent structure [s]
- u Fluctuating part of velocity in x-direction [m/s]
- u^* Friction velocity [m/s]
- v Normal velocity [m/s]
- w Transverse velocity [m/s]
- W_k Frequency spectrum of k [m^2/s]
- W_β Frequency spectrum of β [1/m^{2}s]
- x Longitudinal coordinate [m]
- y Normal coordinate [m]
- z Transverse coordinate [m]

Greek Symbols

- α Velocity gradient at the wall in x-direction [1/s]
- β Velocity gradient at the wall in y-direction [1/ms]
- γ Velocity gradient at the wall in z-direction [1/s]
- δ Thickness of Nernst diffusion layer [m]
- λ Laterial dimension of coherent structure [m]
- ν Kinematic viscosity [m^2/s]

Superscripts

$^+$ Denotes variable made dimensionless with u^* and ν

Background

In 1904, Nernst suggested a model for relating turbulent mass transfer at a boundary to the convective flow. He pictured a stagnant layer of thickness y existing near a boundary and a well mixed region outside this layer. If the concentration of diffusing species is C_B in the well mixed region and is zero at the wall then the rate of mass transfer per unit area, N, is given as

$$N = D\,\frac{C_B}{\delta}\ ,$$

where D is the molecular diffusion coefficient. Hydrodynamics controls the mass transfer rate by controlling the thickness δ.

A number of attempts have been made to offer alternates to the Nernst diffusion layer concept which recognize that the flow close to the wall is not stagnant. These include surface renewal models, the assumption that the eddy diffusion coefficient is proportional to the eddy viscosity and various pseudosteady state eddy models. All of these theories are inconsistent with recently obtained measurements of flow close to a wall and of the fluctuations in the local mass transfer rate to a wall.

Studies of the velocity field have revealed the existence of coherent structures close to the wall which are approximately homogeneous in the flow direction, and which have a lateral dimension of $\lambda^+ \cong 100$ and a period of $T_B^+ \cong 100$ [4]. Studies of the fluctuations in mass transfer rate reveal fluctuations of very large amplitude which have a period of at least an order of magnitude larger than T_B^+ at a Schmidt number of 1000 [6,7,9]. This period increases with increasing Schmidt number. As with the fluctuating velocity field close to a wall, the fluctuations in the rate of mass transfer reveal an elongated structure.

At the Second Symposium on Turbulent Shear Flows [3] a model for the mass transfer process was explored which used structural information on the velocity field obtained in recent experiments to relate mass transfer rates to the flow. The time dependent mass balance equation was solved for a regular periodic flow field with period $T_B^+ = 100$ and with a lateral wavelength of $\lambda^+ = 100$. The implication of this analysis is that the observed low frequency concentration fluctuations close to the wall are not contributing to the transport of mass. The Reynolds transport coefficient is assumed to be associated with concentration fluctuations of period T_B^+ which would exist a distance away from the wall.

This calculation was not successful in explaining available measurements. Consequently we have explored a different approach. The results of this new work are reported in this paper. The chief difference is that the assumption of a velocity field periodic in time is abandoned. Rather, the mass balance equations are solved using actual measurements of the fluctuating velocity field close to the wall. Of particular interest is the solution using filtered velocity signals since this reveals whether high frequency velocity fluctuations are playing any role in controlling the mass transfer rate. New measurements of the structure of the fluctuating mass transfer field and of the structure of the low frequency velocity fluctuations close to the wall are also presented.

Experiments

The experiments were conducted in a vertical five level flow system which provides a straight entrance length of 15 m for a 20 cm circular test section [7]. The test fluid consists of an aqueous solution of potassium iodide and iodine.

The velocity gradients at the wall were measured in a Plexiglas test section which had an array of 20 pairs of rectangular electrodes mounted flush with the wall around the circumference of the pipe. Each pair was in a chevron arrangement with the two electrodes making an angle of 15° to each other. The electrodes formed the cathodes of an electrolysis cell at which the following reaction occurred:

$$I_3^- + 2e^- \rightarrow 3I^- \quad . \tag{2}$$

At large enough cathode voltages the measured current can be related to the velocity gradient at the wall. The sum of the signals from two of these electrodes gave the component of the velocity gradient in the direction of mean flow, α, and the difference gave the component in the spanwise direction, γ [4, 5, 7]. At Re = 20,000 the group of electrode pairs covered a distance of $\Delta z^+ = 100$ and the distance between two pairs was $\Delta z^+ = 5$.

In the mass transfer studies, the test section was a nickel pipe plated with platinum. This constituted the cathode of an electrolysis cell. At large enough voltages the current flowing in the electrolysis circuit is proportional to the mass transfer to the test section. A series of circular wires 0.4 mm in diameter were mounted flush with the wall of the test section in an array in the circumferential direction. These were insulated from the test section by a thin layer of epoxy. By measuring the current flowing to one of these electrodes the local fluctuations in the mass transfer rate could be determined.

Analysis

From an order of magnitude analysis, *Sirkar* and *Hanratty* [8] have argued that molecular diffusion is important only in the y-direction and that the concentration field is approximately homogeneous in the flow direction. The time dependent mass balance equation can therefore be written as

$$\frac{\partial C}{\partial t} + w\frac{\partial C}{\partial z} + v\frac{\partial C}{\partial y} = D\frac{\partial^2 C}{\partial y^2} \tag{3}$$

For large Schmidt numbers the concentration boundary-layer is so thin that w and v may be approximated by a Taylor series expansion so that

$$w = \gamma\,(z, t)\,y \tag{4}$$

and

$$v = \beta\,(z, t)\,y^2 \quad . \tag{5}$$

From the continuity equation the following relation between $\gamma\,(z, t)$ and $\beta\,(z, t)$ is obtained if the flow is assumed approximately homogeneous in the flow direction:

$$\frac{\partial \gamma}{\partial z} + 2\beta = 0 \tag{6}$$

Consequently the mass transfer rate is related to the velocity field either through the specification of $\gamma\,(z, t)$ or of $\beta\,(z, t)$.

As a first approximation the influence of span-wise mixing and of the variation of the flow field in the z-direction were ignored. The equation

$$\frac{\partial C}{\partial t} + v\,(t)\frac{\partial C}{\partial y} = D\frac{\partial^2 C}{\partial y^2} \qquad \begin{array}{ll} C = 0 & y = 0 \\ C = C_B & \text{large } y \end{array} \tag{7}$$

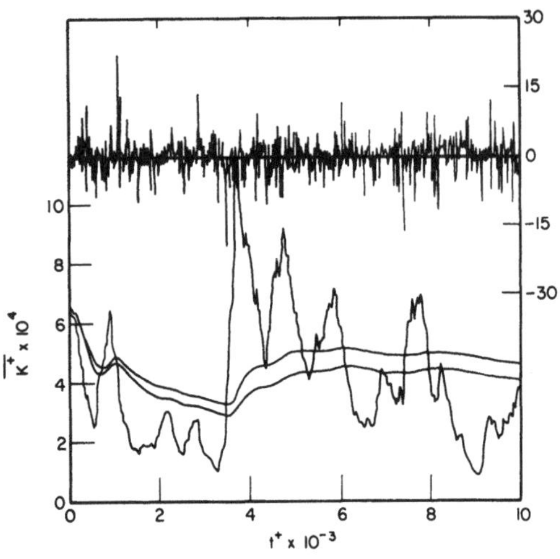

Fig. 1. Dimensionless mass transfer coefficient, K/u^*, and dimensionless velocity, v/u^*, versus dimensionless time, tu^{*2}/v, for $t_t^* = 10{,}150$, as Sc = 1000

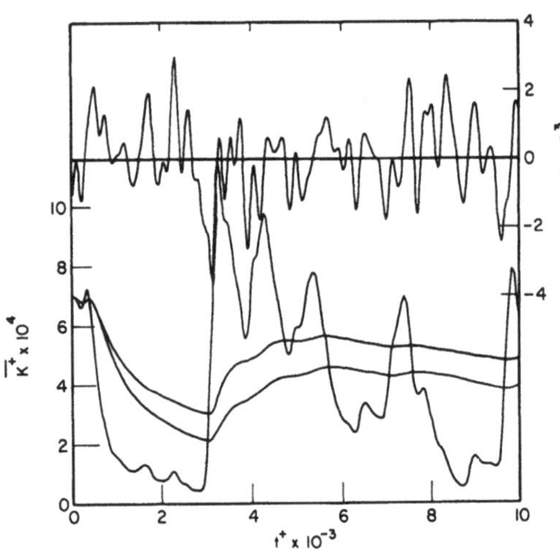

Fig. 2. K/u^* versus tu^{*2}/ν for velocity low-pass filtered with $n_{co}^+ = .0041$, $t_t^+ = 10{,}150$, at Sc = 1000

was solved using approximate measurements of $\beta(t)$. These were obtained from the determination of $\partial\gamma/\partial z$ using two pairs of flush mounted mass transfer probes in chevron configurations. The calculated variation of the mass transfer rate, N, with time was found to be similar at Schmidt number = 1000 for an unfiltered $\beta(t)$ and for a $\beta(t)$ for which ~90% of the energy was removed by using a low pass filter (see Figs. 1 and 2). This suggests (a) that the fluctuations in the mass transfer rate are being caused by low frequency velocity fluctuations and (b) that the rate of mass transfer is being controlled by velocity fluctuations of much lower frequency than the elongated flow structures that dominate the wall region. Similar conclusions have recently been reached by Campbell [2] who presented solutions to a simplified version of Eq. (3) which is linear in the fluctuating quantities.

Solutions of (7) give approximately the same dependency with Schmidt number as found in measurements of the average mass transfer coefficient, $\overline{K}$, and the mean-squared value of the fluctuations in K, $\overline{k^2}$. However, the calculated values of $\overline{k^2}$ are of incorrect magnitude. This suggests that spanwise mixing must be taken into account. The influence of spanwise mixing was assessed by measuring the fluctuations in K simultaneously

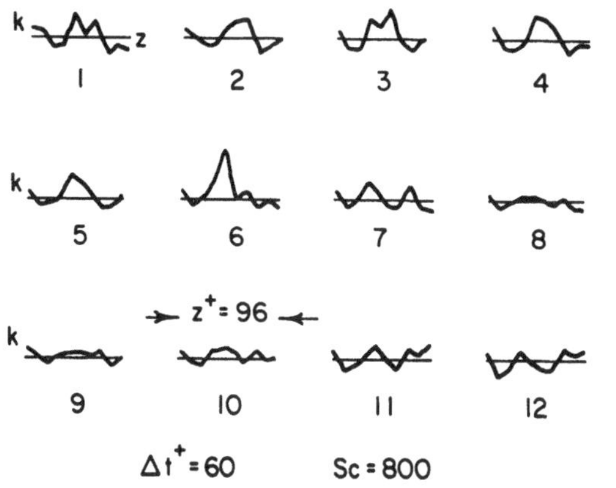

Fig. 3. Instantaneous values of the local fluctuating component of the mass transfer coefficient versus spanwise distance at several consecutive times

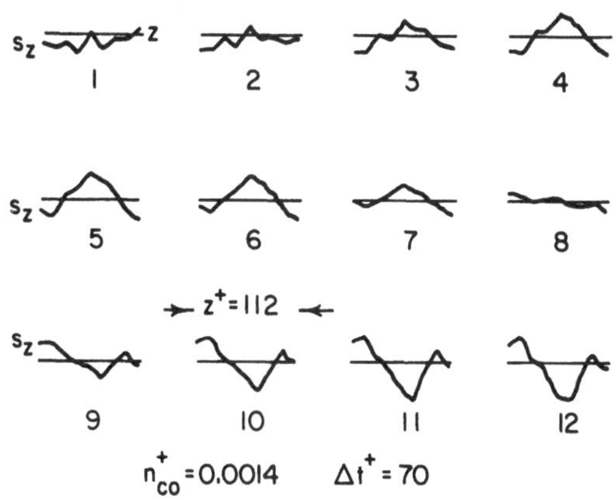

Fig. 4. Instantaneous values of the low-pass filtered spanwise component of the wall velocity gradient versus spanwise direction for several consecutive times

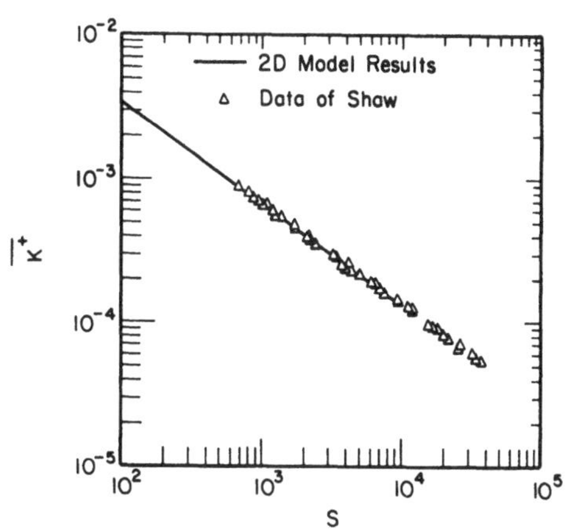

Fig. 5. Comparison of experimental data and results of 2D model for the Schmidt number variation of the average mass transfer coefficient

at multiple wall locations. These revealed a slowly varying pattern with a wavelength approximately equal to $\lambda^+ = 100$ (see Fig. 3). In order to see whether this result is consistent with the finding that the mass transfer is being controlled by low frequency velocity fluctuations, measurements were made of s_z at multiple wall locations. These were treated with a low pass filter so that ~90% of the energy was removed. Variations

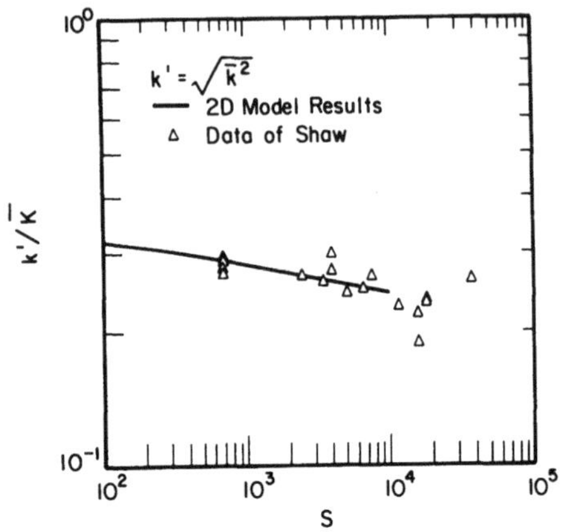

Fig. 6. Comparison of experimental data and results of 2D model for the Schmidt number variation of the relative intensity of the mass transfer coefficient

of the filtered signals in the spanwise direction had a characteristic wavelength on the same order as that observed for the mass transfer fluctuations (see Fig. 4).

In order to include the effect of spanwise variation of the flow, Eq. (2) was solved with

$$v = \beta(t) \cos \frac{2\pi z}{\lambda^+} y^2 \quad .$$

The calculation is not very sensitive to the selection of λ^+ if $\beta(t)$ is normalized so as to give a value of $\overline{v^2}$ that agrees with measurements. A comparison of calculated values of $\bar{K}$ and of $\overline{k^2}$ with experiment showed good agreement (see Figs. 5 and 6).

Method of Calculation

The spectral density function for β is represented by W_β and is defined as

$$\overline{\beta^{+2}} = \int_{-\infty}^{\infty} W_\beta^+(n^+)\, dn^+ \quad , \tag{8}$$

where all quantities have been made dimensionless using wall parameters. It is found that for large Schmidt numbers only the low frequency portion of $W_\beta^+ (n^+)$ for which it is constant, $W_\beta^+(0)$, is important.

Solutions of (7) have revealed good agreement between the calculated spectral function for k at large frequencies and predictions based on the solution to the linear equation presented by *Sirkar* and *Hanratty* [8]

$$W_k^+(n) = \frac{4\overline{K^{+2}}\, W_\beta^+(0)}{(2\pi n^+)^3} \qquad n \to \infty \quad . \tag{9}$$

Therefore, a comparison of the measurements of $W_k(n)$ by *Shaw* and *Hanratty* [8] with Eq. (9) allowed the evaluation of $W_\beta(0)$ as

$$W_\beta^+(0) = 0.010 \quad . \tag{10}$$

The measured $\beta(t)$ used to solve (3) and (7) had a $W_\beta(0)$ smaller than given by (10). This suggests that the measurements are in error because of the averaging of small scale motions and because of the neglect of $\partial u/\partial x$ in (6). This was corrected by multiplying the measured $\beta(t)$ with a constant factor selected to give a value of W_β (0) in agreement with (10). Equations (3) and (7) were solved using finite difference techniques which employed upwinding of the advection terms. The one dimensional equation used an implicit method. The two dimensional study used an alternating-direction-implicit method. The calculation was carried out on the University of Illinois Cyber *CDC* 170/175 Computer. The average computation time for (4) with 10,000 times steps and 20 grid points in the normal direction was 12 s. The consistency of the finite difference schemes was tested by varying the number of grid conditions. Both numerical schemes have been shown to be unconditionally stable [1].

Conclusions

The goal of a theory for mass transfer between a turbulent fluid and a solid is to identify properties of the fluctuating velocity field which are important and to develop a relation between $\bar{K}$ and these properties. The results of the calculations and experiments summarized in this paper present a picture of the process quite different from classical theory.

It is found that at large Schmidt numbers, the concentration boundary-layer close to the surface acts as a filter so that velocity fluctuations of much lower frequency than the most energetic velocity fluctuations are causing turbulent transport. As the Schmidt number increases, smaller and smaller fractions of the turbulence energy are effective in determining the magnitude of $\bar{K}$.

The filtering of measurements of $\gamma(z, t)$ made at multiple locations on the wall reveal no difference in the lateral scale of the low frequency components of $\gamma(z, t)$ and the unfiltered signal.

We therefore find that the mass transfer process is related to the velocity field through W_β (0) and a length characterizing the lateral mixing.

Acknowledgements. This work is being supported by the National Science Foundation under Grant *NSF CPE* 79–20990.

References

1. Campbell, J.A.: "The Use of a Regular Eddy Model to Describe Turbulent Mass Transfer to a Wall", M.S. Thesis, University of Illinois, Urbana, 1979
2. Campbell, J.A.: "The Relation of the Turbulent Velocity Field to Mass Transfer at High Schmidt Numbers," Ph.D. Thesis, University of Illinois, Urbana, 1981
3. Campbell, J.A., Hanratty, T.J.: "Influence of Turbulent Structure on Mass Transfer at a Solid Surface", Second Symposium on Turbulent Shear Flows, London, 1971
4. Hogenes, J.H.A.: "Identification of the Dominant Flow Structure in the Viscous Wall Region of a Turbulent Flow," Ph.D. Thesis, University of Illinois, Urbana, 1979
5. Lau, K.L.: "Study of Turbulent Structure Close to a Wall Using Conditional Sampling Techniques", Ph.D. Thesis, University of Illinois, Urbana, 1980
6. Shaw, D.A.: "Mechanism of Turbulent Mass Transfer to a Pipe Wall at High Schmidt Number", Ph.D. Thesis, University of Illinois, Urbana, 1976
7. Sirkar, K.K.: "Turbulence in the Immediate Vicinity of a Wall and Fully Developed Mass Transfer at High Schmidt Numbers", Ph.D. Thesis, University of Illinois, Urbana, 1969
8. Sirkar, K.K., Hanratty, T.J.: Relation of Turbulent Mass Transfer Rates to a Wall at High Schmidt Numbers to the Velocity Field. J. Fluid Mech., **44**, 589–603 (1970)
9. Van Shaw, P., Hanratty, T.J.: Fluctuations in the Local Rate of Turbulent Mass Transfer to a Pipe Wall. AIChEJ., **10**, 475–482 (1964)

Measurements in the Heated Turbulent Boundary Layer on a Mildly Curved Convex Surface

M.M. Gibson and C.A. Verriopoulos
Department of Mechanical Engineering, Imperial College, London, SW7 2BX, England

Y. Nagano
Nagoya Institute of Technology, Gokiso-Cho, Showa-Ku-Nagoya, Japan

Abstract

Measurements of mean velocity and temperature, and of surface heat flux, have been made in a turbulent boundary layer on a heated convex surface of modest curvature ($\delta/R \sim 0.01$). The results show surprisingly large curvature effects on heat transfer: at the end of the curved plate the Stanton number fell by 18% of the predicted flat plate value; the corresponding fall in the skin friction coefficient was 10%.

Profile measurements of temperature and velocity were obtained well into the viscous sublayer.

Nomenclature

- c_f skin friction coefficient, [Eq. (4)]
- c_p static pressure coefficient, $2(p-p_{ref})/\rho U_{pw}^2$, specific heat
- $\dot{q}''$ heat flux
- R radius of curvature of plate
- s distance measured along curved wall
- T temperature
- T^+ dimensionless temperature [Eq. (6)]
- U velocity
- U^+ dimensionless velocity U/u_τ
- u_τ friction velocity, $\sqrt{\tau_w/\rho}$
- y distance measured normal to the wall
- y^+ dimensionless distance, yu_τ/ν
- Δ_2 enthalpy thickness, [Eq. (2)]
- δ_2 momentum thickness, [Eq. (1)]
- δ boundary layer thickness
- ρ density
- τ shear stress
- ν kinematic viscosity

Subscripts

- p potential flow
- w wall
- ∞ free stream

Introduction

The effects of longitudinal curvature on the turbulence in boundary layers are well known. When, for example, a boundary layer develops on a convex curved surface, such as the suction surface of a wing or turbomachine blade, the angular momentum of the flow increases in the direction of the radius of curvature with the result that the shear stress and turbulence intensity are reduced to values often substantially less than those in an equivalent plane wall flow. The reverse occurs in the flow on a concave surface where the turbulence is increased by curvature. In contrast to laminar flow, where the fractional change in shear stress is of the same order as the ratio of the boundary layer thickness to the radius of curvature, turbulent flow measurements show fractional changes an order of magnitude greater. These effects on the turbulent velocity field are now well documented particularly for convex-wall flow. The survey paper by *Bradshaw* [1] contains details of research up to 1973; since that date a number of detailed hot-wire measurements in isothermal curved flow have been reported [2–6].

In marked contrast to this activity in the study of the hydrodynamics of curved flow the effects of curvature on turbulent heat transfer have received little attention. The supersonic flow measurements of *Thomann* [7], and the more recent results from an incompressible boundary layer reported by *Mayle* et al. [8], show that the turbulent heat flux is affected in the same way as the shear stress: depressed in flow on a convex surface and augmented in flow on a concave surface. Since, however, the corresponding variation of shear stress is not given in these papers it is impossible to judge whether the heat flux changed by a greater or lesser amount. There is no reason to suppose a priori that the changes in these two quantities are equal; on the contrary, in the analogous [9] case of density stratified flow the differences are appreciable. Measurements [10] in the atmospheric surface layer show heat transfer to be significantly more affected by buoyancy than momentum transfer or, stated alternatively, that the turbulent Prandtl number is a strong function of the stratification and increases with increasing stability. The curvature and buoyancy generation terms appear differently in the stress and heat flux equations in each case and suggest that changes of the same order might appear in curved flow [11].

An estimate of the relative changes may be made from the mean temperature and surface heat flux measurements of *Simon* and *Moffat* [12] because details of the turbulent velocity field are reported by *Gillis* et al. [13][1]. Measurements were made in the boundary layer on a ninety degree bend followed by a flat plate recovery flow. The test wall was constructed of individually heated segments which allowed steady state measurement of the spanwise averaged wall heat flux by energy balance. In two experiments maximum values of δ/R were 0.051 and 0.077. In each case the Stanton number at the end of the curved section fell by approximately 34% of the calculated plane-flow value while the skin friction coefficient [13] showed only a 30% fall.

The experiment now to be described, which was conceived before we became aware of *Simon* and *Moffat*'s work, stems from *Gibson*'s [11] speculations based on modelled equations for Reynolds stress and heat flux. The model, which gives good results for heat transfer in density stratifed flow [14], suggested that the effects of curvature on heat transfer are less than on shear stress, with the turbulent Prandtl number in local equilibrium flow decreasing with increasing stability. Coincidentally *So* [15] obtained the contrary result from a different model[2] and has since [16] used the data of [12, 13] to support his hypothesis. We are not, however, concerned here with the details of turbulence modelling. In the present paper we report measurements of mean velocity, mean temperature and surface heat flux made in the boundary layer growing on a moderately curved convex plate for which $\delta/R \cong 0.01$. Previously published data [4, 5] show that this degree of curvature is sufficient to produce measurable effects and it was chosen in order to minimise problems with secondary flow which might be exaggerated by heat transfer. Measurements of the mean and turbulent velocity fields have been described elsewhere [6]; these exhibit the characteristic features of convex wall flow: a diminished rate of layer growth and a progressive damping of turbulence in the layer.

---

[1] Earlier measurements [3] referred to in [12] were made in a nearly but not quite identical wind tunnel and show slight but significant differences in flow conditions and results.

[2] In which the mean strain contributions to the difficult pressure scrambling terms, included by *Gibson* [11], are omitted.

Experimental Arrangement and Procedures

The measurements were made in the boundary layer on the lower, heated, convex wall of an open circuit wind tunnel (Fig. 1). The boundary layer developed first on a flat heated plate 1.22 m long to a thickness, δ, of 17 mm at the start of the curved section, also heated, of constant nominal radius of curvature, R, 2.44 m and length 1.22 m. The tunnel cross section in the plane approach flow was constant at 305 x 102 mm^2 but at the start of the curved plate the section height was increased to 105 mm in order to reduce the negative pressure gradient on the test wall. Figure 1 shows the positions of the traverse stations; the reference free stream velocity measured at station 1.705 mm upstream of the curved plate, was nominally 22.6 ms^{-1} throughout the experiments. The boundary layer was tripped at the start of the flat plate and the momentum thickness Reynolds number at inlet to the curved section was 3300. The maximum value of δ/R, measured at the exit, was 0.011 and the free stream turbulence level was less than 0.4% throughout. Preliminary pressure tube measurements showed that the mean flow was closely two-dimensional up to 150 mm from the curved section exit; in the exit plane the two sides of the mean flow integral momentum equation differed by no more than 10 per cent when the skin friction coefficient was evaluated by the Clauser chart method.

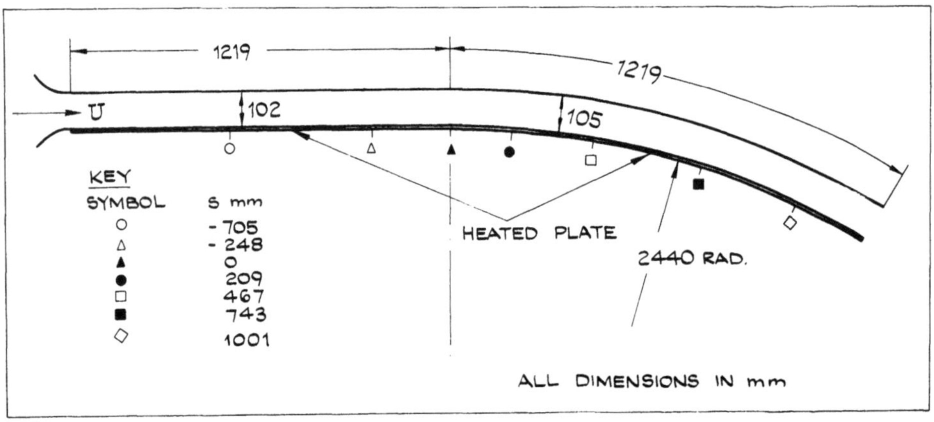

Fig. 1. The wind tunnel working section showing traverse stations and symbols

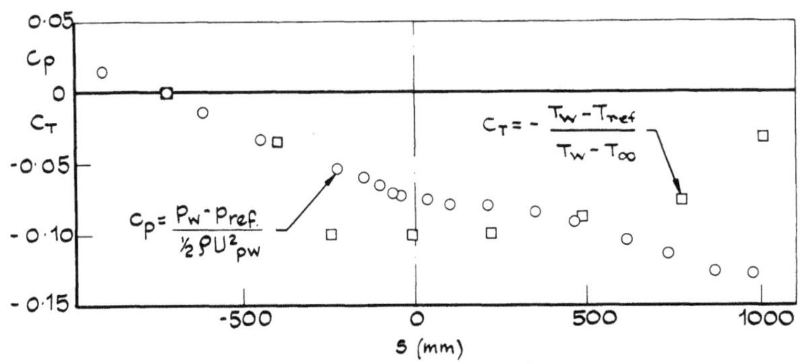

Fig. 2. Streamwise variation of wall static pressure and temperature

The 12.5 mm thick aluminum test plate was heated to 14°C above ambient with an electric blanket. Sets of three copper-constantan thermocouples spaced at 50 mm in the spanwise direction were located at intervals of 120 mm along the flat and curved sections. The variation in plate temperature along the measurement length was found to be within ± 0.75°C from the mean, the greatest temperatures being recorded around the start of the curved section (Fig. 2). Provision was made for inserting miniature heat flux meters at ten locations spaced unevenly (~150 mm) on the centre line in the downstream half of the flat plate and in the curved section. These meters are of the Schmidt-Boelter multiple thermocouple type, 9.4 mm diameter, and were specially made by the Medtherm Corporation in aluminum so as to be compatible with the plate material. Their sensitivity is 40 $mV/W^{-1}\ cm^2$ and their accuracy is estimated at ± 2%.

Mean velocity measurements in the boundary layer were made with a flattened Pitot tube except very close to the wall ($y^+ < 25$) where a constant temperature 5 μm tungsten hot wire was employed. Corrections for wall proximity have not been made in view of the evidence [17] that these are significant only for $y^+ < 4$, or roughly the lower limit of the measurements. Mean temperatures at the same traverse stations were measured with a 12 μm chromel-alumel thermocouple of length-to-diameter ratio 580. Errors associated with conduction to the thermocouple supports and radiation from the heated plate were estimated [18, 19] to be less than 0.03°C. The thermocouple output was amplified (× 1000) in a Precision-Monolithic OPO7A low noise amplifier to be digitised and linearised in an Apple II microcomputer. The hot wire and thermocouple were positioned when close to the plate with a micrometer traverse accurate to 10 μm. The wall temperature recorded by the thermocouple probe differed by no more than ± 0.1°C from the temperatures recorded by the thermocouples set in the plate.

Results and Discussion

The pressure distribution obtained from static pressure tappings in the test plate before the electric blanket was fitted is shown in Fig. 2. The pressure coefficient c_p is defined with respect to reference conditions at the first traverse station and the local potential velocity at the wall, U_{pw}. The integral momentum and enthalpy thicknesses evaluated from mean velocity and temperature traverses on the plate centreline are plotted against streamwise distance, s, in Fig. 3. These quantities are defined as

$$\delta_2 \equiv \int_0^\infty \frac{U}{U_{pw}} \left(1 - \frac{U}{U_p}\right) dy \tag{1}$$

$$\Delta_2 \equiv \int_0^\infty \frac{U}{U_{pw}} \left(1 - \frac{T - T}{T_\infty - T_w}\right) dy \tag{2}$$

The observed higher rate of growth of Δ_2 is consistent with values of Stanton number

$$\mathrm{St} = \frac{-\dot{q}''_w}{\rho c_p U_{pw} (T_\infty - T_w)} \tag{3}$$

which are plotted in Fig. 4. These are greater than values of one half the skin friction coefficient defined by

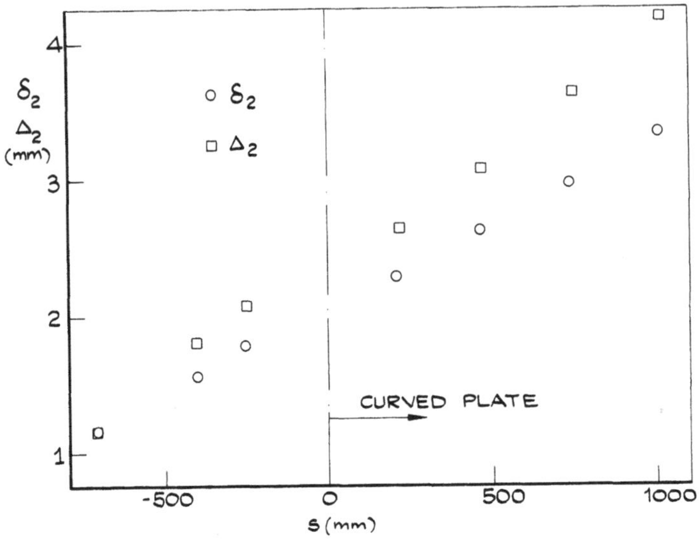

Fig. 3. Growth of boundary layer momentum and enthalpy thicknesses

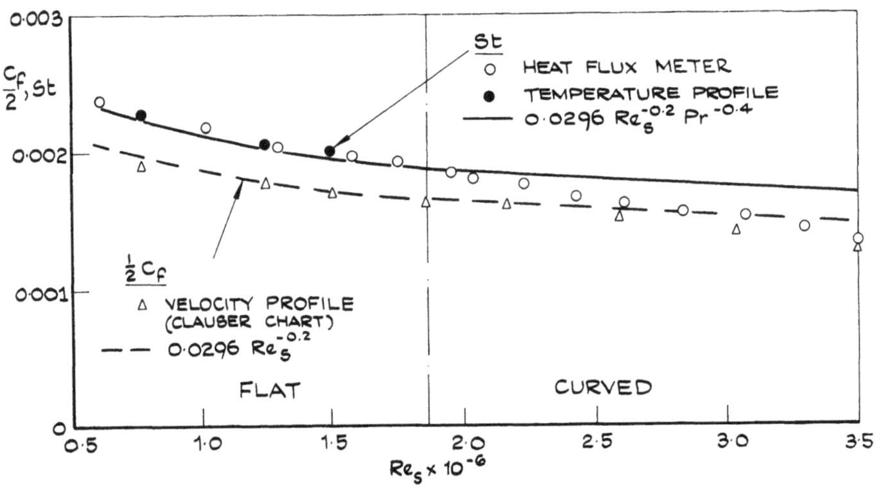

Fig. 4. Streamwise variation of skin friction coefficient and Stanton number plotted against length Reynolds number

$$\frac{c_f}{2} = \frac{\tau_w}{\rho U_{pw}^2} \tag{4}$$

over the entire measurement length. Fluid property values used in (3) and (4) have been evaluated at the mean, or film temperature $0.5\ (T_\infty + T_w)$. In the plane approach flow values of the wall heat flux $\dot{q}_w''$ recorded by the heat flux meters are supplemented by and checked against the local Stanton number obtained by plotting temperature profiles on the thermal equivalent of a Clauser chart. The results are entirely consistent and agree with plane wall values given by the correlation recommended in [20]. The Clauser chart method has been used throughout to calculate c_f from the velocity profiles. The most striking feature of the results plotted in Fig. 4 is that the Stanton number in the curved

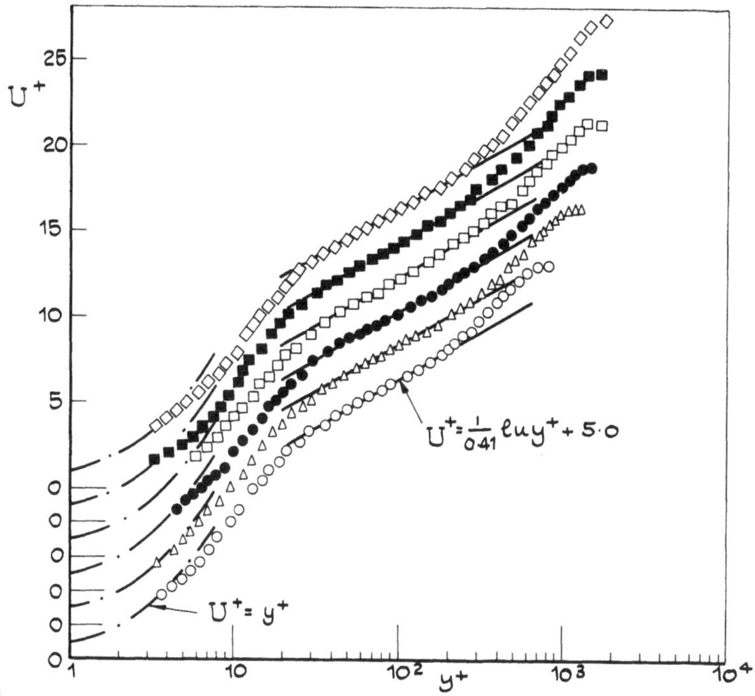

Fig. 5. Measured velocity profiles in universal wall co-ordinates. (Symbols defined in Fig. 1)

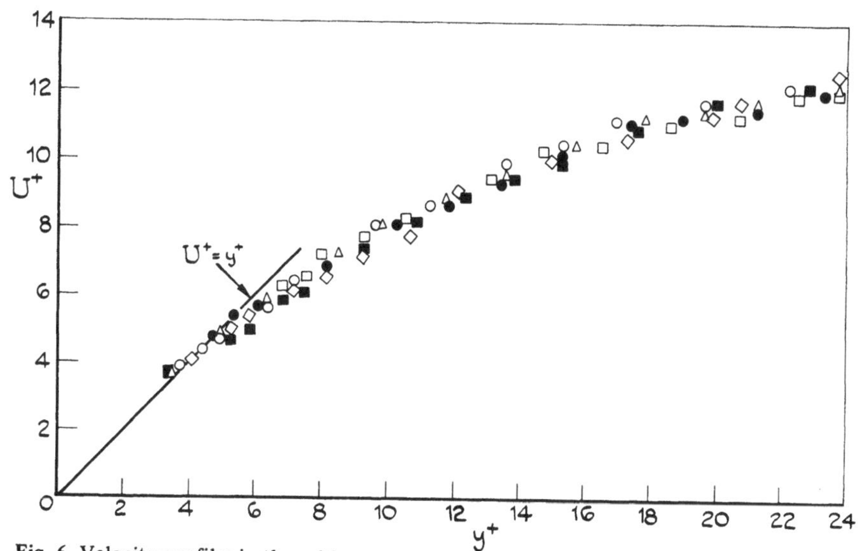

Fig. 6. Velocity profiles in the sublayers

flow falls more rapidly from flat plate values than does the skin friction coefficient. Thus at the last traverse station, $s = 1$ m, St has fallen to 82% and c_f to only 90% of their respective plane flow values. These disproportionate effects are even more marked upstream where in the first half of the curved section a significant drop in St is observed and c_f is little affected. The results are broadly consistent with those of *Simon* and *Moffat* [12]; the milder curvature of the present experiment produces smaller falls in

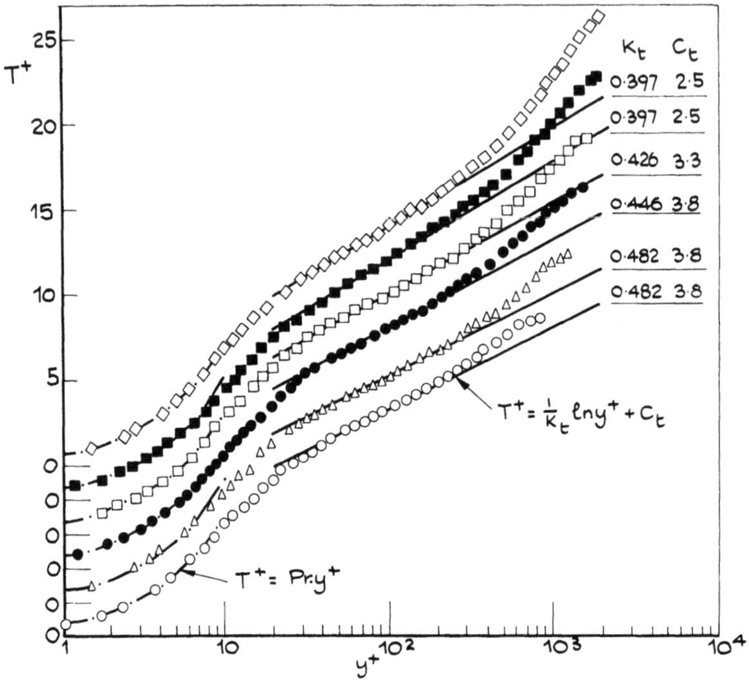

Fig. 7. Measured temperature profiles in universal wall co-ordinates

St and c_f but not proportionately so. The fall in St relative to c_f is apparently much greater than that reported in [12] and [13] and the effects on c_f in the present case of the mild favourable pressure gradient (Fig. 2) must be considered. The pressure gradient term contributes up to 15% of the mean flow momentum balance; this is offset to some extent by the lower growth rate of the accelerated boundary layer (the present measurements of $d\delta_2/dx$ are fully 20% lower than those quoted by *Coles* [21] for uniform pressure flow) so that the net effect on c_f is relatively small, as is suggested by the excellent agreement with the flat plate correlation in the plane approach flow.

Figure 5 is a semi-logarithmic plot of the mean velocity profiles in universal wall co-ordinates ($U^+ \sim y^+$). These all show clearly defined logarithmic layers which are well fitted in the plane and curved flow by

$$U^+ = \frac{1}{K} \ln y^+ + C \tag{5}$$

with $K = 0.41$ and $C = 5.0$. Profiles in the sublayers ($y^+ < 24$) are also shown in the linear plot of Fig. 6. Here the data collapse satisfactorily on a single curve; the scatter appears to be random and within the margins of experimental uncertainty including hot wire positioning error (< 10 μm, unit y^+ corresponds typically to 15 μm). The departure from the $U^+ = y^+$ asymptote at $y^+ = 4$ is attributed to the wall proximity effects documented in [17] for which no correction has been made.

The corresponding plots of T^+ against y^+ are presented in Figs. 7 and 8. In contrast to the Simon-Moffat data for stronger curvature these all show well defined logarithmic regions fitted by the expression

$$T^+ \equiv \frac{\rho u_\tau c_p \,(T - T_w)}{-\dot{q}''_w} = \frac{1}{K_t} \ln y^+ + C_t \tag{6}$$

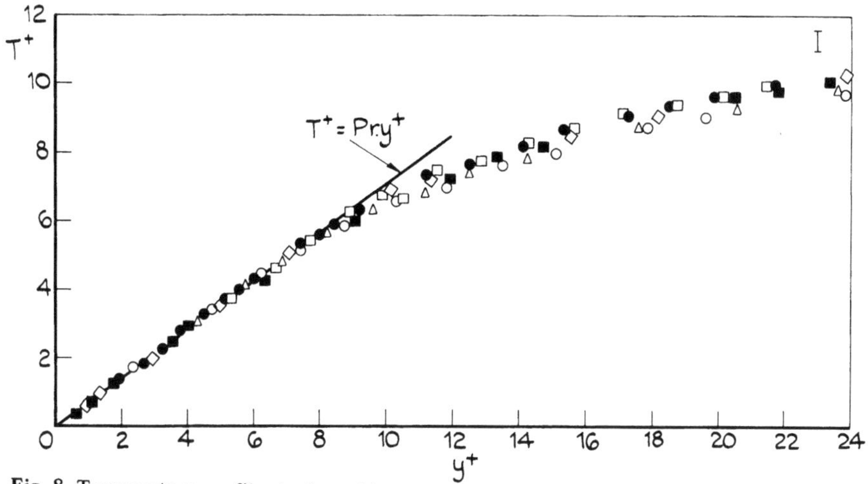

Fig. 8. Temperature profiles in the sublayers

A significant difference from the velocity profiles is that the 'constants' K_t and C_t change in the curved flow as indicated on the graph, the slope increasing (K_t decreasing) with distance along the curved plate. The first two profiles of the plane flow are fitted by [6] when $K_t = 0.48$, $C_t = 3.8$ to give an approximate value[3] of the turbulent Prandtl number near the wall $\text{Pr}_t \cong K/K_t = 0.85$ consistent with values estimated in the *Kader* and *Yaglom* [22] survey of flat plate data. Towards the end of the curved plate, however, the profiles are best fitted using $K_t = 0.4$ so that the turbulent Prandtl number rises to approximately 1.03. This result, which is at variance with our model predictions [11], is not as surprising as it might have been had we not seen the Stanford results [12, 13], which according to *So*'s [16] analysis of the data, exhibit the same trend. Sublayer profiles of T^+ are presented in Fig. 8. It was possible to get much closer to the wall with the thermocouple than with the hot wire probe and, within the estimated uncertainty margins indicated by the bars, the profiles collapse satisfactorily on a single curve with asymptote $T^+ = \text{Pr}\, y^+$ as $y^+ \to 0$. There is a slight but possibly significant difference from the sublayer velocity profiles of Fig. 6 in that scatter is apparently not random but shows a consistent trend for T^+ to increase at a given y^+ as the curved flow develops. It is, however, impossible to be quite certain about this trend in view of the experimental uncertainty.

In a recent paper *Simon* and *Moffat* [23] have used more data [24] from the Stanford experiments to examine the effects of initial boundary layer thickness, free stream velocity, free stream acceleration, an unheated starting length and the maturity of the momentum boundary layer. The principal result to emerge from this study was that, regardless of the initial state, curvature without acceleration eventually forced the boundary layer into the same asymptotic condition: $\text{St} \propto \text{Re}_{\Delta_2}^{-1}$ where Re_{Δ_2} is the enthalpy thickness Reynolds number $U_{pw}\Delta_2/\nu$. It was also found that strong acceleration with curvature changed the Reynolds number exponent from -1 to -2. In the experiments analysed in [23] the boundary layer thickness was varied to give values of δ/R equal to 0.10, 0.05 and 0.02 at the start of the curved plate. The data of the case $\delta/R = 0.02$, which are most nearly comparable to the present measurements, are said to be quite different from the strong curvature results in that the asymptotic state was reached much more slowly. Simon and Moffat note that the -1 exponent is the same as that for a flat

[3] Assuming uniformity of shear stress and heat flux in the logarithmic layer.

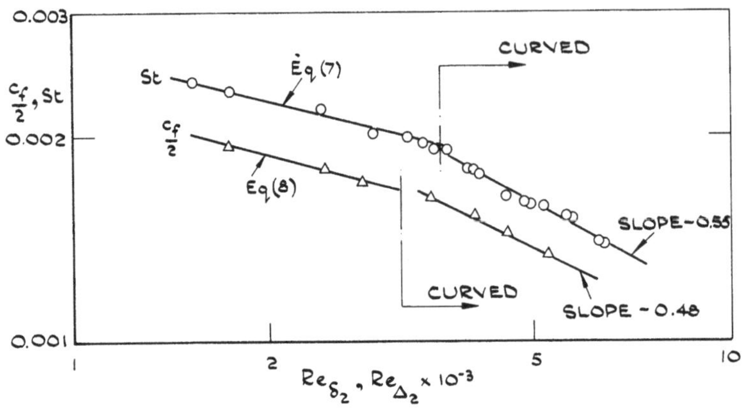

Fig. 9. Stanton number and skin-friction coefficient plotted against integral-thickness Reynolds number

plate laminar boundary layer and suggest that convex curvature causes something similar to "relaminarisation", a term used in connexion with strongly accelerated turbulent boundary layers.

The present measurements neither support nor contradict this hypothesis. The Stanton number is plotted against Re_{Δ_2} in Fig. 9 where the $c_f \sim Re_{\delta_2}$ variation is also shown for comparison. In the approach flow St and c_f are closely fitted by the established flat-plate correlations [20]:

$$St = 0.0125\ Pr^{-0.5}\ Re_{\Delta_2}^{-0.25} \tag{7}$$

$$c_f = 0.025\ Re_{\delta_2}^{-0.25} \tag{8}$$

On the curved plate St and c_f fall more rapidly and the exponents on Re_{Δ_2} and Re_{δ_2} change to -0.55 and -0.48 respectively. The effects of curvature expressed in this way are thus much less severe than those observed even in the weak curvature case of Simon and Moffat. The probable reason for this different behaviour is that in the present experiment the curvature was neither sufficiently strong nor prolonged for the postulated asymptotic date to be reached.

Conclusion

The results of this experiment on a mildly curved heated boundary layer show that convex curvature depresses the surface heat flux by a greater proportionate amount that it does the skin friction. Mean velocity and temperature profiles both exhibit clearly defined logarithmic regions in the curved flow but the gradient of temperature increases as the flow develops. Approximate estimates of the turbulent Prandtl number obtained from the gradients in the logarithmic layer show this quantity to increase from 0.85 at the start of the curved plate to a little over unity at the end. Measurements in the sublayers, in the case of temperature very close to the wall, confirm the internal consistency of the results. Heat transfer on the curved wall did not attain the asymptotic state postulated in [23].

The work presented here is still continuing. Hot-wire measurements of the four non-zero components of Reynolds stress have been made [6] and in the near future we intend to complete our investigation of the velocity field by measuring triple correlations and energy spectra. The mean temperature and surface flux data presented here will be supplemented by profile measurements of the cross-stream and longitudinal components the the heat flux, the temperature variance and the triple correlations.

Acknowledgement. This study is supported by the Science Research Council under Research Grant No. ER/A76684.

References

1. Bradshaw, P.: Effect of streamwise curvature on turbulent flows. AGARDograph **169** (1973)
2. Smits, A. J., Young, S. T. B., Bradshaw, P.: The effect of short regions of high surface curvature on turbulent boundary layers. J. Fluid Mech. **94**, 209 (1979)
3. Gillis, J.C., Johnston, J.P.: "Experiments on the Turbulent Boundary Layer over Convex Walls and its Recovery to Flat-Wall Conditions. In *Turbulent Shear Flows II* ed. by L.J.S. Bradbury, F. Durst, B.E. Launder, F.W. Schmidt, J.H. Whitelaw (Springer, Berlin, Heidelberg, New York 1980) pp. 116–128
4. Hoffmann, P.H., Bradshaw, P.: "Turbulent Boundary Layers on Surfaces of Mild Longitudinal Curvature". Imperial Coll. Aero. Rpt. 78–104 (1978)
5. Ramaprian, B. R., Shivaprasad, B. G.: The structure of turbulent boundary layers along mildly curved surfaces. J. Fluid Mech. **85**, 273 (1978)
6. Gibson, M.M., Verriopoulos, C.A., Vlachos, N.S.: "Measurements in the Turbulent Boundary Layer on a Convex Wall", Int. Symp. on Applications of Fluid Mechanics and Heat Transfer to Energy and Environmental Problems, Patras, Greece (1981)
7. Thomann, H.: Effect of streamwise wall curvature on heat transfer in a turbulent boundary layer. J. Fluid Mech., **33**, 283 (1968)
8. Mayle, R.E., Blair, M.F., Kopper, F.C.: Turbulent boundary layer heat transfer on curved surfaces. ASME J. Heat Transfer **101**, 521 (1979)
9. Bradshaw, P.: The analogy between streamline curvature and buoyancy in turbulent shear flow. J. Fluid Mech. **36**, 177 (1969)
10. Businger, J. A., Wyngaard, J. C., Izumi, Y., Bradley, E. F.: Flux-profile relationships in the atmospheric surface layer. J. Atmos. Sci. **28**, 181 (1971)
11. Gibson, M.M.: An algebraic stress and heat flux model for turbulent shear flow with streamline curvature. Int. J. Heat Mass Transfer **21**, 1609 (1978)
12. Simon, T.W., Moffat, R.J.: "Heat Transfer Through Turbulent Boundary Layers – the Effects of Introduction of and Recovery from Convex Curvature". ASME paper 79–WA/GT–10 (1979)
13. Gillis, J.C., Johnston, J.P., Kays, W.M., Moffat, R.J.: "Turbulent Boundary Layer on a Convex Curved Surface". Rpt. No. HMT-31, Thermosciences Div., Mech. Eng. Dept., Stanford University (1980)
14. Gibson, M. M., Launder, B. E.: Ground effects on pressure fluctuations in the atmospheric boundary layer. J. Fluid Mech. **86**, 491 (1978)
15. So, R.M.C.: "The Effects of Streamline Curvature on Reynolds Analogy". ASME paper 77–WA/FE–17 (1977)
16. So, R. M.C.: Heat transfer modelling for turbulent shear flows. Z. Angew. Math. Phys. **32**, 541 (1981)
17. Oka, S., Kostic, Z.: Influence of wall proximity on hot wire velocity measurements. DISA Inf. **13**, 29 (1972)
18. Blackwell, B. F., Moffat, R. J.: Design and construction of a low-velocity boundary-layer temperature probe. ASME J. Heat Transfer **97**, 313 (1975)
19. Hishida, M., Nagano, Y.,Nakamura, Y.: Temperature distribution in the turbulent boundary layer in a circular pipe. Bull. JSME **21**, 157, 1175 (1978)
20. Kays, W.M., Crawford, M.E.: *Convective Heat and Mass Transfer* (McGraw-Hill, New York 1980)
21. Coles, D.: The law of the wake in the turbulent boundary layer. J. Fluid Mech. **1**, 191 (1956)
22. Kader, B. A., Yaglom, A. M.: Heat and mass transfer laws for fully turbulent wall flows. Int. J. Heat Mass Transfer **15**, 2329 (1972)
23. Simon, T.W., Moffat, R.J.: "Turbulent Boundary Layer Heat Transfer Experiments: a Separate Effects Study on a Convexly Curved Wall". ASME paper 81–HT–78 (1981)
24. Simon, T.W., Moffat, R.J., Johnston, J.P., Kays, W.M.: "Turbulent Boundary Layer Heat Transfer Experiments: Convex Curvature Effects Including Introduction and Recovery", Rpt. No. HMT–32, Thermosciences Div., Mech. Eng. Dept., Stanford University (1980)

Part II

Scalar Transport

Introductory Remarks

K.N.C. Bray
Department of Aeronautics and Astronautics, University of Southampton, Southampton S09 5NH, England

The subject of heat and mass transfer in turbulent flows, that is, the transport of scalar properties such as composition or temperature differences, is important in a number of fields. Examples may be found in chemical and mechanical engineering, meteorology and pollutant dispersion. A particularly challenging problem is its application to the phenomenon of turbulent combustion in engines and furnaces. Designers of chemical reactors, combustion engines, furnaces, etc., as well as meteorologists and oceanographers, require increasingly accurate theoretical models of turbulent heat and mass transfer processes. The five papers in this section on scalar transport may all be seen as steps towards this goal.

The two previous volumes in the present series, *Turbulent Shear Flows* 1 and 2, both contain much material relevant to the scalar transport problem, but the book edited by *Bradshaw* [1] provides an easier entry to the large literature on the subject. Recent reviews of topics in turbulent reacting flows, with emphasis on combustion, may be found in the volume edited by *Libby* and *Williams* [2].

In many applications, it is heat and mass transfer to or from a solid wall that is of interest. The transport then takes place through a boundary layer. The papers by *Campbell* and *Hanratty* in the present volume, and by *Belorgey* et al. [3] in *Turbulent Shear Flows 2* provide examples of this type of study. On the other hand, the papers in the present section all deal with turbulent flows that are well removed from boundary layers on solid surfaces. In these circumstances interest is focused on turbulent transport from one region of the flowfield to another.

In classical treatments, the time-averaged turbulent flux of a scalar c in direction x_α, $\overline{c'v'_\alpha}$, is assumed to be proportional to $-\partial\bar{c}/\partial x_\alpha$. The use of this gradient transport or eddy diffusivity approximation cannot be justified, except in highly restrictive circumstances, but is often difficult to avoid in engineering calculations. One way of doing so is to use a second-order moment method, *Vandrome* and *Kollmann* (this volume), in which Reynolds stress components and flux components such as $\overline{c'v'_\alpha}$ are determined from their own transport equations. Even here, however, it is a common practice, *Lewellen* et al. [4], to invoke a gradient transport closure for third-order moments such as $\overline{c'v'_\alpha v'_\beta}$.

Although the five papers in this section deal with a wide spectrum of topics two common strands may be found. The first is the gradient transport approximation: when it will fail, and why, and how its use may be avoided in modelling calculations. The second is the role of coherent structures in turbulence and, implicitly, the question of whether empirical closure of the time-averaged equations can provide an adequate description of turbulent flow phenomena.

Sreenivasan, *Tavoularis* and *Corrsin* have tested the gradient transport approximation by measuring both the mean scalar flux and the gradient of the corresponding mean scalar property in a variety of turbulent flows. They find that the gradient transport model is adequate for homogeneous turbulent flows but cannot handle inhomogeneous flows even qualitatively. The explanation proposed is that large eddy structures predominate in the inhomogeneous flows, and are then responsible for a major portion of the turbulent transport.

The work of *Kewall* and *Keffer* also identifies important effects of large eddy structures. In a uniform flow, the wake of a heated circular cylinder approaches a so-called self-preserving state, in which time-averaged flow properties can be expressed uniquely in terms of a single independent similarity variable. Gradient transport is usually found to be an acceptable approximation in the self-preserving regime. *Kewall* and *Keffer* allow their wake to develop in a distorted free stream and find that the flow is no longer self-preserving. This is attributed to coherent structures in the wake which are amplified by the distortion.

Chemical reaction is introduced as an additional factor in the work of *Sherikar* and *Chevray*. Their initially separate reactant species mix together and react in a two-dimensional mixing layer whose structure is dominated by large coherent eddy strutures. When the reactants meet, the chemistry is sufficiently fast for the progress of the reaction to be limited by diffusional mixing. However, the reactants are diluted with inert gas so the temperature rise is much smaller than in the genuine diffusion flame of *Eickhoff* and *Grethe.* Flow visualisation allows the scalar transport mechanism to be described qualitatively in terms of the engulfment of fresh material into the mixing layer by the coherent eddy structures. The growth of the mixing layer is identified with the pairing process whereby two coherent eddies combine to form a single larger coherent structure.

A seemingly different picture of scalar transport emerges from *Pope's* analysis of a very similar problem: transport of a chemically inert species in a plane two-dimensional mixing layer. It is time-averaged quantities that are predicted and the formulation contains no explicit recognition of the existence of coherent structures. *Pope's* statistical analysis of the self-preserving plane mixing layer is based on a numerical solution of the transport equation for the joint probability density function (pdf) for velocity and scalar concentration. This joint pdf contains all the necessary single-point information from which time-averaged properties of the flow can be reconstructed. A great potential virtue of the formulation is that it can be applied to variable density reacting flows, where the terms describing the processes of turbulent convection, chemical reaction, buoyancy and mean pressure gradient effects all appear in closed form. That is, these terms can be treated exactly, without invoking any empirical closure approximations. The terms that do require empirical closures describe various dissipation and pressure fluctuation effects. An interesting question for the future is the extent to which *Pope's* predictions will be sensitive to the details of this empiricism.

Finally, turbulent combustion is studied in the work of *Eickhoff* and *Grethe*. Among the additional complications introduced by combustion are the following. On the one hand, heat release causes very large density variations and expansion of fluid elements, leading to modifications in the turbulent flow structure. On the other hand the turbulent flow strongly influences mixing at the molecular scale, which often controls the time-average rate of heat release. The mutual interaction between turbulence and combustion can lead to new turbulence phenomena which are absent in constant density turbulent flows. Examples include a flame-generated turbulence mechanism and turbulent mass transport in a time mean pressure gradient, leading to counter-gradient diffusion, *Bray* et al. [5].

Eickhoff and *Grethe* explore a different kind of coherent structure from that studied by *Sherikar* and *Chevray*. They assume that the combustion and heat release in their turbulent jet diffusion flames occur in laminar diffusion flamelets, which are embedded in the turbulent flow. This is an important simplification, because it allows the complex chemical kinetics of hydrocarbon oxidation to be computed once and for all in a laminar flame calculation. It thus avoids the formidable task of modelling the interaction between the fluctuating thermochemical state and the hydrocarbon kinetics. As *Liew* et al. [6] have demonstrated, even the laminar flame calculation can be avoided if experimental profiles of composition and temperature through a laminar flame are available.

Williams [7] has shown that laminar flamelets can exist in a turbulent flow so long as the laminar flame thickness is less than the Kolmogorov microscale of the turbulence. Within this regime of validity the laminar flamelets form 'coherent microstructures' which influence other properties of the flow, *Libby* and *Bray* [8], in addition to reaction rates. The laminar flamelet microstructure is sharply in contrast with the random turbulent microstructure implicit in the work of *Pope* [9].

The turbulent flame calculations of *Eickhoff* and *Grethe* use their laminar flamelet results to determine mean reaction rates. They also employ a $k-\epsilon$ turbulence model and therefore assume the validity of the gradient transport approximation.

The next few years can hardly fail to provide interesting developments arising from the work reported in this section. We must hope that these developments will include a resolution of the apparently conflicting points of view that the papers describe.

References

1. Bradshaw P. (ed): *Turbulence*, Topics in Applied Physics, Vol. 12 (Springer, Berlin, Heidelberg, New York)
2. Libby P.A., Williams F.A. (eds): *Turbulent Reacting Flows*, Topics in Applied Physics Vol. 44 (Springer, Berlin, Heidelberg, New York 1980).
3. Belorgey M., Nguyen A.D., Trinite M.: 'Diffusion From a Line Source in a Turbulent Boundary Layer With Transfer to the Wall, in *Turbulent Shear Flows 2*, ed. by L.J.S. Bradbury, F. Durst, B.E. Launder, F.W. Schmidt, J.H. Whitelaw (Springer, Berlin, Heidelberg, New York 1980)
4. Lewellan W.S., Teske M.E., Sheng Y.P.: 'Micrometeorological Applications of a Second-Order Closure Model of Turbulent Transport', in *Turbulent Shear Flows 2*, ed by L.J.S. Bradbury, F. Durst, B.E. Launder, F.W. Schmidt, J.M. Whitelaw (Springer, Berlin, Heidelberg, New York 1980)
5. Bray K.N.C., Libby P.A., Masuya G., Moss J.B.: Turbulence production in premixed turbulent flames. Combust. Sci. Technol. **25**, 127–140 (1981)
6. Liew S.K., Bray K.N.C., Moss J.B.: A flamelet model of turbulent non-premixed combustion. Combust. Sci. Technol. **27**, 69–73 (1981)
7. Williams F.A.: 'A review of Some Theoretical Considerations of Turbulent Flame Structure', in Analytical and Numerical Methods for Investigation of Flow Fields with Chemical Reactions, Especially Related to Combustion, AGARD Conf. Proc. 164, NATO, Paris, 1975, pp. II 1-1 to II 1–25
8. Libby P.A., Bray K.N.C.: Implications of the laminar flamelet model in premixed turbulent combustion. Combust. Flame **39**, 33–41 (1980)
9. Pope S.B.: Monte Carlo calculations of premixed turbulent flames. 18th Intern. Symp. on Combustion, Combustion Institute (1981) pp. 1001–1010

A Test of Gradient Transport and Its Generalizations

K.R. Sreenivasan
Applied Mechanics, Mason Laboratory, Yale University, New Haven, CT 06520, USA

S. Tavoularis
Mechanical Engineering Department, University of Ottawa, Ontario, Canada K1N 6N5

S. Corrsin
Chemical Engineering Department, Johns Hopkins University, Baltimore, MD 21218, USA

Abstract

Heat flux measurements in several specially designed turbulent flows with inhomogeneous temperature field are presented and analyzed with the purpose of evaluating the performance of the gradient transport models (GTM) as well as several of the generalizations, and of understanding the circumstances in which the GTM works in spite of the a priori conclusions to the contrary. One of the flows considered is a uniform grid-generated flow; two others are shear flows with transverse homogeneity and a constant mean velocity gradient. The fourth is the wake of a circular cylinder in which an asymmetric temperature field is created by heating a combination of thin wires located off-axis. A GTM with a constant turbulent diffusivity adequately describes the turbulent heat flux in all three homogeneous flows, while in the inhomogeneous flow, forcing gradient transport to the measured heat flux results in negative diffusivity over a part of the flow. Direct evaluation from the present and other measurements shows that none of the generalizations of the GTM is adequate. It is suggested that the large eddy transport in inhomogeneous shear flows is the possible cause for the failure of the GTM, and a simple criterion set forth here shows that its apparent success in symmetrically heated free shear flows is largely a happenstance caused by the imposed boundary conditions.

Nomenclature

- d = diameter of the cylinder
- D = turbulent diffusivity
- h = height of the wind-tunnel test section
- l_θ = half the distance between half-maximum θ' points
- l = characteristic length scale for turbulent diffusion, $= V\tau$
- L_f = longitudinal integral length scale
- L_g = transverse integral length scale
- L_θ = half the distance between $T_{max}/2$ points
- M = mesh size of the grid
- q' = $(\overline{u_i u_i})^{1/2}$
- T = mean temperature rise above the ambient
- u_i = velocity fluctuation in the direction i
- U_i = mean velocity in the direction i
- V = characteristic velocity scale for turbulent diffusion
- V_c = characteristic bulk convection velocity
- w = wake defect velocity
- x_i = coordinate axes; $i=1$ is along the flow and $i=2$ is along the (transverse) direction of maximum shear
- θ = temperature fluctuation
- τ = characteristic time scale for turbulent diffusion

Suffixes

0 = centreline value
max = maximum value
′ = root-mean-square value

Introduction

Virtually all the "classical turbulent theories" model turbulent transport of momentum, heat or a passive contaminant by linear mean gradient models. This hypothesis, probably attributable to *de St. Venant* or to *Boussinesq*, has appeared in different forms – usually incorporating an ad hoc estimate of the proportionality coefficient, the "eddy viscosity" or "eddy diffusivity" (for example, [1, 2]).

In spite of the enormous development in turbulence modelling witnessed over the last two decades (e.g., [3]), the simplicity of the gradient transport models (GTM) – often with enough adjustable parameters – appears to be responsible for their persistent use in some areas of engineering practice, especially in meteorology and oceanography. The GTM has been used in modelling such diverse physical phenomena as the growth and dispersal of populations [4], the solar radiation in dynamical models of terrestrial climate (for an extensive review, see [5]), etc. It is thus no surprise that a recurring concern has been voiced on the general conditions for the validity of the GTM [6–8]. Among other things, gradient transport models require that the characteristic scale of the turbulent transporting mechanism must be small compared with the dimension characteristic of the inhomogeneity of the mean transported quantity. It has been pointed out several times [6–8] that nearly all turbulent flows violate this basic a priori requirement, and yet a reasonable degree of success has been claimed for the GTMs, especially in free shear flows. Therefore, a study directed towards determining the reasons for this (apparent) success of the GTMs appeared to be worthwhile; this is one of the major goals of the present study.

It is perhaps apt to quote the following from *Saffman* [9] here because it reflects our current concern rather well: "The continual preaching against the eddy diffusivity hypothesis... has not served any useful purpose. The effort would have been better spent trying to understand the reasons for the apparent success and the circumstances in which the hypothesis must (not ought to) fail...". We hoped that some light would be shed on this issue if measurements were made in the three classes of flows in which a priori arguments would suggest that GTM

a) 'ought to' and does work,
b) ought not to but does work,
c) ought not to and does not work.

Clearly, there are no turbulent flows in which GTM 'ought to' work, but homogeneous flows offer the best possible chance of success. (We shall briefly return to this point somewhat later.) Consequently, we set up some experiments in homogeneous shear flows. Surprising success has often been claimed for the GTMs in free shear flows, and so we sought to deliberately create conditions in a free shear flow in which GTM would not work. Together, these constituted our experimental effort in this study. Presentation of the experimental data in these flows with a view to spur possible development of more sophisticated turbulence modelling is a secondary purpose of our study.

Several instances can now be quoted which, over the years, have demonstrated the inadequacies of the GTMs [10–19]. Mindful of these inadequacies to turbulence, several formal generalizations have been proposed [8, 20–22]. Alternatively, it has been thought that formal generalizations do not address themselves to the heart of the problem, and so some ad hoc 'corrections' (which could even be drastic), based on a totally different physical mechanism, have also been proposed (for example [23]). However, the validity of any of these models in more than one situation remains to be tested. This is the second major purpose of the present study. We tested these various generalizations not only for our flow in which GTM does not work, but also in another similar flow [23] which was selected for this purpose chiefly on the basis of the reasonable degree of detail to which the appropriate data have been published.

In this paper, we shall be concerned almost exclusively with the turbulent heat transport. The appropriate form of the diffusivity concept can then be written as

$$-\overline{u_i\theta} = D_{ij}\,(\partial T/\partial x_j)$$

where D_{ij} is the second-order diffusivity tensor. (The diffusivity tensor is not diagonal, or even symmetric; using a combination of experiment in homogeneous shear flows, theory and informed speculation, *Tavoularis* and *Corrsin* [24] have evaluated several components of D_{ij}.) In the simplest case with the dominant mean temperature gradient essentially in the direction x_2, the GTM gives

$$-\overline{u_2\theta} = D_{22}(\partial T/\partial x_2) \equiv D\,(\partial T/\partial x_2)\ , \tag{1}$$

where D is the diffusivity. It is in this simple form that the GTM will concern us here.

Flow Configuration

Figure 1 shows a schematic representation of the experimental configurations used here. The wind tunnel was of the open-return type with a nominal test section 30 cm × 30 cm, and about 3.65 m long. Air flow was created by two axial fans in tandem. An essentially constant pressure field was created by adjusting the vertical walls.

Three homogeneous flows and one inhomogeneous shear flow were studied. One of the homogeneous flows was the uniform flow produced behind a square-mesh biplane cylindrical-rod grid with 2.54 cm mesh and solidity 0.36. The mean velocity was nominally uniform across the test section. The temperature field was created by electrically heating each of the horizontal rods of the grid separately, so that a desired inhomogeneous mean temperature field could be created (Fig. 1a).

The two other homogeneous flows with nominally uniform positive or negative mean velocity gradient were created by placing a shear generator 20 mesh sizes downstream of the turbulence-generating grid (see Fig. 1b,c). The shear generator is an array of horizontal non-uniformly spaced cylindrical rods [25] and could be inserted in the wind-tunnel such that dU_1/dx_2 could be positive (Fig. 1b) or negative (Fig. 1c). The temperature field was created exactly as in the uniform flow case.

Lastly, the inhomogeneous free shear flow studied here is the wake of a circular cylinder ($d = 1.1$ cm) in which the temperature field was produced by heating three thin parallel wires (diameter 0.127 mm) mounted asymmetrically with respect to the cylinder on a wooden frame which could be inserted (see Fig. 1d) at any of the three locations ($x_1/d = 1.2$, 2.3, and 46) downstream of the cylinder. Also, the spacing be-

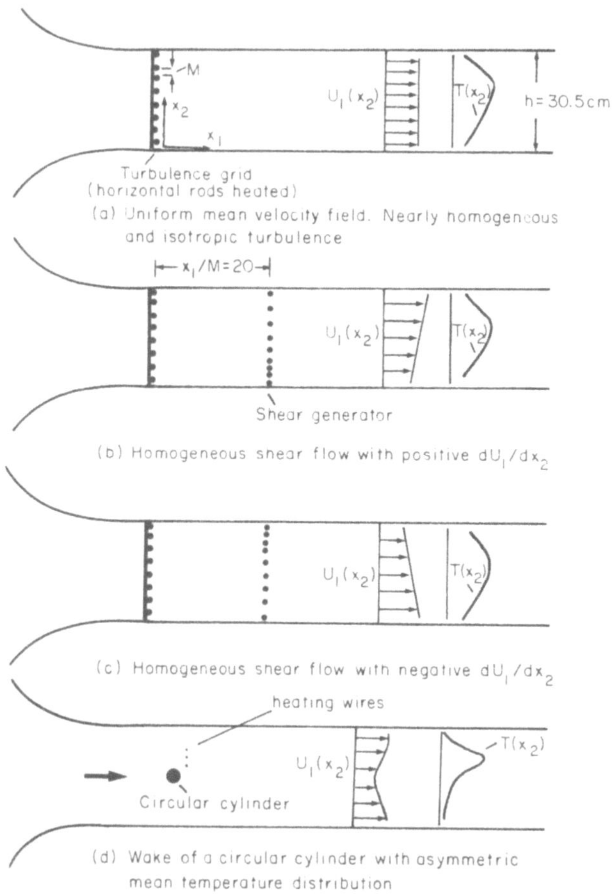

Fig. 1. Schematic of experimental set-up

tween the wires, their transverse location with respect to the cylinder and the heating current in each of the wires could be adjusted independently to obtain within limits any desired mean temperature distribution.

Instrumentation

Mean velocity U_{10} along the tunnel centreline was measured with a pitot-static tube. The mean velocity profile $U_1(x_2)$ and the velocity fluctuations u_1 and u_2 were measured with a DISA 55P51 gold-plated X-wire probe, with sensing elements 5 μm in diameter and 1 mm in length, powered by two DISA 55D01 constant temperature anemometers; dc power supplies were used to minimize the noise level. The mean temperature profile $T(x_2)$ and the reference temperature upstream of the grid were measured with two Fenwal Electronics GC32M21 thermistor probes. The temperature fluctuation θ was measured with a DISA 55P31 platinum wire probe with the wire length of 0.4 mm and diameter 1 μm. The temperature wire was positioned vertically at a distance of about 0.5 mm from the nearest wire of the X-wire probe, and was operated at a constant current of 0.3 mA on a home-made constant current source [26]. The operating current was low enough to render the velocity sensitivity of the temperature wire negligible. The temperature contamination of the velocity signals was eliminated by correcting them with the instantaneous local temperature measured with the temperature wire [27]. The vertical position of the probes was adjusted with a variable speed motor and a gear mechanism.

All signals were amplified and low-pass filtered at an upper cut-off frequency of 5 kHz. The signals were also corrected for noise assuming that the noise is statistically independent of the signal. The signals were digitized and processed on a DEC PDP 11/40 digital computer.

Results in Homogeneous Flows

Mean Velocity and Mean Temperature Distributions

Measurements were made at $x_1/M = 60$ and 128 but, in most cases, only those made at $x_1/M = 60$ are reported here. In all the cases, there was a significant region of two-dimensionality in the mean quantities.

Figure 2 shows the distribution of mean velocity and mean temperature rise for the three homogeneous flows. For the uniform flow experiment, the mean velocity was uniform across the test section to within about 2% of the centreline velocity of 17.2 m s^{-1}. The grid mesh Reynolds number was about 29,100. The two other flows had roughly linearly varying mean velocity profiles (except for the last point in each case on the low velocity side). The centreline velocity U_{10} in both cases was 16.0 m s^{-1}, and $|dU_1/dx_2|$ was 17.9 s^{-1}.

In all three cases, the mean temperature distributions were quite similar. The maximum mean temperature rise of about 2.7°C was low enough to consider heat as a passive scalar. This was also verified by noting that the measured root-mean-square velocity intensities with and without heating were essentially the same. A measure of the inhomogeneity of the temperature field is given by the parameter $(dT/dx_2)\, L_g/T_{max}$. In the present experiments, the highest value assumed by this parameter was about 0.5, signifying a sizeable inhomogeneity.

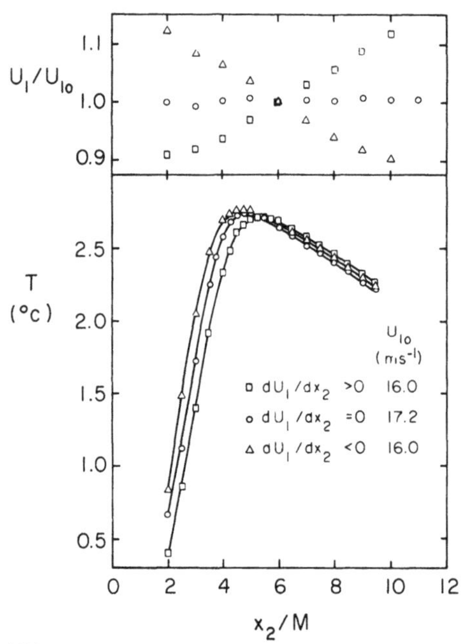

Fig. 2. Transverse distributions of mean velocity and mean temperature

Root-Mean-Square Intensities

Figure 3 shows the transverse distribution of the normalized root-mean-square velocity fluctuations in the central two-thirds of the tunnel hight. For the uniform grid flow, both u_1' and u_2' are uniform to within ±5%. The ratio $u_1'/u_2' \cong 1.15$, quite comparable to that in other similar flows [28]. For the shear flows, on the other hand, there is a ±12% variation in u_1'/U_{10} and about ±8% variation in u_2'/U_{10}. For the present purposes, these distributions were considered sufficiently homogeneous in the transverse direction. On the average, $u_1'/u_2' = 1.23$. At $x_1/M = 128$, the turbulence intensity distributions are homogeneous to a somewhat better accuracy. Figure 4 shows the transverse distribution of the normalized root-mean-square temperature fluctuation.

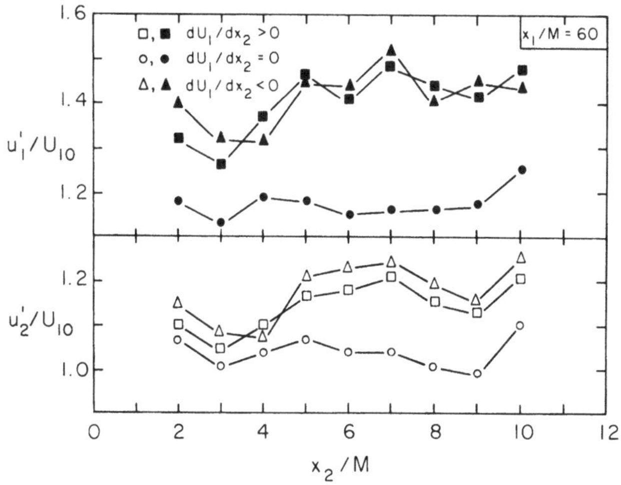

Fig. 3. Transverse variation of turbulence intensities

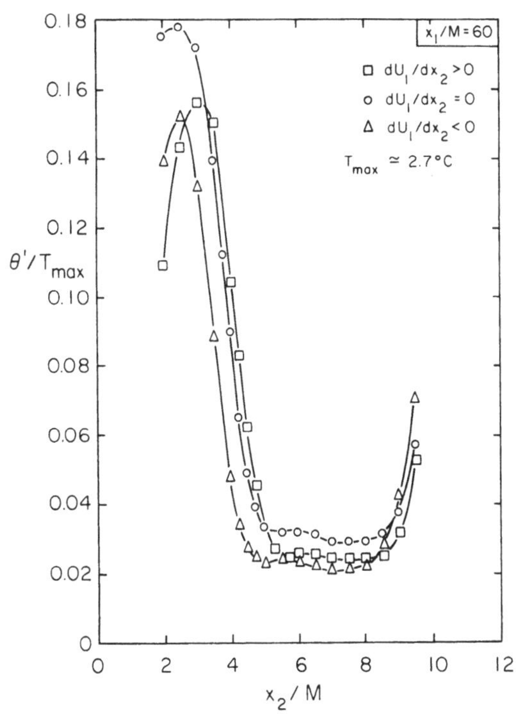

Fig. 4. Transverse distribution of the root-mean-square temperature fluctuation

Heat Transport in the x_2-Direction

Figure 5 shows how at $x_1/M = 60$ the correlation coefficient $-\overline{u_2\theta}/u_2'\theta'$ varies with respect to x_2/M. It is seen in comparison with Fig. 2 that $\overline{u_2\theta}$ changes sign essentially where the mean temperature gradient vanishes. Figure 6 in which $\overline{u_2\theta}$ is plotted against the corresponding total values of dT/dx_2 shows this more clearly.

Data are presented for both $x_1/M = 60$ and 128. A 6th order polynomial was fitted to the measured mean temperature distribution to obtain dT/dx_2. The estimated error bounds for both quantities are shown in the figure. It is clear that in all three cases $\overline{u_2\theta}$ is zero when dT/dx_2 is zero, and a gradient transport model with constant diffusivity is quite satisfactory.

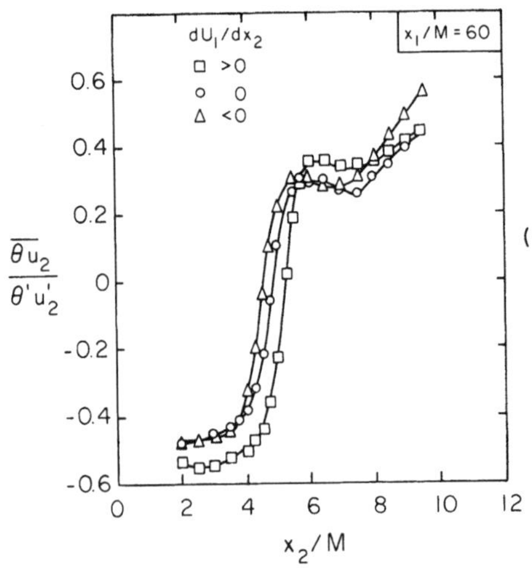

Fig. 5. The coefficient of correlation between transverse velocity and temperature fluctuations

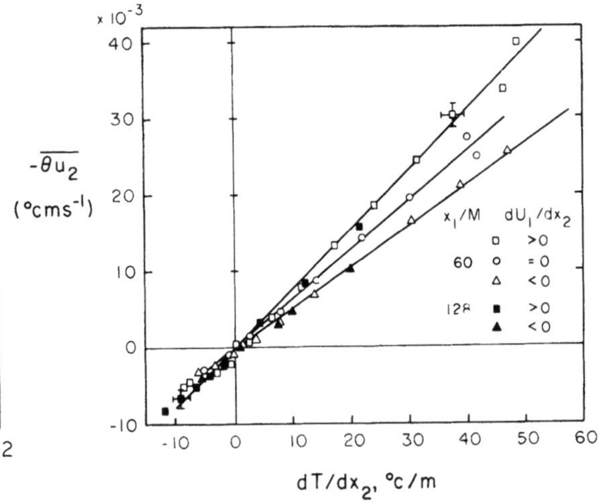

Fig. 6. The variation of the turbulent heat flux with mean temperature fluctuations

Table 1. The turbulent diffusivity, and the characteristic length and time scales of turbulent diffusion in the present homogeneous flows

dU_1/dx_2	D/γ	l [cm]	τ [ms]	l/L_g
> 0	36.6	0.42	22.8	0.11
= 0	30.5	0.36	20.1	0.35
< 0	25.5	0.29	15.8	--

Table 2. The turbulent diffusivity, and the characteristic length and time scales of turbulent diffusion in the *Tavoularis-Corrsin* flow [27]

x_1/h	D/γ	l [cm]	τ [ms]	l/L_g
6	59.5	0.44	15.4	0.11
7.5	72.5	0.48	15.2	0.11
9.5	93.0	0.53	14.6	0.10
11.0	114.0	0.59	14.5	0.10

The diffusivities evaluated from slopes of the straight lines in Fig. 6 are given in Table 1 (the molecular diffusivity $\gamma = 0.21 \times 10^{-4}$ m² s⁻¹). Interestingly, the turbulent transport in the case with $dU_1/dx_2 > 0$ is more efficient, and that in the case with $dU_1/dx_2 < 0$ is less efficient, than that in the uniform velocity flow. The reason for this is not immediately apparent (for example, from the equation for $\overline{u_2\theta}$, since the mean velocity gradient does not appear explicitly in such an equation).

For the case of constant diffusivity, we can write [29]

$$D = Vl = V^2\tau \quad . \tag{2}$$

Presumably, V and τ must respectively be the root-mean-square velocity and integral time scale of velocity following a material point. However, we shall assume that $V = u_2'$ (this is strictly true only for stationary homogeneous turbulence), and compute the values of l and τ from the measured diffusivity. These are also shown in Table 1.

In the present flow with $dU_1/dx_2 > 0$, the measured (Eulerian) transverse integral scale of turbulence L_g at $x_1/M = 70$ was about 40 mm. (The integral length scales here and elsewhere in this paper were obtained by evaluating the area up to the first crossing under the appropriate correlation curve. For a discussion of this point, see [30]). It should be remarked that the integral length scale in this flow is one order of magnitude larger than the length scale l characteristic of turbulent diffusion. No measurements of L_g were made for the flow with $dU_1/dx_2 < 0$, but the same conclusion can be expected to hold here also. The implication is that for homogeneous uniformly sheared flows, the eddies most effective in turbulent transport are *not* of the order of the integral scale, but an order of magnitude smaller. On the other hand, for the homogeneous non-sheared turbulence ($dU_1/dx_2 = 0$), L_g estimated from the relation [31]

$$L_g/M = 0.065\,(x_1/M - 3)^{0.4}$$

is about 8.3 mm, which is only about $2^{1/2}$ times as large as l.

Asymmetrically Heated Wake

Measurements were made for several configurations of the heating wires and the wake generator, but results are presented here only for that shown in Fig. 7. Measurements were made only at one station ($x_1/d \cong 100$). It would have been desirable to have made the measurements further downstream (where the flow would be self-preserving to a better degree of approximation), but the limitations of accuracy in heat transport measurements for small T led to this choice. Figure 7 shows transverse profiles of normalized velocity defect w/w_0 ($w_0 = 1.78$ m s⁻¹), u_2'/w_0, T, θ'/T_{max} and L_f/d; here L_f was obtained by evaluating the area up to the first crossing under the auto-correlation function of u_1 and converting the resulting integral time scale to a length scale via Taylor's "frozen field" approximation. It is seen that even at $x_1/d \cong 100$, the presence of the wires in the wake results in a slight asymmetry in the velocity near the maximum defect region, and stronger asymmetries in u_2' and L_f profiles.

Figure 8a shows a plot of the heat flux $-\overline{u_2\theta}$ and the mean temperature gradient $\partial T/\partial x_2$ at several points across the wake. Clearly, there is a small but finite region (the shaded region in the figure) in which the heat flux and the mean temperature gradient are of opposite sign implying negative diffusivity, or heat transport against the mean temperature gradient.

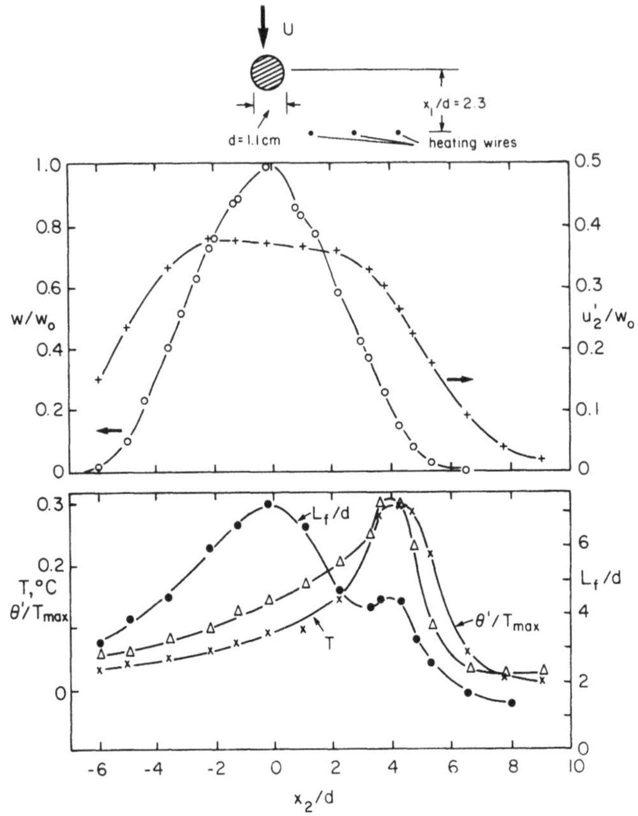

Fig. 7. A representative experimental configuration of the circular cylinder/heating-wire combination, and the resulting transverse distributions of mean defect velocity and temperature rise

A more direct demonstration of the inadequacy of the GTMs is given in Fig. 8b which shows that $-\overline{u_2\theta}$ when plotted against $\partial T/\partial x_2$ forms a closed loop. (If a GTM were applicable, the loop would collapse on to a single curve through the origin; if the diffusivity were also constant, the curve would be a straight line.) For the discussion to follow, the rough correspondence between the various key points of this loop and their physical location is indicated by the use of the same letters A, B, C, D and E in Fig. 8a, b, and the inset to Fig. 8b. Large negative x_2/d (say, around A in the inset) correspond to the vicinity of the origin in Fig. 8b. As x_2/d increases (algebraically), both $\partial T/\partial x_2$ and $-\overline{u_2\theta}$ increase until at B the (positive) maximum value of $\partial T/\partial x_2$ is reached. Beyond B, $\partial T/\partial x_2$ decreases but $-\overline{u_2\theta}$ does not keep pace with $\partial T/\partial x_2$ and is finite and large even in the vicinity of C where $\partial T/\partial x_2$ is small. For even larger x_2/d, $\partial T/\partial x_2$ is negative (path CD) until the negative maximum of the temperature gradient is attained at D; around D, $-\overline{u_2\theta}$ is still decreasing (see Fig. 8a), however. The path DE constitutes the return to $\partial T/\partial x_2 = 0$ as x_2/d approaches large positive values.

Corresponding regions in which the turbulent momentum transport $-\overline{u_1 u_2}$ occurs against the direction of mean velocity gradient have been observed in many different flow situations [10–20]. These regions have been called regions of "energy reversal" [13] or, more commonly in the later literature, as regions of "negative production", although the appropriateness of either term has been questioned. For example, it has been pointed out [32] that the total production terms are given by $-\overline{u_i u_j}\;\partial U_i/\partial x_j$, and that in the regions where $-\overline{u_1 u_2}\;(\partial U_1/\partial x_1)$ is negative, the other production terms $-(\overline{u_1^2} - \overline{u_2^2})\,\partial U_1/\partial x_1 - \overline{u_1 u_2}\;\partial U_2/\partial x_1$ are positive and (though small) of the right

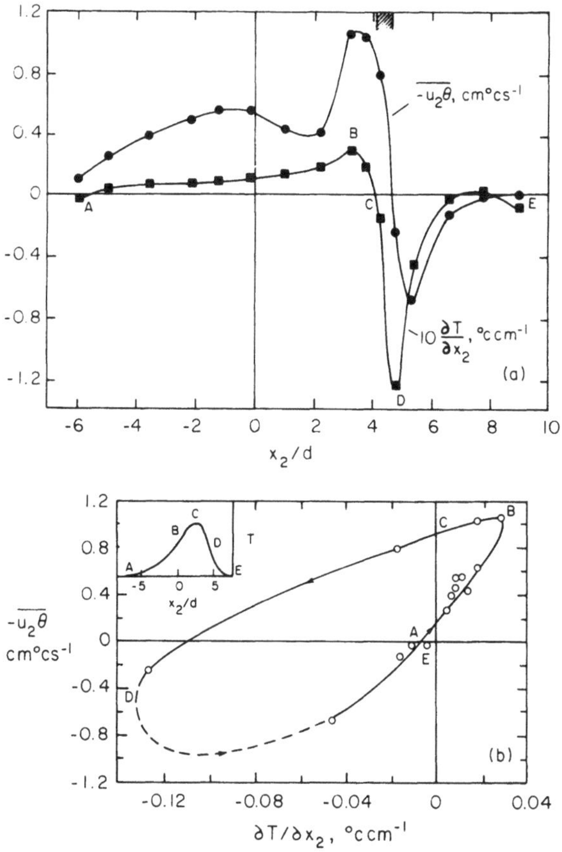

Fig. 8 a,b. The variation of turbulent heat flux (**a**) across the flow width and (**b**) with mean temperature gradient for the asymmetrically heated flow

magnitude to counteract locally the negative values of $-\overline{u_1 u_2}\ \partial U_1/\partial x_2$. However, this conclusion is negated by other measurements [14, 33] in which all the production terms except $-\overline{u_1 u_2}\ (\partial U_2/\partial x_1)$ were measured. (However, in these two latter cases, the magnitude of the negative values of the total production is very small.) Lastly, we mention *Hinze*'s [20] conclusion that even if the sum-total of production terms is negative, it does not imply energy transfer back to the mean flow.

In the heat transfer case, there are fewer measurements of the θ^2-production terms [18, 23, 34], but the balance of evidence does suggest that the total thermal production terms

$$-\overline{u_2\theta}\ \partial T/\partial x_2 + \overline{u_1\theta}\ \partial T/\partial x_1$$

add up to negative values in some small flow region. In the present wake, mean temperature profiles were not measured sufficiently closely to evaluate $\partial T/\partial x_1$ accurately, but rough estimates suggest that the inclusion of $-\overline{u_1\theta}\ (\partial T/\partial x_1)$ does not alter the sign of total production in any substantial way.

'Generalizations' of GTM

One obvious way to generalize the GTM for inhomogeneous situations is to replace the constant diffusivity by a space-dependent diffusivity. *Corrsin* [8] has noted that this is not a self-consistent generalization to adopt. Using a heuristic analogy based on generalized one-dimensional random walks, he inferred an alternative expression for the mean flux $\overline{F}$ of a passive scalar contaminant (concentration Γ) with a dominant mean concentration gradient $(\partial\Gamma/\partial x_2)$. *Corrsin*'s analysis is based essentially on a Taylor expansion for inhomogeneities and unsteadiness in the transport, with (effectively) the characteristic length scale l as the small parameter. His analysis shows that to O (l^2) in steady transport

$$-\overline{F} = D\,(\partial\Gamma/\partial x_2) + l\Gamma\,(\partial V/\partial x_2)\ . \tag{3}$$

The expression is valid for unsteady transport also, but to O (l).

Another generalization of simple GTM has been proposed by *Lumley* [21], based on the power series expansion of a continuous probability density function for displacement of tagged particles. The formulation appears to be in much the same spirit as a "Kramers-Moyal expansion". Keeping only the first two terms of this infinite series, *Lumley* showed that the net integrated flux is given by:

$$\overline{F} = \overline{\Gamma}\,\frac{d\,(\overline{\Delta x_2})}{dt} - \frac{1}{2}\,\frac{\partial}{\partial x_2}\left[\Gamma\,(x_2)\,\frac{d\,(\overline{\Delta x_2})^2}{dt}\right]\ . \tag{4}$$

Lumley argued that for an almost homogeneous situation

$$\frac{d(\overline{\Delta x_2})}{dt} = \frac{1}{4}\,\frac{\partial}{\partial x_2}\,\frac{d(\overline{\Delta x_2})^2}{dt}$$

so that Eq. (4) becomes

$$-\overline{F} = D\,\frac{\partial\overline{\Gamma}}{\partial x_2} + \frac{1}{2}\overline{\Gamma}\,\frac{\partial D}{\partial y}\ , \tag{5}$$

where we have used the simple result from the Brownian motion theory that the diffusivity D is given by

$$D = \frac{1}{2}\,\frac{d(\overline{\Delta x_2})^2}{dt}\ .$$

For a homogeneous situation, $d\,(\overline{\Delta x_2})/dt = 0$, so that Eq. (4) reduces to the classical GTM.

Kranenburg [22] derived an alternative expression based on the Lagrangian formulation of particle dispersion. If one neglects viscosity, the concentration of a passive scalar quantity is conserved along a fluid particle trajectory, and one can then write:

$$\Gamma\,(\boldsymbol{x}, t) - \Gamma\,(\boldsymbol{x} - \int_{t-\tau_K}^{t} V(t'\,|\,\boldsymbol{x}, t)\,dt', t - \tau_K) = 0$$

where $V(t'|\boldsymbol{x}, t)$ is the time (t') dependent Lagrangian velocity of the fluid particle passing through $\boldsymbol{x}$ at time t. *Kranenberg* applied the mean value theorem to the integral in the above equation, and took the analysis through the following steps. First, he assumed that τ_K was *small* enough to permit the particle trajectories to be approximated by straight lines (with the stochasticity entering into the problem through the stochasticity in τ_K). Then, using the Eulerian conservation equation for the scalar function Γ evaluated in the preceding steps, he obtained, in the limit of *large* diffusion times, the result that

$$-\bar{F} = D\,(\partial\Gamma/\partial x_2) + l\,\frac{\partial}{\partial x_2}\,(D\partial\Gamma/\partial x_2)\ . \tag{6}$$

The self-consistency of the assumptions is not immediately obvious.

All these three models share a few things in common. They introduce no additional adjustable constants into the respective expressions and seek to obtain (in a more or less rational framework) corrections necessitated by inhomogenities in a process that is basically gradient-transport type; in some sense, the departures from the GTM are perceived to be small.

There is general awareness that in turbulent shear flows, transport occurs not only by the diffusive motion of the small eddies (as envisaged in the GTMs) but also by the motion of large eddies comparable in size to the flow width. It appears unlikely that the transport due to these bulk motions will be proportional to the local concentration gradient; if the large-scale motion is potent in its transport characteristics, it appears equally unlikely that the total transport can be successfully evaluated by applying small corrections to the gradient transport type mechanism. Motivated by these considerations, *Townsend* [35] made a rough model for the total transport. In its simplest sense, the model can be represented by

$$-\bar{F} = D\,(\partial\Gamma/\partial x_2) + V_c\Gamma_{max} \tag{7}$$

where V_c is a bulk convection velocity (to be determined externally) and Γ_{max} is the maximum concentration difference within the flow field.

Motivated essentially by these considerations, *Beguier* et al. [23] proposed a somewhat more specific model for the bulk convection. Their model is no more than a straightforward analogy to *Beguier*'s [36] momentum transport model, whose brief description is therefore approximate. Let A and B be the two points in the flow where $\partial U_1/\partial x_2 = 0$ and $\overline{u_1u_2} = 0$, respectively, and l_m be the distance between A and B. Region AB is the negative production region. Since this region is generally thin, the mean concentration gradient at B can be written as

$$\left(\frac{\partial U_1}{\partial x_2}\right)_B = l_m\left(\frac{\partial^2 U_1}{\partial x_2^2}\right)_A\ .$$

Beguier then models the momentum transport due to bulk motion by

$$D_m\left(\frac{\partial U_1}{\partial x_2}\right)_B = D_m l_m\left(\frac{\partial^2 U_1}{\partial x_2^2}\right)_A\ ,$$

where D_m is the momentum eddy diffusivity which itself is assumed to be given by

$$D_m = K_m W^2\left(\frac{\partial q'}{\partial x_2}\right)_A$$

with W as the 'width' of the flow and K_m as a free constant. In analogy to this model, the total heat transport can be written as

$$-\overline{u_2\theta} = D\,(\partial T/\partial x_2) + K_\theta l_\theta^3\,(\partial q'/\partial x_2)\,(\partial^2 T/\partial x_2^2) \tag{8}$$

where K_θ is a constant and $(\partial^2 T/\partial x_2^2)$ is to be evaluated at the point where $(\partial T/\partial x_2) = 0$. This is effectively the model given in [23].

Test of Generalized Models

All expressions for the models briefly discussed above are given to the first order of 'correction'. A knowledge of the Lagrangian root-mean-square velocity and integral time scales across the shear flows under consideration is now needed, but is at the moment not available. For the specific purposes of testing these models, we shall replace l by the Eulerian integral scale L_f, V by the Eulerian root-mean-square velocity u'_2 and the diffusivity D by $L_f u'_2$. A more appropriate definition based on Lagrangian scales would presumably differ from the present one, but would not lead to qualitatively different conclusions. Particularizing the various models to heat transport (i.e., replacing the concentration Γ by the temperature T), we have:

$$-\overline{u_2\theta} = D\,(\partial T/\partial x_2) \equiv R_G \tag{9a}$$

$$-\overline{u_2\theta} = D\,(\partial T/\partial x_2) + L_f T\,(\partial u'_2/\partial x_2) \equiv R_c \tag{9b}$$

$$-\overline{u_2\theta} = D\,(\partial T/\partial x_2) + (T/2)\,(\partial D/\partial x_2) \equiv R_L \tag{9c}$$

$$-\overline{u_2\theta} = D\,(\partial T/\partial x_2) + L_f\,\partial/\partial x_2\,(D\,\partial T/\partial x_2) \equiv R_K \tag{9d}$$

$$-\overline{u_2\theta} = D\,(\partial T/\partial x_2) + K_\theta l_\theta^{\,3}\,(\partial q'/\partial x_2)\,(\partial^2 T/\partial x_2^{\,2}) \equiv R_B \tag{9e}$$

$$-\overline{u_2\theta} = D\,(\partial T/\partial x_2) + V_c T_{max} \equiv R_T \tag{9f}$$

Referring now to Fig. 8b, any improved alternative to the GTM must be capable of collapsing the closed loop on to a single curve or nearly so. This is the simple criterion

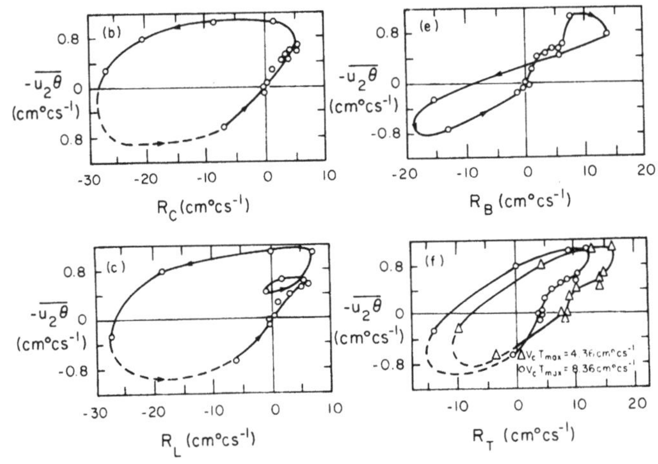

Fig. 9. A test of generalizations of GTM for the present asymmetrically heated flow

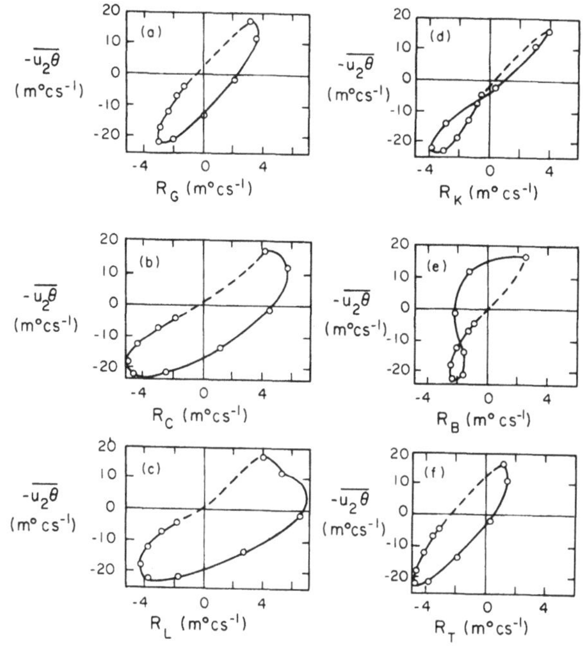

Fig. 10. A test of generalizations of GTM for asymmetrically heated mixing layer [23]

that we shall adopt as a measure of success of any given model. From both Figs. 9 and 10, the models due to *Corrsin* [8] and *Lumley* [21] appear to aggravate rather than improve the situation. The *Kranenburg* model appears to work reasonably well for the mixing layer data [23], but has poor performance in the present flow. For testing Eq. (9e) using the present flow data, we have replaced q' by $[3(u_1^2 + u_2^2)/2]$. This and other assumptions implied in Eq. (8) introduce an uncertainty in the precise value of K_θ. We tried several values of K_θ, and the best value (Fig. 9e) was found to be 0.015. Note that in contrast to the suggestion of [23], this constant assumed a positive value: the significance of a model whose constant changes sign from one flow to another is not clear. In the simple version of *Townsend*'s model, the correction due to the bulk transport is a constant number on the right-hand side, and this can only translate the closed loop without either shrinking or enlarging it. The effect is shown in Figure 9f for two values of the correction, and in Figure 10f for one.

Discussion

In the homogeneous shear flows examined here, the length scale characteristic of the eddies most effective in transporting heat were found to be small compared with the integral scale of turbulence (Table 1), or the scale of inhomogeneity of the temperature field. The transport is thus mainly a diffusive effect due to these small eddies and it comes as no surprise that the GTM works as well as it does. This can be shown also formally by noting that all the criteria listed by *Corrsin* [8] as being necessary for the GTM to work are satisfied in these flows. The only one requiring some comment is the condition that

$$\frac{l^2}{24}\,\frac{(\partial^3 T/\partial x_2^3)}{(\partial T/\partial x_2)} \ll 1 \ . \tag{10}$$

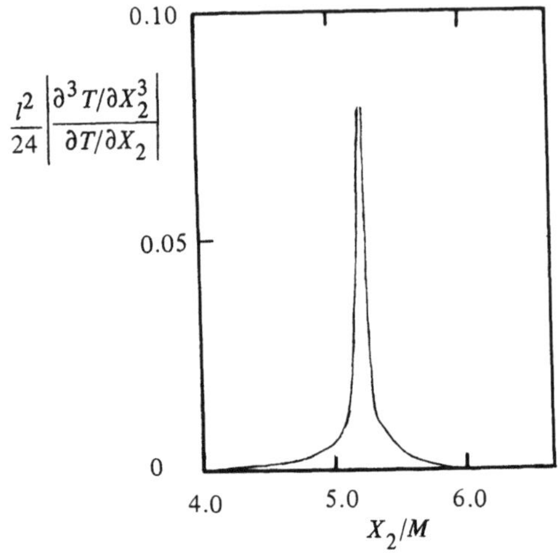

Fig. 11. A test of validity of the GTM for turbulent transport in the homogeneous flow with $dU_1/dx_2 = 0$. Note that the required condition (10) is violated in a very localized region, which is thus of no consequence

This is effectively a size limitation on the "mean free path" l of the diffusion process. Clearly, there is always a region in the flow where the above inequality is not satisfied, but this is such a local aberrant (see Fig. 11) that its effects do not manifest significantly.

Our experiments have emphasized that the GTM which appears perfectly adequate with a constant diffusivity for the homogeneous flows cannot handle inhomogeneous flows even qualitatively. The chief difference between the homogeneous and inhomogeneous shear flows is the dominance of the large structures in the latter; it appears that the large structures are responsible for a sizeable fraction of the transport process, and it is in the modelling of these effects that none of the generalizations of the GTM discussed earlier has had reliable success. One of the currently held views is that a turbulent shear flow is essentially a consequence (and not the cause) of these transport-efficient large structures and their mutual interactions. If this view is correct, an altogetherly different approach, which does not even invoke a mean field, is necessary before the problem can be resolved.

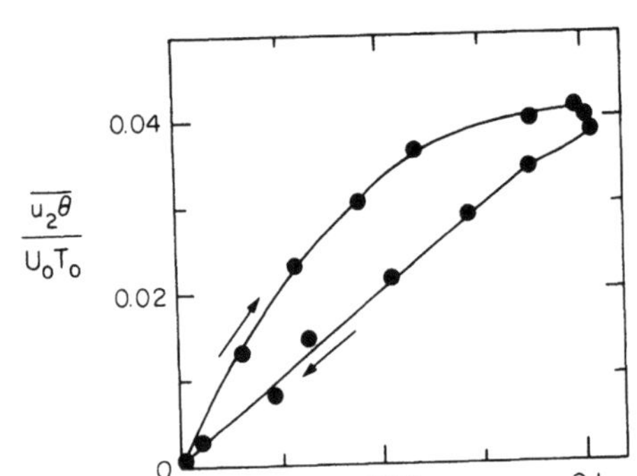

Fig. 12. The variation of turbulent heat flux with mean temperature gradient in a symmetrically heated co-flowing jet [37]

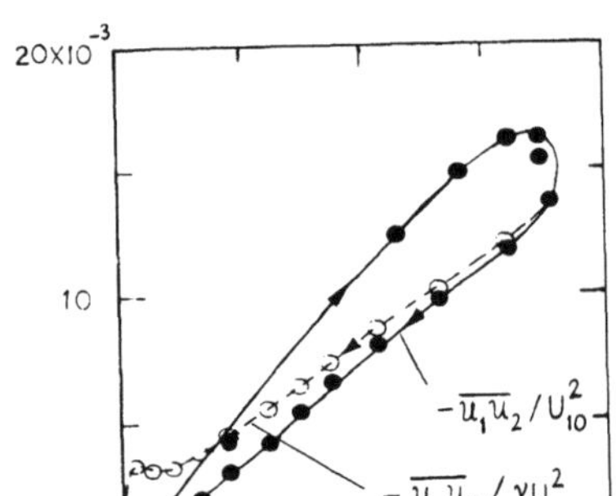

Fig. 13. The variation of turbulent momentum flux with the mean velocity gradient in an axisymmetric jet [38]. Notice how the introduction of the intermittency factor 'over-corrects' the heat flux values

Finally, we examine briefly the reason why GTM has often been claimed to work *well* in some free shear flows. Figure 12 shows a plot of $\overline{u_2\theta}$ vs $(\partial T/\partial x_2)$ – now appropriately normalized – for one half of a symmetrically heated co-flowing jet [37]. It is again seen that the data form a loop instead of a single curve. However, the flow symmetry about the centreline forces both $\overline{u_2\theta}$ and $\partial T/\partial x_2$ to be zero at the same point, so that the return part of the loop is now constrained to go through the origin. This results in a much smaller loop than would be the case if asymmetries existed. This is the reason for the apparent success of the GTM in symmetric flows; the large eddies are no less effective in transporting heat in these flows.

So far, we have concentrated on the heat transport, but the present conclusions should be relevant also for the momentum transport. We may now briefly consider the momentum transport in an axisymmetric jet [38], and examine the applicability of the GTM in the context of the smallness or otherwise of the appropriate loop. Figure 13 shows that a plot of $\overline{u_1 u_2}$ as a function of $\partial U_1/\partial x_2$ indeed forms a loop that is not too wide, again for reasons of imposed boundary conditions; GTM should thus work reasonably, as indeed it does. If one took account of the outer intermittency of the jet, it is seen from Fig. 13 that no qualitative changes in the conclusions result. Towards the outer edge, the intermittency 'over-corrects' the expectation from the GTM.

Acknowledgement. This research was supported by the Atmospheric Science Division of the National Science Foundation.

References

1. Monin, A.S., Yaglom, A.M.: *Statistical Fluid Mechanics: Mechanics of Turbulence,* ed. by J.L. Lumley. Vol. 1 M.I.T. Press, Cambridge (1971)
2. Hinze, J.O.: *Turbulence* 2nd ed. McGraw-Hill, New York (1976)
3. Launder, B.E.: "Heat and Mass Transport", in *Turbulence*, ed. by P. Bradshaw, Topics in Applied Physics, Vol. 12 Springer, Berlin, Heidelberg, New York 1978 pp. 231–287
4. Cohen, D.S., Murray, J.D.: A generalized diffusion model for growth and dispersal in a population. J. Math. Biol. **12**, 237–249 (1981)
5. Saltzman, B.: A survey of statistical dynamical models in the terrestrial climate. Adv. Geophys. **20**, 183–304 (1978)
6. Batchelor, G.K.: Note on free turbulent flows, with special reference to the two-dimensional wake. J. Aeronaut. Sci. **17**, 441–445 (1950)
7. Corrsin, S.: Heat transfer in isotropic turbulence. J. Appl. Phys. **23**, 113–118 (1952)
8. Corrsin, S.: Limitations of gradient transport models in random walks and in turbulence. Adv. Geophys. **18**A, 25–60 (1974)
9. Saffman, P.G.: "Problems and Progress in the Theory of Turbulence", in *Structure and Mechanism of Turbulence* II, ed. by H. Fiedler, Lecture Notes in Physics, Vol. 7 (Springer, Berlin, Heidelberg, New York 1978), pp. 183–304
10. Eskinazi, S., Yeh, H.: An investigation of fully developed turbulent flows in a curved channel. J. Aeronaut. Sci. **23**, 23–31 (1956)
11. Mathieu, J., Tailland, A.: Etude d'un jet plan dirigé tangentiellement à une paroi. C.R. Acad. Sci. **256**, 2768–2771 (1963)
12. Gee, M.T., Bradshaw, P.: Turbulent wall jets with and without external stream. NPL Bull. **3252** (1960)
13. Eskinazi, S., Erian, E.F.: Energy reversal in turbulent flows. Phys. Fluids **12**, 1988–1998 (1969)
14. Erian, E.F.: Influence of pressure gradient on turbulent flow with asymmetric mean velocity. ASME-J. Appl. Mech. **91**, 901–904 (1969)
15. Hanjalic, K. Launder, B.E.: Fully developed asymmetric flow in a plane channel. J. Fluid Mech. **51**, 301–335 (1972)
16. Palmer, M.D., Keffer, J.F.: An experimental investigation of an asymmetrical turbulent wake. J. Fluid Mech. **53**, 593–610 (1972)
17. Fabris, G.: "Conditionally Sampled Turbulent Thermal and Velocity Fields in the Wake of a Warm Cylinder and Its Interaction with an Equal Cool Wake"; Ph.D. dissertation, Illinois Institute of Technology (1979)

18. Charney, G., Schon, J.P., Alcaraz, E., Mathieu, J.: "Thermal Characteristics of a Turbulent Boundary Layer with Inversion of Wall Heat Flux", Proc. of the Symposium on Turbulent Shear Flows, Penn. State University, University Park, 1977, pp. 15, 47–15.55
19. Morel, R., Awad, M., Schon, J.P., Mathieu, J.: "Experimental Study of an Asymmetric Thermal Wake", in *Structure and Mechanism of Turbulence* I, ed. by H. Fiedler. Lecture Notes in Physics, Vol. 75 (Springer, Berlin, Heidelberg, New York, 1978) pp. 36–45
20. Hinze, J.O.: Turbulent flow regions with shear stress and mean velocity gradient of opposite sign. Appl. Sci. Res. **22**, 168–175 (1970)
21. Lumley, J.L.: Modelling turbulent flux of passive scalar quantities in inhomogeneous flows. Phys. Fluids **18**, 619–621 (1975)
22. Kranenburg, C.: On the extension of gradient-type transport to turbulent diffusion in inhomogeneous flows. Appl. Sci. Res. **33**, 163–175 (1977)
23. Beguier, C., Fulachier, L., Keffer, J.F.: The turbulent mixing layer with an asymmetrical distribution of temperature. J. Fluid Mech. **89**, 561–587 (1978)
24. Tavoularis, S., Corrsin, S.: "Theoretical and Experimental Determination of the Turbulent Diffusivity Tensor in Homogeneous Turbulent Shear Flow", Proc. of the 3rd Symposium on Turbulent Shear Flows, University of California, Davis, 1981, pp. 15.24–15.27
25. Rose, W.G.: Results of an attempt to generate a homogeneous turbulent shear flow. J. Fluid Mech. **25**, 97–120 (1966)
26. Tavoularis, S.: A Circuit for the measurement of instantaneous temperature in heated turbulent flows. J. Phys. E **11**, 21–23 (1978)
27. Tavoularis, S., Corrsin, S.: Experiments in nearly homogeneous turbulent shear flow with a uniform mean temperature gradient. Part 1, J. Fluid Mech. **104**, 311–347 (1981)
28. Comte-Bellot, G., Corrsin, S.: The use of a contraction to improve the isotropy of grid-generated turbulence. J. Fluid Mech. **25**, 657–682 (1966)
29. Corrsin, S.: "Some Current Problems in Turbulent Shear Flow", in *Naval Hydrodynamics,* Proc. of the 1st Symp. on Naval Hydrodynamics, ed. by F. Sherman Publication No. 515,(National Academy of Science, National Research Council, Washington, D.C. 1957) pp. 373–400
30. Comte-Bellot, G., Corrsin, S.: Simple Eulerian time correlation of full and narrow-band velocity signals in grid-generated, isotropic' turbulence. J. Fluid Mech. **48**, 273–337 (1971)
31. Sreenivasan, K.R., Tavoularis, S., Henry, R., Corrsin, S.: Turbulent fluctuations and scales in grid-generated turbulence. J. Fluid Mech. **100**, 597–621 (1980)
32. Wilson, J.D.: Turbulent transport of mean kinetic energy in countergradient shear stress regions. Phys. Fluids **17**, 674–675 (1974)
33. Beguier, C.: Mesures de tensions de Reynolds dans un écoulement dissymétrique en régime turbulent incompressible. J. Méc. **4**, 319–334 (1965)
34. Fulachier, L., Keffer, J.F., Beguier, C.: Production negative de fluctuations turbulentes de température dans le case d'un cré neau de chaleur s'épanouissant dans une zone de mélange", C.R. Acad. Sci. **280**, pp. 519–522 (1975)
35. Townsend, A.A.: *The Structure of Turbulent Shear Flows,* (University Press, Cambridge 1956)
36. Beguier, C.: Ecoulements plans dissymétriques en régime turbulent incompressible, par. C.R. Acad. Sci. A**268**, 69–72 (1969)
37. Antonia, R.A., Prabhu, A. Stephenson, S.E.: Conditionally sampled measurements in a heated turbulent jet. J. Fluid Mech. **72**, 455–580 (1975)
38. Wygnanski, I.J., Fiedler, H.E.: Some measurements in the self-preserving jet. J. Fluid Mech. **38**, 577–612 (1969)

Calculations of Velocity-Scalar Joint pdf's

S.B. Pope

Massachusetts Institute of Technology, Department of Mechanical Engineering,
Cambridge, MA 02139, USA

Abstract

A joint pdf transport equation has been solved to calculate the properties of the self-similar, plane, turbulent jet. The joint probability density function (pdf) is that of the three velocity components and of the nozzle-fluid concentration, which is a conserved, passive scalar. The advantage of basing a turbulence closure on the joint pdf equation is that convective transport appears in closed form. Consequently, the gradient-diffusion assumption is avoided. For self-similar shear flows, the joint pdf is a function of five variables. Its transport equation is solved by a Lagrangian method with stochastic modelling of the unclosed terms. The calculated Reynolds stresses and triple velocity correlations agree well with the experimental data. Calculated pdf's and joint pdf's are also reported.

Nomenclature

C_R, C_ϕ	model constants
D	velocity dissipation term
E	conditionally-expected value
$f(V,\psi)$	velocity-scalar joint pdf
G	scalar dissipation term
$p(\psi)$	scalar pdf
$\langle p\rangle, p'$	mean and fluctuating pressure
R	redistribution term
r, θ, z	polar-cylindrical coordinates
$t, \Delta t$	time, time interval
$\boldsymbol{U}, \boldsymbol{u}$	velocity, fluctuating velocity
V	independent velocity variables
x, y, z	axial, lateral and spanwise coordinates
$y_{1/2}$	jet half width

Greek Symbols

Γ	molecular diffusion coefficient
ϵ	dissipation rate of turbulent kinetic energy
η	cross-stream variable $y/y_{1/2}$
μ	molecular viscosity
ϕ, ϕ'	scalar, fluctuating scalar
ψ	independent scalar variable
ω, ω^*	turbulent frequency, normalised frequency

Introduction

A modelled joint pdf equation [1] has been solved to calculate the one-point statistical properties of the self-similar plane turbulent jet. The equation solved is for the joint prob-

Permanent address: Cornell University, Mechanical and Aerospace Engineering, Ithaca NY 14853, USA

ability density function (pdf) of the three velocity components $\boldsymbol{U}(\boldsymbol{x},t)$ and a conserved passive scalar $\phi(\boldsymbol{x},t)$. With θ being the non-dimensional, cross-stream variable, the joint pdf is $f(\boldsymbol{V},\psi;\theta)$, where $\boldsymbol{V} \equiv V_1, V_2, V_3$ and ψ are the independent variables corresponding to $\boldsymbol{U}$ and ϕ. Since $f(\boldsymbol{V},\psi;\theta)$ is a function of five variables, the solution of its transport equation by a finite-difference method is impracticable. Instead, the transport equation is solved by a Lagrangian method with stochastic modelling of the unclosed terms. The simplest possible models are used – not the most accurate – the emphasis being on demonstrating the use of the joint pdf equation and the performance of the solution procedure.

Transport equations for pdf's are useful in modelling turbulent flows because non-linear one-point processes (such as convection and reaction) can be treated without approximation [1–3]. The transport equation for the joint pdf of a set of scalars $p(\psi;\boldsymbol{x},t)$ is particularly useful in reactive flows since the term pertaining to reaction appears in closed form, irrespective of the complexity and non-linearity of the reaction scheme. *Pope* [4] and *Janicka* et al. [5] have reported accurate calculations of premixed and diffusion flames based on the scalar pdf equation. For more than one or two scalars, finite-difference solutions are impracticable, but a Monte Carlo method [6] has been developed for the general case.

The great advantage of considering the joint pdf of velocity and scalars is that convective transport appears in closed form. Consequently, the assumption of turbulent transport by gradient diffusion is avoided. In fact, for a variable-density reactive flow, the terms pertaining to convection, reaction, buoyancy and the mean pressure gradient all appear in closed form. Only the effects of molecular mixing and the fluctuating pressure gradient need to be modelled. Models for these processes have been given in a recent paper [1] which contains a derivation and discussion of the joint pdf equation.

Self-similar free shear flows are good test cases for the modelled joint pdf equation. The condition of self-similarity can be used to reduce the number of spatial dimensions to one – namely, the normalized cross-stream coordinate θ. And for self-similar flows there is a wealth of experimental data. The plane jet was chosen for this initial study because of the simple coordinate system and boundary conditions.

In the following three sections, the joint pdf transport equation is presented, the solution procedure is described, and the results of the calculations are reported. The results include profiles of the mean velocity, Reynolds stresses and triple velocity correlations, all of which are compared with the experimental data of *Heskestad* [7]. Calculated scalar pdf's and joint pdf's are also reported but no data are available for comparison.

Joint pdf Equation

The joint pdf $f(\boldsymbol{V},\psi;\boldsymbol{x},t)$ contains all the one-point statistical information about the velocity $\boldsymbol{U}(\boldsymbol{x},t)$ and the conserved, passive scalar $\phi(\boldsymbol{x},t)$ in a constant-density turbulent flow. If $Q(\boldsymbol{U},\phi)$ is any function of $\boldsymbol{U}$ and ϕ, then its mean value (at any $\boldsymbol{x},t$) can be determined from the joint pdf by

$$\langle Q(\boldsymbol{U},\phi)\rangle = \int\int Q(\boldsymbol{V},\psi)\, f(\boldsymbol{V},\psi)\, d\boldsymbol{V}\, d\psi \quad . \tag{1}$$

Here and below, $\int d\boldsymbol{V}$ represents integration over the whole of the velocity space and, similarly, $\int d\psi$ represents integration over all possible values of ϕ. Substituting $Q(\boldsymbol{U},\phi) = U_j$ in Eq. (1) shows the mean velocity to be

$$<U_j> = \iint V_j f(V,\psi)\, dV\, d\psi \quad . \tag{2}$$

Similarly the Reynolds stresses are

$$<u_i u_j> = \iint (V_i - <U_i>)\,(V_j - <U_j>)\, f(V,\psi)\, dV\, d\psi \quad , \tag{3}$$

and the scalar fluxes are

$$<u_i\phi'> = \iint (V_i - <U_i>)\,(\psi - <\phi>)\, f(V,\psi)\, dV\, d\psi \quad . \tag{4}$$

Thus, any one-point statistic can be obtained from $f(V,\psi)$.

The pdf of the scalar alone $p(\psi; x, t)$ is obtained by integrating $f(V,\psi)$ over velocity space

$$p(\psi) = \int f(V,\psi)\, dV \quad , \tag{5}$$

and the joint pdf of velocity $g(V; x, t)$ is

$$g(V) = \int f(V,\psi)\, d\psi \quad . \tag{6}$$

Each of the pdf's satisfies the normalization condition

$$\int p(\psi)\, d\psi = \int g(V)\, dV = \iint f(V,\psi)\, dV\, d\psi = 1 \quad . \tag{7}$$

A transport equation for $f(V,\psi;x,t)$ can be derived from the conservation equations for U and ϕ [1]:

$$\begin{aligned} \frac{\partial f}{\partial t} + V_i \frac{\partial f}{\partial x_i} - \frac{\partial f}{\partial V_i}\frac{\partial <p>}{\partial x_i} &= \frac{\partial}{\partial V_j}\,[f\,\mathrm{E}(\partial p'/\partial x_j \,|\, U = V, \phi = \psi)] \\ &- \frac{\partial}{\partial V_j}\,[f\,\mathrm{E}(\mu\nabla^2 U_j \,|\, U = V, \phi = \psi)] \\ &- \frac{\partial}{\partial \psi}\,[f\,\mathrm{E}(\Gamma\nabla^2\phi \,|\, U = V, \phi = \psi)] \quad . \end{aligned} \tag{8}$$

(The fluid density ρ has been set to unity.) The terms on the left-hand side represent convection and the effect of the mean pressure gradient. These terms are in closed form and therefore require no modelling. The terms on the right-hand side of the equation contain (as unknowns to be modelled) conditionally-expected values.

The first term on the right-hand side

$$R(V,\psi;x,t) = \frac{\partial}{\partial V_j}\,[f\mathrm{E}(\partial p'/\partial x_j \,|\, U = V, \phi = \psi)] \quad , \tag{9}$$

represents the effects of the fluctuating pressure gradient. The term can be decomposed into three parts [1]: a transport term and two redistribution terms. As in Reynolds-stress models [8], the redistribution terms do not affect the mean velocity or the turbulent kinetic energy, but they redistribute the energy in velocity space. The *rapid* part of the redistribution is due to pressure fluctuations caused by mean-velocity gradients: the *Rotta* part [9] is due solely to the turbulence. Models for all three contributions to

$R(V,\psi;x,t)$ are available [1], but for this initial study only the Rotta term is included.

In Reynolds-stress models, the Rotta term corresponds to a linear return to isotropy, i.e.

$$\frac{\partial}{\partial t} <u_i u_j> = \ldots - C_R \, \omega (<u_i u_j> - 1/3 <u_l u_l> \delta_{ij}) \quad , \tag{10}$$

where C_R is a constant and $\omega(x,t)$ is the turbulent frequency (dissipation rate/turbulent kinetic energy). In the joint pdf equation, this term is modelled by a stochastic process that randomly reorientates the energy in velocity space. The effect on the Reynolds stresses is just that given by Eq. (10).

The second term on the right-hand side of Eq. (8)

$$D(V,\psi;x,t) = -\frac{\partial}{\partial V_j} \, [f \, \mathrm{E}(\mu \nabla^2 U_j \,|\, U = V, \phi = \psi)] \quad , \tag{11}$$

corresponds to dissipation – it does not affect the mean velocity but decreases the turbulent kinetic energy. Neglecting low-Reynolds-number terms, D is related to the dissipation [1] by

$$\epsilon = \int\!\!\int 1/2 \; V_i \; V_i \; D(V,\psi;x,t) \, dV \, d\psi \quad . \tag{12}$$

In the joint pdf equation, *Curl*'s coalescence/dispersal model [10] is used to represent this process. (Because the turbulent frequency ω is defined in terms of ϵ, no additional model constant arises).

The final term in Eq. (8)

$$G(V,\psi;x,t) = -\frac{\partial}{\partial \psi} \, [f \, \mathrm{E}(\Gamma \nabla^2 \phi \,|\, U = V, \phi = \psi)] \quad , \tag{13}$$

corresponds to scalar dissipation and is analogous to the velocity dissipation term D. The effect of the term is to reduce the scalar variance $<\phi'^2>$ without affecting the mean $<\phi>$. Again, *Curl*'s model [10] is used to represent the process. The model involves a constant C_ϕ which is the ratio of the velocity to scalar turbulent time scales. Following conventional modelling [11] we take $C_\phi = 2.0$.

The calculations for the self-similar plane jet are performed in the polar-cylindrical coordinates r, θ, z. The distance from the virtual origin is r, the angle to the plane of symmetry is θ, and z is the spanwise distance. Some of the results are reported in the conventional Cartesian coordinate system (x,y,z) in which x is the axial distance ($x = r \cos\theta$) and y is the lateral distance ($y = r \sin\theta$). The half-width of the jet is $y_{1/2}$. It is assumed that the turbulent frequency $\omega(r,\theta,z)$ does not vary across the jet – i.e. that ω is independent of θ. Then, the conditions for self-similarity require that the normalized frequency

$$\omega^* \equiv \omega y_{1/2} \,/<U>_0 \quad , \tag{14}$$

be a constant. ($<U>_0$ is the center-line velocity.) By transforming Eq. (8) into polar-cylindrical coordinates and applying the conditions of self-similarity, a transport equation is obtained for the joint pdf $f(V_r, V_\theta, V_z, \psi;\theta,t)$. This equation is obtained without invoking the boundary-layer assumptions and without neglecting any terms. It is solved for $0 \leqslant \theta \leqslant \theta_{max} = 0.5$, with symmetry conditions applied at $\theta = 0$ and with the boundary conditions at θ_{max} corresponding to potential flow with $\phi = 0$.

The values of the constants $\omega^* = 0.165$ and $C_R = 4.5$ are chosen to produce approximately the correct spreading rate ($dy_{1/2}/dx \cong 0.1$) and center-line turbulent kinetic energy ($1/2 \langle u_i u_i \rangle_0 / \langle U \rangle_0^2 \cong 0.065$). The value of C_R is greater than the usual value $C_R = 1.5$ [8] because the modelled redistribution term has to account for both the Rotta and rapid terms.

To summarize, in the joint pdf equation, Eq. (8), convective transport appears in closed form. Consequently, there are no Reynolds stresses, scalar fluxes or other velocity correlations to be modelled: the gradient-transport assumption is avoided. There are three unclosed terms representing, respectively, redistribution of kinetic energy in velocity space, dissipation of velocity fluctuations, and dissipation of scalar fluctuations. These three terms are modelled by simple stochastic processes, each of which proceeds at a rate proportional to the turbulent frequency ω. For the self-similar plane jet, the turbulent frequency is assumed to be uniform across the jet. A fuller description and analytic expressions for the models can be found in [1].

Solution Procedure

Since the joint pdf $f(V_r, V_\theta, V_z, \psi; \theta, t)$ is a function of five independent variables and time, it is obvious that its transport equation cannot be solved by a standard numerical method. The solution is obtained as the steady state of a transient calculation. The calculation advances from time t to time $t + \Delta t$ in two fractional steps. In the first step, the exact terms are treated by solving the equation

$$\frac{\partial f}{\partial t} + V_i \frac{\partial f}{\partial x_i} - \frac{\partial f}{\partial V_i} \frac{\partial \langle p \rangle}{\partial x_i} = 0 \quad , \tag{15}$$

by a Lagrangian method. In the second step, stochastic models are used to solve the equation

$$\frac{\partial f}{\partial t} = R + D + G \quad . \tag{16}$$

After the second step, the result is the pdf at time $t + \Delta t$ (according to Eq. 8) plus a truncation error of order Δt^2. The truncation error can be reduced at will by reducing Δt.

In the numerical procedure, the pdf is represented indirectly by a large number (N) of elements in the solution domain $0 \leqslant \theta \leqslant \theta_{max} = 0.5$. The nth element is located at $\theta^{(n)}$ and has the properties $U_r^{(n)}$, $U_\theta^{(n)}$, $U_z^{(n)}$, $\phi^{(n)}$. The elements are approximately uniformly distributed. Consequently, the number M of elements in the small sub-interval $\theta - 1/2\Delta\theta < \theta^{(n)} \leqslant \theta + 1/2\Delta\theta$ is

$$M \approx \Delta\theta N / \theta_{max} \quad , \tag{17}$$

Let $n_1, n_2, ..., n_M$ be the M elements in the sub-interval. Then, at location θ, an approximation to the mean of any function $Q(\boldsymbol{U},\phi)$ is

$$\langle Q(\boldsymbol{U},\phi) \rangle \cong \frac{1}{M} \sum_{m=1}^{M} Q(\boldsymbol{u}^{(n_m)}, \phi^{(n_m)}) \ . \tag{18}$$

Similarly, the joint pdf at θ can be approximated by the histogram formed from the M elements.

In the approximation of the mean, Eq. (18), there is a truncation error (of order $\Delta\theta^2$) and a statistical sampling error of order $M^{-1/2}$. These errors can, in principle, be reduced at will by decreasing $\Delta\theta$ and increasing $N\Delta\theta$. In practice, for realistic values of N and $\Delta\theta$, the error is large, and a more sophisticated method of determining means based on least-square cubic splines is used. But the principle is the same.

The solution of Eq. (15) is facilitated by the observation that it is the pdf transport equation for the Lagrangian system:

$$\frac{d\boldsymbol{x}}{dt} = \boldsymbol{U} \;, \quad \frac{d\boldsymbol{U}}{dt} = -\nabla \langle p \rangle \;, \quad \frac{d\phi}{dt} = 0 \;. \tag{19}$$

Thus, for the first of the fractional steps, the properties of each of the N elements evolve according to Eq. (19) for a time Δt.

In the second fractional step, stochastic models are used to solve Eq. (16). Pairs of elements that are close to each other in physical space interact to cause redistribution and velocity and scalar dissipation – the terms R, D and G. The solution domain is divided into K sub-intervals, of width $\Delta\theta = \theta_{max}/K$, and the ensembles of elements within each sub-interval are treated separately.

In Curl's model for the velocity dissipation (D), each ensemble is treated in turn and the following procedure is repeated n_D times, where

$$n_D = \omega \Delta t M \;. \tag{20}$$

Two elements, denoted by n and m, are selected at random from the ensemble, and their velocities are replaced with the values

$$\boldsymbol{U}^{(n)} = \boldsymbol{U}^{(m)} = 1/2(\boldsymbol{U}_0^{(n)} + \boldsymbol{U}_0^{(m)}) \;, \tag{21}$$

where $\boldsymbol{U}_0^{(n)}$ is the value of $\boldsymbol{U}^{(n)}$ before the process. This has the effect of decreasing the energy $\langle \boldsymbol{U} \cdot \boldsymbol{U} \rangle$ without affecting the mean $\langle \boldsymbol{U} \rangle$.

The implementation of Curl's model for the scalar dissipation (G) is precisely analogous. A pair of elements (n and m) is selected at random and the values of ϕ are replaced with

$$\phi^{(n)} = \phi^{(m)} = 1/2\,(\phi_0^{(n)} + \phi_0^{(m)}) \;. \tag{22}$$

This process is repeated

$$n_G = C_\phi\, \omega \Delta t M \;, \tag{23}$$

times, which has the effect of decreasing the scalar variance $\langle \phi'^2 \rangle$ without affecting the mean $\langle \phi \rangle$.

In the stochastic model of redistribution (R), a pair of elements is selected at random from the ensemble. The velocities of the two elements (n and m) are then replaced with the values

$$\boldsymbol{U}^{(n)} = 1/2\,(\boldsymbol{U}_0^{(n)} + \boldsymbol{U}_0^{(m)}) + 1/2\,\xi\, |\boldsymbol{U}_0^{(n)} - \boldsymbol{U}_0^{(m)}| \;, \tag{24}$$

and

$$\boldsymbol{U}^{(m)} = 1/2\,(\boldsymbol{U}_0^{(n)} + \boldsymbol{U}_0^{(m)}) - 1/2\,\xi\, |\boldsymbol{U}_0^{(n)} - \boldsymbol{U}_0^{(m)}| \;, \tag{25}$$

where ξ is a random vector of unit length, uniformly distributed on the unit sphere. This transformation from $U_0^{(n)}$, $U_0^{(m)}$ to $U^{(n)}$, $U^{(m)}$ corresponds to a random rotation of the elements in velocity space. Since neither their mean position nor their separation is altered, the transformation conserves both momentum and energy; the effect of the random rotation is to decrease the anisotropy.

This process is repeated n_R times, where

$$n_R = C_R \, \omega \Delta t M \quad . \tag{26}$$

The two fractional steps advance time by Δt. Starting from a specified initial condition, this sequence of operations is repeated until the statistically steady state is reached. The steady state is independent of the initial condition.

For the calculations reported in the next section, there were $N = 7{,}200$ elements, and $K = 48$ sub-intervals in the solution domain $0 \leqslant \theta \leqslant \theta_{max} = 0.5$. With a time step of $\Delta t = 0.2\, y_{1/2}/\langle U \rangle_0$, 400 time steps were taken. (The sampling error is decreased by time-averaging: hence the large number of steps). The total CPU time on a VAX computer was 27 min, at a cost of $ 16.

Results

The calculations reported in this section were obtained from the numerical solution of the modelled joint pdf equation. Calculated profiles of mean velocity, Reynolds stresses and triple correlations are compared with the hot-wire data of *Heskestad* [7]. Scalar pdf's and a joint velocity pdf are also reported, but there are no data for comparison. All the results are normalised with the center-line values $\langle U_0 \rangle$ and $\langle \phi_0 \rangle$.

Figure 1 shows the mean axial velocity profile. Although the calculations are performed in polar-cylindrical coordinates, the results are transformed to cartesian coordinates for comparison with the data. Thus Fig. 1 shows the mean velocity in the x-direction against the normalised y-coordinate,

$$\eta \equiv y/y_{1/2} \quad . \tag{27}$$

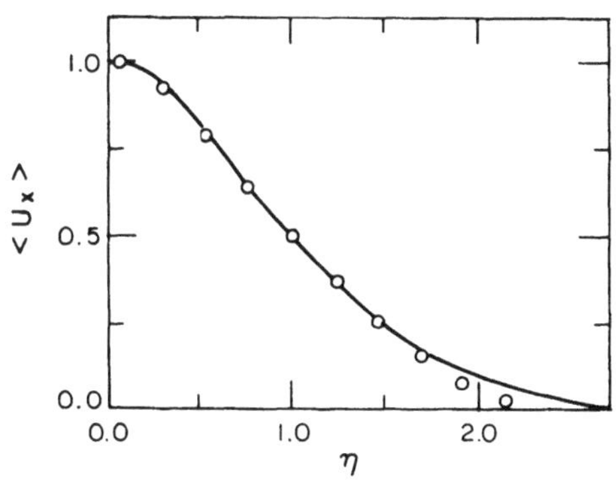

Fig. 1. Mean axial velocity against lateral distance: (——) calculation; (○) experiment [7]

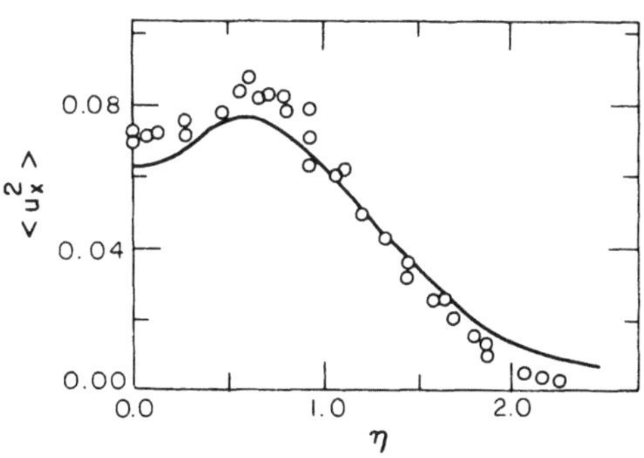

Fig. 2. Mean square axial velocity fluctuations against lateral distance: (——) calculation; (○) experiment [7]

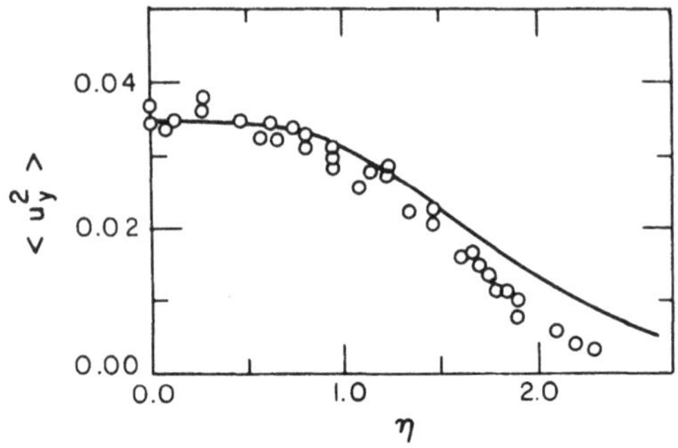

Fig. 3. Mean square lateral velocity fluctuations against lateral distance: (——) calculation; (○) experiment [7]

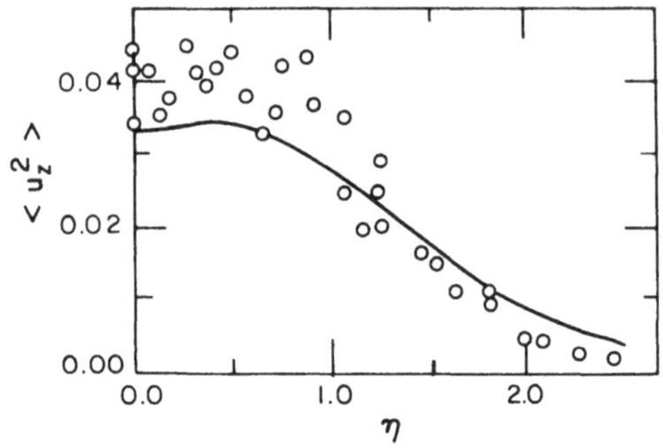

Fig. 4. Mean square spanwise velocity fluctuations against lateral distance: (——) calculation; (○) experiment [7]

There is good agreement between the calculated and measured profiles, but it should be remembered that the normalisation forces agreement at $\eta = 0$ and $\eta = 1$.

Figures 2–4 show the normal stress profiles $<u_x^2>$, $<u_y^2>$ and $<u_z^2>$. The profile shapes are calculated quite accurately as are the relative magnitudes, i.e.

$$<u_x^2> \approx 2<u_y^2> \approx 2<u_z^2> \quad . \tag{28}$$

It may be recalled that the constants ω^* and C_R were chosen so that the calculated spreading rate and the center-line kinetic energy agree with the measured values. It is not completely surprising, therefore, that the magnitude of the calculated normal stresses agrees well with the data. One feature of interest is the maximum in $<u_x^2>$ away from the plane of symmetry. This occurs because the production of $<u_x^2>$ is relatively small at $\eta = 0$.

Profiles of the shear stress $<u_x u_y>$ are shown in Fig. 5. It may be seen that the measured values are, typically, 20% lower than the calculations. Using the measured

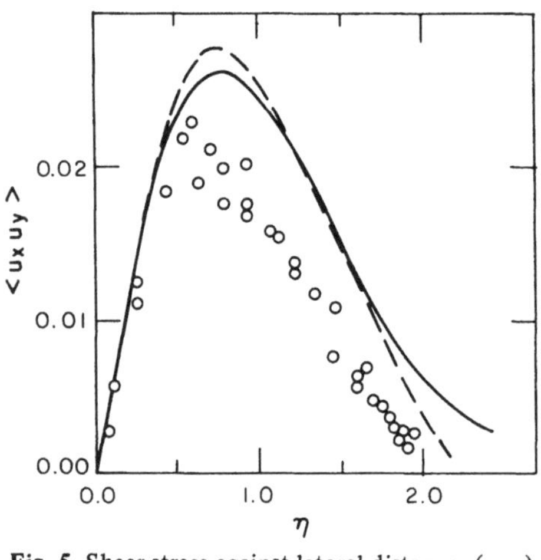

Fig. 5. Shear stress against lateral distance: (——) calculation; (○) experiment [7]; (– – –) deduced from measured mean velocity and momentum balance [7]

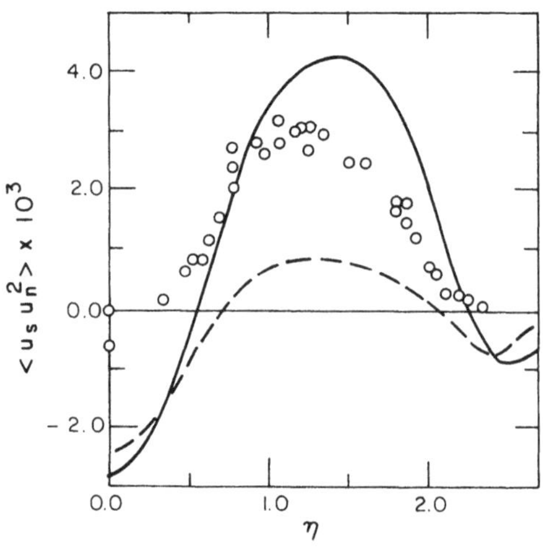

Fig. 6. Triple velocity correlation against lateral distance: (——) calculation; (○) experiment [7]; (– – –) gradient model [8] evaluated from calculated Reynolds stresses

mean velocity and normal stresses, Heskestad used the axial momentum equation to obtain an independent estimate of the shear stresses. This estimate (indicated by the broken line) is in good agreement with the calculation. This supports Heskestad's conclusion that the measured values of $<u_x u_y>$ are subject to error.

Heskestad's results include triple velocity correlations, measured in a local mean-streamline coordinate system. In the x-y plane, the fluctuating velocity components parallel and normal to the mean velocity vector are u_s and u_n. Figure 6 shows measured and calculated values of the triple correlation $<u_s u_n^2>$. Also shown on the figure is the value of the triple correlation obtained from Launder, Reece and Rodi's [8] gradient diffusion model

$$<u_i u_j u_k> = -\frac{0.11}{\omega} <u_i u_l> \frac{\partial}{\partial x_l} <u_j u_k> + <u_j u_l> \frac{\partial}{\partial x_l} <u_k u_i> + <u_k u_l> \frac{\partial}{\partial x_l} <u_i u_j> \quad (29)$$

Near to the plane of symmetry $0 \leqslant \eta \leqslant 0.5$, both the joint-pdf calculations and the gradient model give significant negative values, while the measured values are close to zero. The maximum value of $<u_s u_n{}^2>$ occurs at $\eta \approx 1$: the maximum value from the joint pdf calculation is 30% greater than that measured, while for the gradient model it is 80% less. The level of agreement between the measurements and the joint-pdf calculations is satisfactory considering the likely measurement error and the simplicity of the closure assumptions.

This completes the comparison of calculated velocity statistics with the experimental data of Heskestad. For the scalar field, the only measurements are for the mean profile, see [12] for example. It is found that for the self-similar plane jet, the spreading rate of the mean scalar is 40% greater than the spreading rate of the mean velocity. In the calculations, however, the scalar spreading rate is 5% less. This disappointing result is undoubtedly due to the simplistic modelling of the fluctuating pressure gradient terms. Future calculations that include the rapid term and the pressure transport term will, hopefully, not suffer from this defect. The normalised profile (not shown) is calculated accurately.

Some pdf's are now reported to illustrate the type of quantities that can be calculated. Figure 7 shows the scalar pdf $p(\psi)$

$$p(\psi) = \int f(V, \psi)\, dV \quad , \quad (30)$$

for different values of η. On the plane of symmetry ($\eta = 0$), the pdf is a bell-shaped curve and some negative skewness is evident. At $\eta = 0.5$ the pdf has the same shape but the mean is shifted to a lower value and the standard deviation is increased. At $\eta = 1.0$, the effect of the lower bound becomes evident. That is, the minimum possible value of ϕ is zero, and hence the pdf is zero for negative values of ψ. As can be seen from the figure, this leads to a discontinuity in $p(\psi)$ at $\psi = 0$. At the furthest location shown, $\eta = 1.5$, the pdf is centered on low values of ψ, corresponding to recently entrained fluid.

Also shown on Fig. 7 are the contributions to $p(\psi)$ of outward flowing fluid. The scalar pdf can be decomposed into two parts

$$p(\psi) = p_+(\psi) + p_-(\psi) \quad , \quad (31)$$

where

$$p_+(\psi) = \int_{-\infty}^{\infty} \int_{0}^{\infty} \int_{-\infty}^{\infty} f(V_r, V_\theta, V_z, \psi)\, dV_r\, dV_\theta\, dV_z \quad , \tag{32}$$

and

$$p_-(\psi) = \int_{-\infty}^{\infty} \int_{-\infty}^{0} \int_{-\infty}^{\infty} f(V_r, V_\theta, V_z, \psi)\, dV_r\, dV_\theta\, dV_z \quad . \tag{33}$$

The contribution to $p(\psi)$ from outward flowing fluid is $p_+(\psi)$, which is shown as the broken line on Fig. 7. On the plane of symmetry, inflowing and outflowing fluid are indistinguishable: hence $p_+(\psi) = 1/2\, p(\psi)$. As η increases, $p_+(\psi)$ is biased towards larger values of ψ [compared with $p(\psi)$] since outward flowing fluid comes from regions of higher mean levels.

Figure 8 shows the joint pdf of the radial and circumferential velocities at $\eta = 1.0$. This joint pdf is defined by

$$f_{r\theta}(V_r, V_\theta) = \iint f(V_r, V_\theta, V_z, \psi)\, dV_z\, d\psi \quad . \tag{34}$$

The figure shows $f_{r\theta}(V_r, V_\theta)$ plotted against V_r for different values of V_θ. At this location ($\eta = 1.0$) the mean circumferential velocity is $\langle U_\theta \rangle = -0.04$, and the mean radial velocity is $\langle U_r \rangle \approx 0.5$. It may be seen that the joint pdf is a maximum close to $V_r = \langle U_r \rangle$, $V_\theta = \langle U_\theta \rangle$. That the curves of $f_{r\theta}(V_r, V_\theta)$ move to higher values of V_r as V_θ

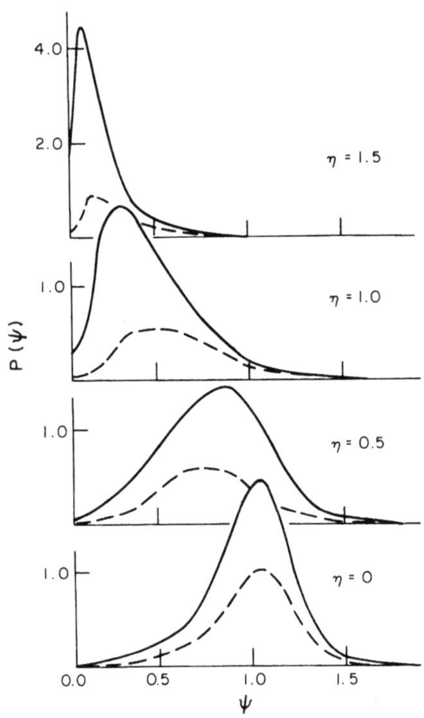

Fig. 7. Scalar pdf's and the contribution from outward-flowing fluid at different lateral positions: (——) $p(\psi)$; (– – –) $p_+(\psi)$

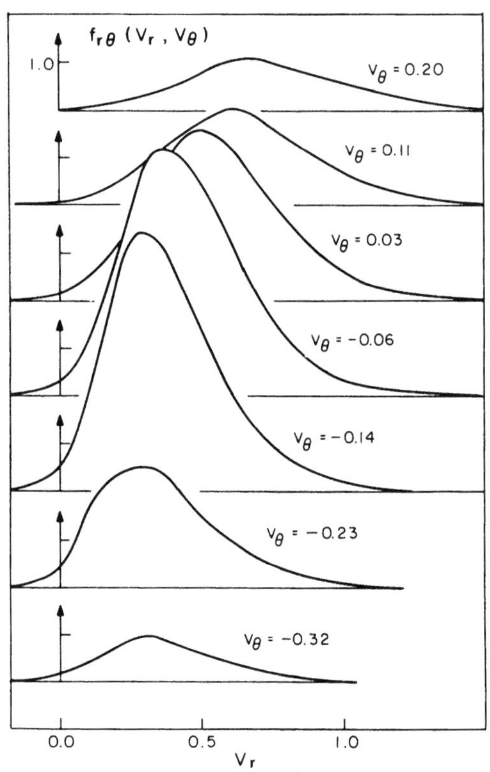

Fig. 8. Joint pdf of radial and circumferential velocity at $\eta = 1.0$

indicates that the shear stress $<u_r u_\theta>$ is positive (cf. Fig. 5). For large V_θ ($V_\theta = 0.20$) it may be seen that the distribution is broader than for small values ($V_\theta = 0.32$). This is because the normal stress $<u_r^2>$ is greater for $\eta < 1$, where outward flowing fluid ($V_\theta > 0$) originates.

Figures 7 and 8 illustrate just two quantities that can be obtained from the calculated joint pdf. Any other one-point statistic can also be deduced – for example, conditionally expected values and conditional pdf's.

Conclusion

The velocity-scalar joint pdf transport equation has been solved for the self-similar plane jet. The solution procedure is not computationally expensive even though the joint pdf is a function of five independent variables. In general, there is good agreement between the calculations and measurements.

Acknowledgement. This work was supported in part by grant number CPE8000026 from the National Science Foundation, Engineering energetics program.

References

1. S.B. Pope: Phys. Fluids **24**, 588 (1981)
2. T.S. Lundgren: Phys. Fluids **12**, 485 (1969)
3. C. Dopazo, E.E. O'Brien: Acta Astronaut. **1**, 1239 (1974)
4. S.B. Pope: 18th International Symposium on Combustion, The Combustion Institute (1981)
5. J. Janicka, W. Kolbe, W. Kollmann: Proc. of the Heat Transfer and Fluid Mechanics Institute. Stanford, 1978, p. 296
6. S.B. Pope: Combust. Sci. Technol. **25**, 159 (1981)
7. G. Heskestad: J. Appl. Mech. **32**, 721 (1965)
8. B.E. Launder, G.J. Reece, W. Rodi: J. Fluid Mech. **68**, 537 (1975)
9. J.C. Rotta: Z. Phys. **129**, 547 (1951)
10. R.L. Curl: AIChE J. **9**, 175 (1963)
11. D.B. Spalding: Chem. Eng. Sci. **26**, 95 (1971)
12. P.E. Jenkins, V.W. Goldschmidt: J. Fluids Eng. **95**, 581 (1973)

Aerosol Formation in a Mixing Layer

S.V. Sherikar and R. Chevray

Department of Mechanical Engineering, State University of New York at Stony Brook, Stony Brook, NY 11794, USA

Abstract

Flow visualization has for long been a powerful tool in experimental studies in fluid mechanics. Here, a novel technique utilizing the NH_3-HCl reaction is described. The product of this reaction appears as a white and dense aerosol which can readily be photographed. Since the reaction itself is fast, information is also provided on local entrainment and mixing. This technique has been applied to advantage to the study of coherent structures in a plane mixing layer the results of which are presented here.

Nomenclature

$F(d)$	total number of particles per unit volume larger than a given number d
H	shape factor for boundary layer (δ_D/δ_M)
N	total number of particles per unit volume
U	mean flow velocity
d	particle diameter
k	rate constant for NH_3-HCl reaction [1.9×10^{-17} cm^3/s (at STP)]
k_{nucl}	rate constant for nucleation of particulate NH_4Cl (2×10^{-229} cm^{48}/s at STP)
k_{cond}	rate constant for condensation of colloidal NH_4Cl
m	arbitrary number
q	a constant
t	time
α	constant = $(13 \pm 3) \times 10^4$ cm^{-1}
$[NH_3]$	concentration of NH_3
[HCl]	concentration of HCl
δ_D	displacement thickness
δ_M	momentum thickness

Subscripts

(g)	refers to gaseous phase
(s)	refers to solid phase
1	refers to fast stream
2	refers to slow stream

Introduction

One of the simplest free shear flows is the turbulent plane mixing layer formed when two parallel streams of different velocities are brought together. Substantial work has been reported on this in the last decade, and, although there is considerable scatter in data obtained by different researchers, owing mainly to different experimental conditions (there is greater deal of agreement on this than on the interpretation of the data itself), some important features of this type of flow have been brought to light. One such feature is the existence of large coherent structures having some kind of orderly behavior and their strong influence on the development of the mixing layer, put in evidence in low Reynolds number flows by, among others, *Browand* and *Laufer* [1]. That this was not a "relic of transition" and that this phenomenon is exhibited also at large Reynolds numbers was shown by *Brown* and *Roshko* [2], and *Dimotakis* and *Brown* [3].

The observations made on coherent structures so far indicate that the development of a mixing layer is largely controlled by two kinds of vortices in the flow – the large, primary vortices which are essentially two-dimensional and oriented spanwise and counter-rotating vortex pairs that are secondary vortices, resulting from secondary instabilities, which are embedded in primary vortices and in the vortex layers which connect the latter [4]. The Reynolds stresses and growth (and hence entrainment) of the mixing layer are controlled mainly by the primary vortices while the secondary vortices contribute to internal mixing and, possibly, modification of Reynolds stresses.

Of interest is also the case of mixing layer with an external disturbance imposed on the initial conditions (importance of which on the initial development of mixing layers has been proved satisfactorily). It has been found that the external disturbance tends to make prominent the scales of the initial instability that scale with the external disturbance thereby resulting in higher coherence in the shear layer [5]. Another scheme of great practical interest is the mixing layer in which chemical reaction is taking place. Although it has been known that chemical kinetics and turbulence influence each other, the interaction is not well-understood. A clear understanding of this then becomes essential before schemes of practical interest generally involving complex turbulent flows and complex chemical kinetics can be investigated.

With such an uphill task for turbulence researchers, and a lack of sufficient knowledge even in simple cases, it is apparent that qualitative description can be useful in understanding the phenomenon. Flow-visualization, which has for long been a powerful tool in experimental studies in fluid mechanics, is of value here. Flow-visualization provides certain advantages since effects such as jitter and loss of phase-coherence do not blur the results as in quantitative measurements. Moreover, it can provide an instantaneous picture of the entire flow-field.

A new technique for flow-visualization is described here and its results for the study of a plane mixing layer are presented. With this it is possible to "observe" the transport of a passive scalar in the flow and, also, reaction taking place in the mixing layer.

Principle of the Flow-Visualization Technique

This technique utilizes the reaction between gaseous ammonia and gaseous hydrogen chloride which produces dense white fumes of ammonium chloride. Little effort has been directed so far in assessing different aspects of this reaction and its applicability for flow-visualization. While the reaction is rapid and irreversible at room temperature, the actual aerosol formation can be attributed only to condensation and growth of particles described by colloidal chemistry of NH_4Cl. The generation of solid particles of NH_4Cl can be written in terms of the following equations:

$$NH_{3\,(g)} + HCl_{(g)} \xrightarrow{k} NH_4Cl_{(g)} \quad , \quad k = 1.9 \times 10^{-17}\ \text{cm}^3/\text{s (at STP)}$$

$$(q+1)\,NH_4Cl_{(g)} \xrightarrow{k_{nucl}} (NH_4Cl)_{q+1\,(s)} \quad , \quad k_{nucl} = 2 \times 10^{-229}\ \text{cm}^{48}/\text{s}$$

$$(NH_4Cl)_{m\,(s)} + NH_4Cl_{(g)} \xrightarrow{k_{cond}} (NH_4Cl)_{m+1\,(s)} \quad , \quad m \geqslant (q+1)$$

The first equation describes a chemical reaction between the species in gaseous phase while the second and third describe a physical process involving homogenous nucleation and definite growth of NH_4Cl particles, possible whenever they have $(q + 1)$ or more monomers. Particles with q monomers are as likely to grow as to disintegrate, and those with less than q monomers tend to revert to molecular state. The value of q has been cited as 7.5 by *Heicklen* [6]. Gravitational settling, wall-deposition and coagulation are unimportant in the condensation of NH_4Cl vapors.

The particle number density can be estimated from the rate law for nucleation given by Heicklen:

$$\frac{dN}{dt} = k_{nucl} \; [NH_3]\,[HCl]^{8.5} \quad , \qquad k_{nucl} = 2 \times 10^{-229} \text{ cm}^{48}/\text{s}$$

The aerosol generated by formation of NH_4Cl is then poly-dispersed, the particle size distribution being dependent on the extent of mixing and time. It is expected that if $F(d)$ is the total number of particles per unit volume that are larger than a given diameter d and $N(t)$ is the total number of particles at time t, then

$$F(d) = N(t)\, e^{-\alpha d} \tag{1}$$

where α is weakly dependent on time (Heicklen). Typical size distributions obtained by *Countess* and *Heicklen* [7] are shown in Fig. 1. Details of the application of this principle and discussion of the results obtained are described below.

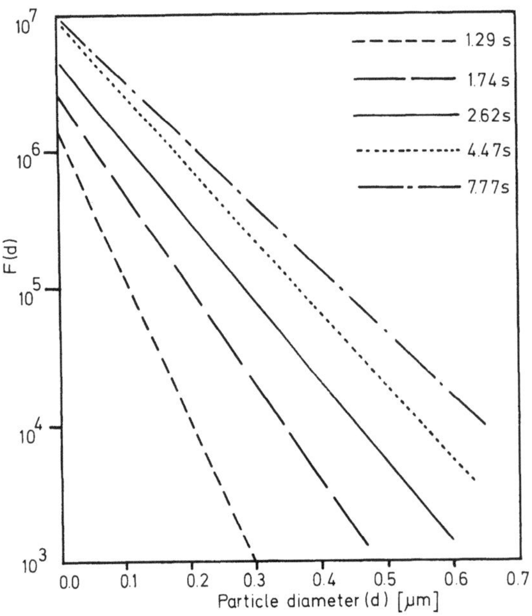

Fig. 1. Particle size distribution: $F(d)$, the number of particles per cc greater than the diameter d, vs d for various reaction times at 25°C, for initial concentrations of 60 ppm of each NH_3 and HCl in 1 atm of N_2 (*Countess* and *Heicklen* [2])

Experimental Facility

Flow-visualization in the mixing layer of two parallel streams of air moving at different velocities was carried out in a specially designed wind-tunnel in which the two streams are separated from each other prior to their mixing in the test-section. The test-section, which is 25 cm x 100 cm long (x 15 cm wide) has glass sides so that visualization is possible. With the velocities used values of Reynolds numbers of the order of 10^6, based on maximum velocity and maximum downstream distance, are attainable. Honeycombs and screens are placed in the flow to reduce the free-stream turbulence. Initial tests revealed that the boundary layer on the splitter plate was laminar ($H = \delta_D/\delta_M = 2.2$) with momentum thicknesses of 0.6 mm and 0.9 mm in fast and slow streams respectively. The splitter plate was made out of 1/16″ thick stainless steel plate and tapered over 25 cm and smoothed to a sharp trailing edge. Velocity profile in each section was flat with a variation of less than 1% when measurements were made with blowers. In addition, an inert gas delivery system was set up to study reactions in which the reactants could react with any of the constituents of air. This system delivers high purity nitrogen from eight bottles simultaneously so that large flow rates can be obtained.

For flow-visualization with NH_4Cl aerosol, it is important that the flow and test-section be illuminated well. In the experiments reported here, the flow was illuminated by three 1500-W straight-line-filament quartz lamps. Considering the initial concentrations of NH_3 and HCl (~0.5%) and time scales of the flow, it is reasonable to estimate that the particle size distribution of NH_4Cl aerosol would be in the range 0.05 μm to 0.4 μm in diameter. The largest particles therefore correspond to relaxation time of 7.4×10^{-5} s, from which one can safely conclude that the aerosol will follow the small scale fluctuations in the flow.

Results and Discussion

Some typical results from the flow-visualization in plane mixing layer using this new technique are illustrated by plates 1–4. These plates show selected sequences from movies corresponding to different conditions.

The sequences in plates 1 and 2 were taken with the chemicals NH_4 and HCl brushed on either side of the splitter plate. In this case, the flow is seeded only near the trailing edge of the plate and the chemical reaction is thus completed very soon thereafter so that the picture is that of a non-reacting flow carrying a passive contaminant. The transport of this contaminant is readily observed since the dense white NH_4Cl aerosol formed near the splitter plate marks its path well. Plate 3 shows a similar sequence for the case of the two streams seeded fully with NH_3 and HCl. The species react as soon as they get in contact at the molecular level in the developing shear layer and produce NH_4Cl which can be observed. Plate 4 shows the corresponding case for two streams seeded with the reacting species introduced 1.25 cm vertically above and vertically below the splitter plate. The large coherent structures marked by NH_4Cl when the mixing layer grew to a certain size and entrained the chemical species within it were observed as expected.

Several observations can be made from these photographs. In plate 1, to the far left (near the trailing edge of the splitter plate) the process of amplification of the Kelvin-Helmholtz instability producing a pair of two-dimensional large vortices (which merge further downstream) can be seen very clearly. Also interesting are "empty" pockets of air in some of the vortices half-way downstream which gives further evidence of the entrainment by a mechanism of engulfment of the surrounding fluid by the vortices

PLANE MIXING LAYER

HCl and NH_4OH brushed on each side of splitter plate

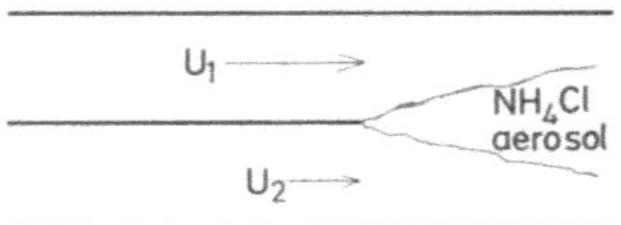

Test-section length L=85 cm
Mixing layer width $\Delta \simeq 15$ cm
$U_1 = 2{\cdot}1$ m/s
$\frac{U_2}{U_1} = 0{\cdot}57$
$\mathbb{R}_\Delta = 2{\cdot}2 \times 10^4$
32 frames/s
Shutter-speed = $1/216$ s

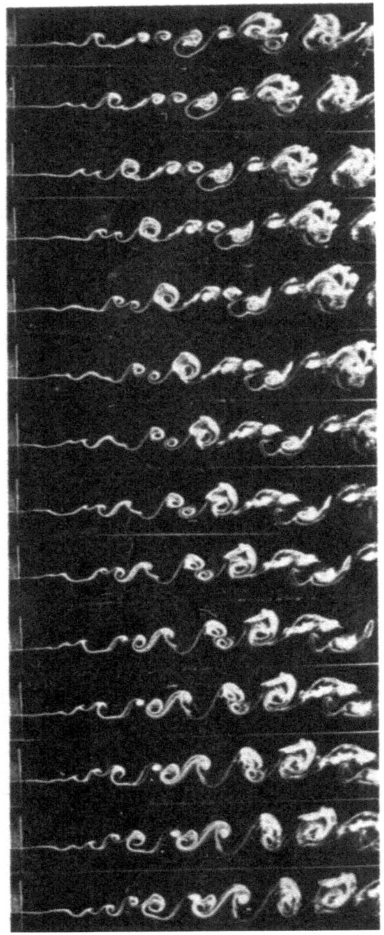

Plate 1

PLANE MIXING LAYER

HCl and NH_4OH brushed on each side of splitter plate

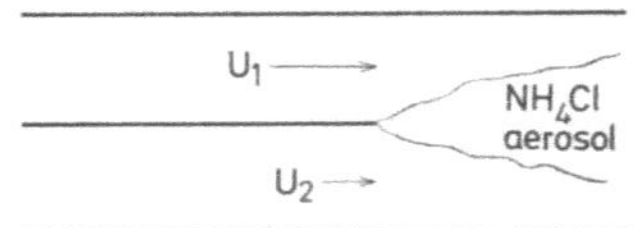

Test-section length L = 85
Mixing layer width $\Delta \simeq 15$
$U_1 = 2{\cdot}1$ m/s
$\frac{U_2}{U_1} = 0{\cdot}76$
$\mathbb{R}_\Delta = 2{\cdot}2 \times 10^4$
32 frames/s
Shutter-speed = $1/216$ s

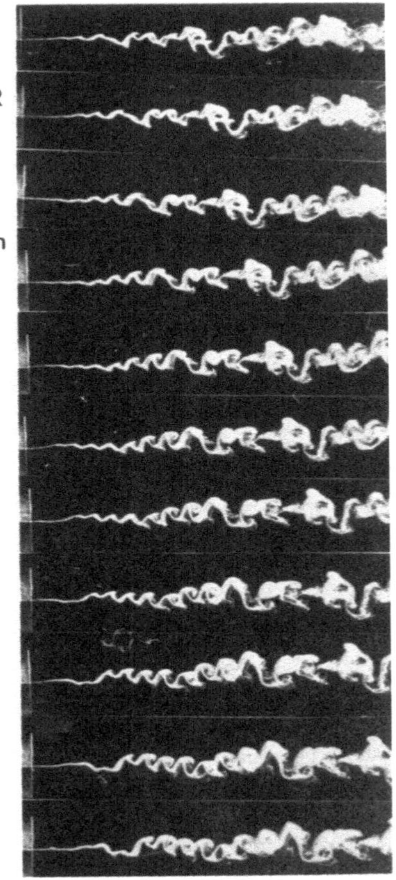

Plate 2

PLANE MIXING LAYER

Both streams uniformly seeded with HCl and NH_4OH

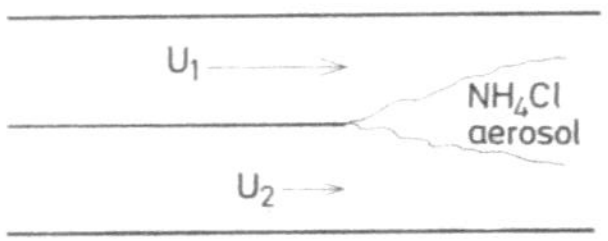

Test-section length L=85 cm
Mixing layer width $\Delta \simeq 15$ cm
$U_1 = 2{\cdot}1$ m/s
$\frac{U_2}{U_1} = 0{\cdot}50$
$\mathbb{R}_\Delta = 2{\cdot}2 \times 10^4$
32 frames/s
Shutter-speed = $1/160$ s

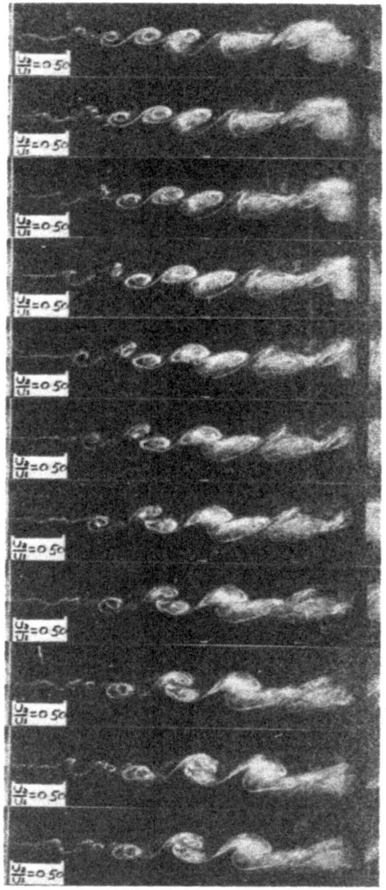

Plate 3

PLANE MIXING LAYER

HCl and NH_4OH brushed as shown

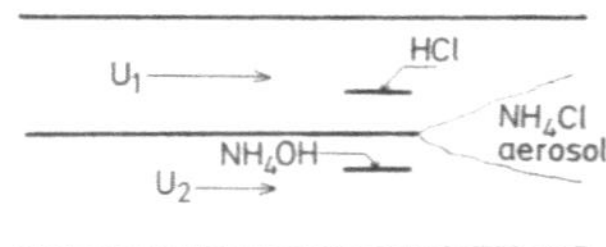

Plate separation = $1{\cdot}25$ cm
Test-section length L = 85cm
Mixing layer width $\Delta = 15$cm
$U_1 = 1{\cdot}6$m/s
$\frac{U_2}{U_1} = 0{\cdot}48$
$R_\Delta = 1{\cdot}6 \times 10^4$
32 frames/s
Shutter-speed = 1/160s

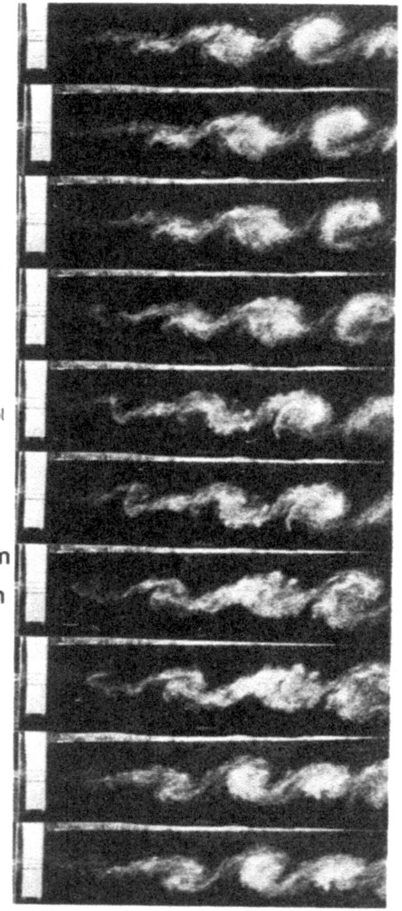

Plate 4

and the interface connecting these. Plate 1 also shows a relative rare event of tripling of vortices.

In plate 3, the reaction can be seen to progress downstream and, in contrast to empty pockets of air as in plate 1, the vortices are filled because the reacting species in the surrounding fluid are engulfed by the vortices and they continue to react and form NH_4Cl aerosol. This clearly indicates that fluid inside these vortices is fully mixed. The dependence of growth of the mixing layer on velocity ratio can be qualitatively judged from plates 1 and 2. Amalgamation (pairing and tripling) seems to be characteristic of the growth of the mixing layer – as can be observed from the smaller growth rate for the case $U_2/U_1 = 0.76$ (plate 2) in which pairing of the vortices is infrequent, and the correspondingly larger growth rate for the case $U_2/U_1 = 0.57$ (plate 1) characterized by pairing and occasional tripling of vortices. The average size of coherent structures vs downstream distance is shown in Fig. 2 for three different velocity ratios. This is representative of dependence of mixing layer growth on velocity ratio. Another feature of the mixing layer is the broad spectrum of scales (vortex spacings, for example) which changes as flow develops downstram and this qualitative behavior is observed. The flow-visualization technique has the advantage here of providing a description of the flow (at least qualitative) because, although correlation measurements can provide information

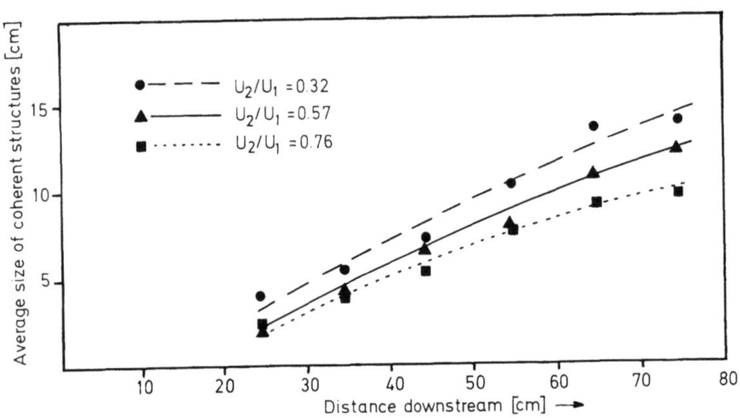

Fig. 2. Growth of coherent structures in plane mixing layer

about the coherent structures and the length scales involved, the data obtained is degenerated because the jitter, phase incoherence and higher order instabilities on the flow. Moreover, visualization of the flow gives an instantaneous picture of the entire field.

Conclusions

A feasible technique for flow-visualization utilizing the reaction between NH_3 and gaseous HCl, resulting in formation of NH_4Cl aerosol, has been developed. Its results in the case of a plane mixing layer give good qualitative information of the flow, both for the transport of a passive scalar as well as for the case in which a reaction is taking place.

Acknowledgement. We would like to thank the National Science Foundation for its support under Grant No. CME 8012961.

References

1. Browand, F.K, Laufer, J.: "The Role of Large Scale Structures in the Initial Development of Circular Jets", Proceedings, 4th Biennial Symposium on Turbulence in Liquids, September 1975, Univ. of Missouri, Rolla, p. 333
2. Brown, G.L., Roshko, A.: On density effects and large structures in turbulent mixing layers. J. Fluid Mech. **64**, 775 (1974)
3. Dimotakis, P.E., Brown, G.L.: The mixing layer at high Reynolds number: Large structure dynamics and entrainment. J. Fluid Mech. **78**, 535–560 (1976)
4. Roshko, A.: "The Plane mixing Layer: Flow Visualization Results and Three Dimensional Effects", *Proceedings*, International Conference on the Role of Coherent Structures in Modelling Turbulence and Mixing, Madrid, Spain, 1980
5. Wygnanski, I., Oster, D., Fiedler, H.: "The Forced, Plane, Turbulent Mixing-Layer: A Challenge for the Predictor, *Proceedings*, Second Symposium on Turbulent Shear Flows, Imperial College, London, 1979
6. Heicklen, J.: *Colloid Formation and Growth* (Academic Press, New York 1976)
7. Countess, R.J., Heicklen, J.: Kinetics of particle growth. II. Kinetics of the reaction of ammonia and hydrogen chloride and the growth of particulate ammonium chloride. J. Phys. Chem. **77** (4), 444–447 (1973)

The Role of Coherent Structures in the Development of a Uniformly Strained Turbulent Wake

J.G. Kawall and J.F. Keffer

Department of Mechanical Engineering, University of Toronto,
Toronto, Canada

Abstract

The development of a uniformly distorted, passively heated cylinder-generated wake is investigated experimentally. Characteristic local length and intensity scales for the flow are determined from the experimental data and compared to theoretical expressions derived on the assumption that the uniformly distorted wake is a self-preserving flow. It is found that, although the scales based on mean (streamwise velocity and temperature) profiles and the half-intermittency point length scale are in reasonable agreement with the predictions, the scales based on root-mean-square profiles and, in particular, the scale characterizing the lateral extent of the large-scale, intermittent turbulent bulges of the flow deviate from the predicted behaviour, signifying that the distorted wake does not really undergo a truly self-preserving development. Evidence is presented indicating that this lack of self-preservation is the result of amplification, by the distortion field, of coherent structures within the flow. These organized motions – which are, in fact, the turbulent bulges – are identified by means of coherence functions, phase spectra and lateral velocity autospectra.

Nomenclature

- a strain constant [mm^{-1}]
- C co-spectrum [s]
- D cylinder diameter [mm]
- E autospectrum [s]
- l_0 length scale based on a mean streamwise velocity profile [mm]
- l'_0 length scale based on an RMS streamwise velocity profile [mm]
- $\hat{l}_0$ length scale based on a mean-square streamwise velocity profile [mm]
- l_θ length scale based on a mean temperature profile [mm]
- l'_θ length scale based on an RMS temperature profile [mm]
- n frequency [Hz]
- Q quadrature spectrum [s]
- T local mean temperature [K]
- T_1 free-stream temperature [K]
- t time [s]
- U local mean streamwise velocity [ms^{-1}]
- U_d mean velocity defect [ms^{-1}]
- U_1 free-stream velocity [ms^{-1}]
- u' RMS streamwise velocity [ms^{-1}]
- u_0 intensity scale based on a mean streamwise velocity profile [ms^{-1}]
- u'_0 intensity scale based on an RMS streamwise velocity profile [ms^{-1}]
- u, v fluctuating velocity components in the x and y directions respectively [ms^{-1}]
- x_s streamwise distance between the cylinder location and the beginning of the distortion [mm]

x, y, z	Cartesian, streamwise, lateral and spanwise coordinates	θ_0	intensity scale based on a mean temperature profile [K]
$\overline{Y}$	half-intermittency point length scale [mm]	θ'_0	intensity scale based on an RMS temperature profile [K]
Γ	coherence function	σ	characteristic lateral length of intermittent turbulent bulges [mm]
γ	intermittency factor	Φ	phase spectrum
Θ	local mean temperature difference [K]		
θ'	RMS temperature [K]		

Introduction

The problem of whether a classical free turbulent shear flow can undergo a simple self-preserving development when a specially aligned strain field is imposed upon it has been considered by relatively few investigators [1–4]. It is, nonetheless, a problem of both practical and intrinsic interest, inasmuch as the reaction of the flow to such a strain field can provide information on the basic structural features of the turbulence.

In the present work, attention is focused upon the spread of a heated cylinder-generated wake subjected to a constant and uniform rate of strain. The strain field was aligned in such a way as to distend the flow in the spanwise direction and simultaneously constrict it laterally – without applying any strain in the streamwise direction. The heat input level was kept sufficiently low so that the temperature could be considered a passive scalar contaminant. Length and intensity scales based upon the lateral distributions of conventional mean and root-mean-square (RMS) streamwise velocities and temperatures and length scales based upon intermittency factor distributions were determined. As well, one point coherence functions and phase spectra of simultaneously-measured streamwise and lateral velocity signals, and the autospectra of these signals were examined. We note that coherence functions and phase spectra are important statistical tools in the study of turbulent shear flows, as they serve as a means of detecting the presence of coherent structures in such flows [5–8].

Description of the Flow

A definition sketch of the flow is shown in Fig. 1. The imposed strain field was effected by means of a distortion duct with exponentially varying sides and a constant cross-sectional area (see Fig. 2). The flow within such a duct is subjected to a pure strain rate in the cross-stream directions but is not accelerated in the streamwise direction.

In the Appendix, it is shown that if the uniformly strained wake is capable of undergoing a self-preserving development, then, in the plane of symmetry of the flow ($z = 0$), all characteristic length and intensity scales should be directly proportional to $\exp[-a(x - x_s)/2]$, where a is a positive strain constant for a given distortion duct and s pertains to conditions at the entrance to the distortion duct.

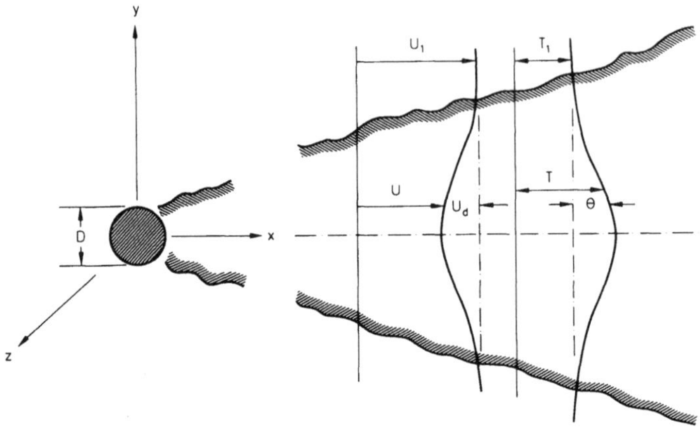

Fig. 1. Definition sketch

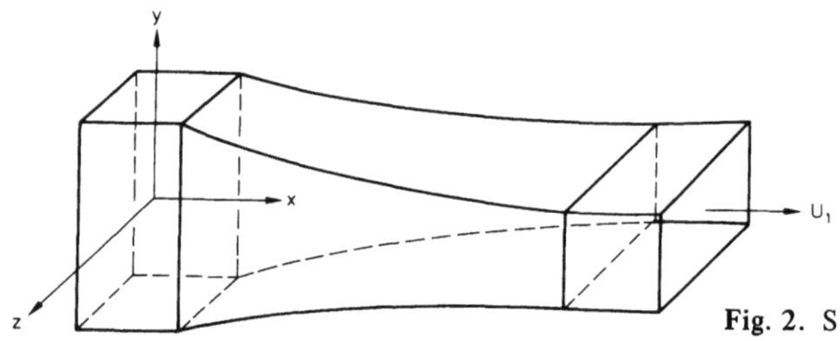

Fig. 2. Schematic diagram of distortion duct

Definitions

The intensity scales based upon the mean and RMS (streamwise velocity and temperature) profiles are given by the centre-line values of U_d, u', Θ and θ'; and the length scales based on these profiles are defined as one-half the distances between the (positive and negative) lateral locations where the various variables assume one-half their centre-line values.

The length scales derived from the intermittency factor γ profiles are (I) the so-called half-intermittency point, $\overline{Y}$, which is the distance between the centre-line of the flow and the lateral location where $\gamma = 0.5$, and (II) the characteristic lateral extent of the large-scale, intermittent turbulent bulges of the flow, σ. This latter is given by:

$$\sigma^2 = \int_{y_\infty}^{y_0} y\,\gamma(y)\,dy - \overline{Y}_\infty^2 \quad ,$$

where y_0 is the lateral position where γ first attains its maximum value of unity and y_∞, the position at which γ becomes zero.

The one-point coherence function for $u(t)$ and $v(t)$, which describes the coherence or degree of correlation between $u(t)$ and $v(t)$ at frequency n, is defined as:

$$\Gamma_{uv}(n) = \frac{C_{uv}^2(n) + Q_{uv}^2(n)}{E_{uu}(n)\,E_{vv}(n)} \quad .$$

Here, $C_{uv}(n)$ is the co-spectrum and represents the correlation between in-phase components of $u(t)$ and $v(t)$ at frequency n; $Q_{uv}(n)$ is the quadrature spectrum and represents the correlation between 90° out-of-phase components at frequency n; $E_{uu}(n)$ and $E_{vv}(n)$ are the normalized autospectra of $u(t)$ and $v(t)$ respectively.

The one-point phase spectrum, which describes the phase relationship between the Fourier components of $u(t)$ and $v(t)$ is given by:

$$\Phi_{uv}(n) = \tan^{-1}\left[\frac{Q_{uv}(n)}{C_{uv}(n)}\right] .$$

Experimental Details

The wake was generated by a circular heating element (referred to as the cylinder) of diameter, D, equal to 6.6 mm, mounted horizontally in the centre of a variable-speed, open-circuit wind tunnel. This tunnel, which is fully described by *Kawall* [9], has three test sections, viz., a pre-strain section, a straining section, similar to the distortion duct described previously, and a post-strain section. The tunnel was operated at a speed, U_1, of about 10 ms^{-1}, which resulted in a cylinder Reynolds number of approximately 4200.

The temperature and velocity signals were measured by means of DISA 55 M anemometer systems in conjunction with cold-wire (resistance thermometer), normal hot-wire and X-wire probes. The temperature data were obtained with the cylinder heated to a sufficiently low level so as to ensure that the velocity field was not affected by the temperature field. The velocity data were obtained with the cylinder unheated in order to simplify the signal processing. All measurements were taken in the vertical centre-plane of the tunnel (z = 0) at x/D values of 65, 81, 89, 96, 135, 154, 173, 188 and 212. The first four values corresponded to positions within the pre-strain region, the other five values, to positions inside the distortion.

The anemometer signals were low-pass filtered and sampled digitally at rates such that the problem of aliasing was effectively eliminated. The digital data were analysed subsequently on an IBM 370–165 digital computer.

Results and Discussion

The normalized mean profiles for the velocity and temperature fields wihtin the pre-strain and strain regions are indicated in Fig. 3, from which it is clear that these profiles are self-similar. Figures 4 and 5 show the streamwise variations of the mean profile length and intensity scales. The agreement between the measured and predicted variations is seen to be quite good, implying that the mean temperature and streamwise velocity distributions are self-preserving within the distortion.

An inherently more sensitive test for self-preservation of the flow is provided by the behaviour of the RMS quantities. The normalized distributions of RMS streamwise velocities and temperatures are depicted in Figs. 6 and 7. It is apparent that these profiles are not strictly self-similar. Thus, the detailed structure of the distorted wake examined here did not attain the self-preserving state (self-similarity being a necessary condition for self-preservation). Figures 8 and 9 show that, within the distortion, the RMS scales decrease at a rate significantly smaller than that required for self-preservation.

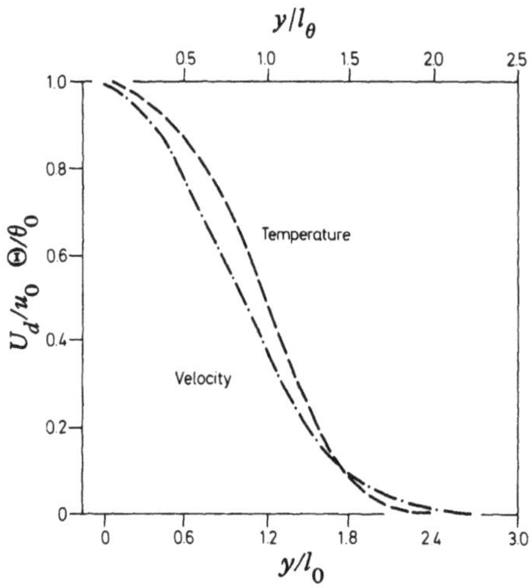

Fig. 3. Normalized mean streamwise velocity and temperature distributions

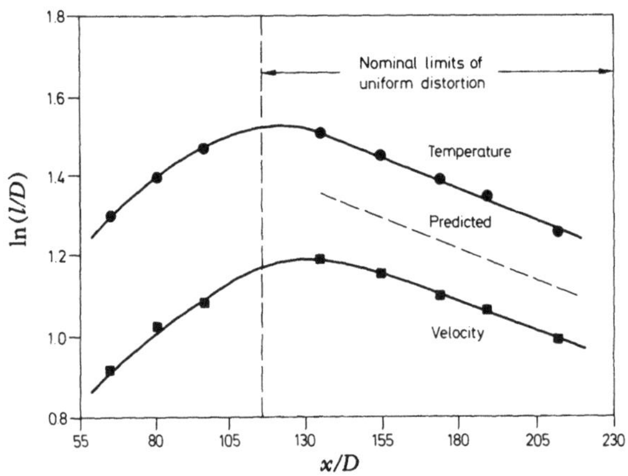

Fig. 4. Mean profile length scales. (■) velocity; (●) temperature; (---) -a/2 slope

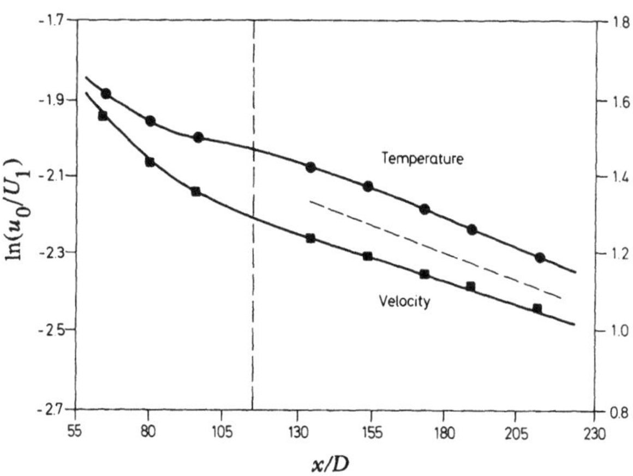

Fig. 5. Mean profile intensity scales. The symbols have the same meaning as in Fig. 4

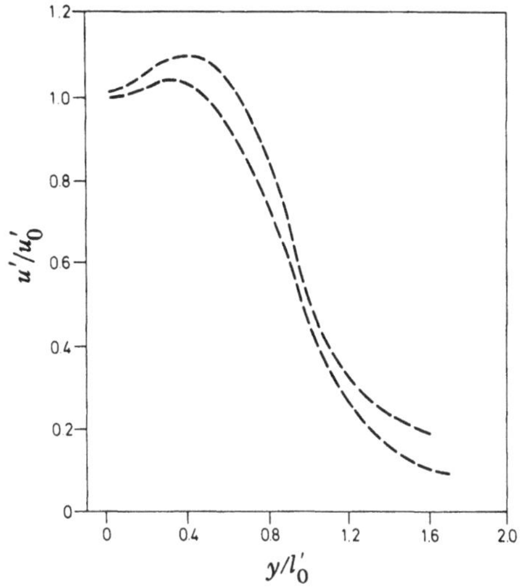

Fig. 6. Normalized RMS streamwise velocity distributions

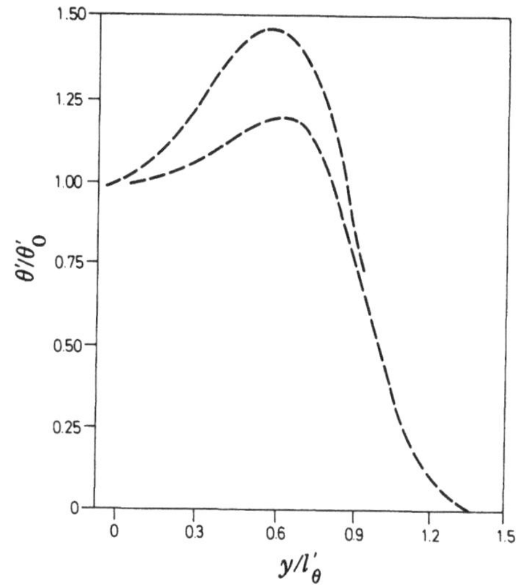

Fig. 7. Normalized RMS temperature distributions

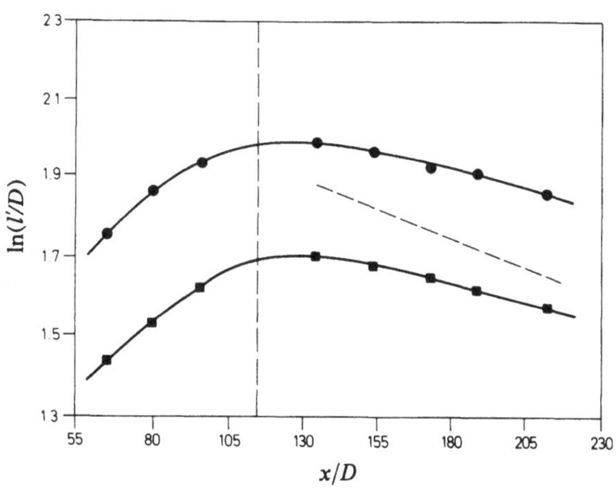

Fig. 8. RMS profile length scales. The symbols have the same meaning as in Fig. 4

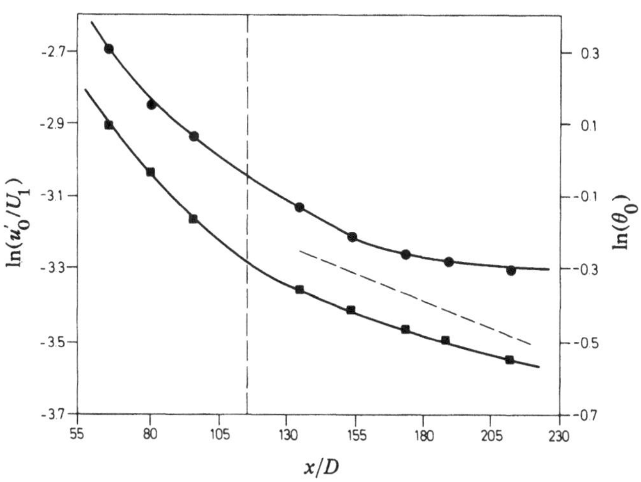

Fig. 9. RMS profile intensity scales. The symbols have the same meaning as in Fig. 4

In Fig. 10, RMS streamwise velocity length scales obtained in this study are compared with those found by *Reynolds* [1] and by *Keffer* [2] for similar flow situations. (Note that $\hat{l}_0$ in this figure represents one-half the distance between the half-maximum mean-square velocity points and is, therefore, not equal to l'_0). It can be seen that at all downstream locations, the present scales are larger than those of the other studies. This is probably because x_s and D were, respectively, about 50% and about 40% greater for the present experiments than they were in the case of the Reynolds-Keffer experiments. Yet, the data obtained here and those obtained by Keffer exhibit much the same trends (with respect to the predicted scale behaviour). On the other hand, Reynolds' data and Keffer's data, which were determined under essentially identical experimental conditions are quite different both in trend and in magnitude. We may infer from this that the detailed self-preservation claimed by Reynolds, as revealed by the turbulence (i.e., RMS) quantities, does not really exist for these flows.

The normalized intermittency factor profiles were found to be self-similar and well-represented by the Gaussian distribution function [9]. The streamwise variations of the lateral length scales ($\overline{Y}$ and σ) that characterize these profiles are presented in Fig. 11. As expected, $\overline{Y}$ (being a measure of the gross width of the flow) conforms well to the "self-preserving" behaviour. In contrast, σ, the characteristic lateral extent of the large-scale, intermittent turbulent bulges, deviates completely from the predicted variation. This behaviour of σ clearly lends support to the view (based on the RMS scale results) that uniformly distorted wakes do not attain a truly self-preserving state.

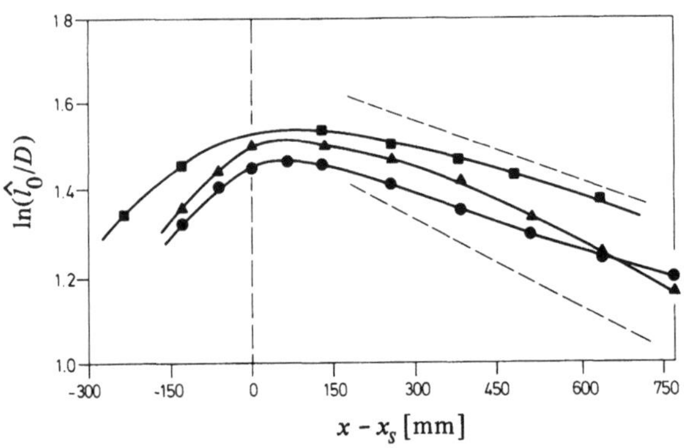

Fig. 10. Comparison of length scales based on mean-square streamwise velocity profiles. (■) present results; (●) *Keffer* [2]; (▲) *Reynolds* [1], (—) -a/2 slope

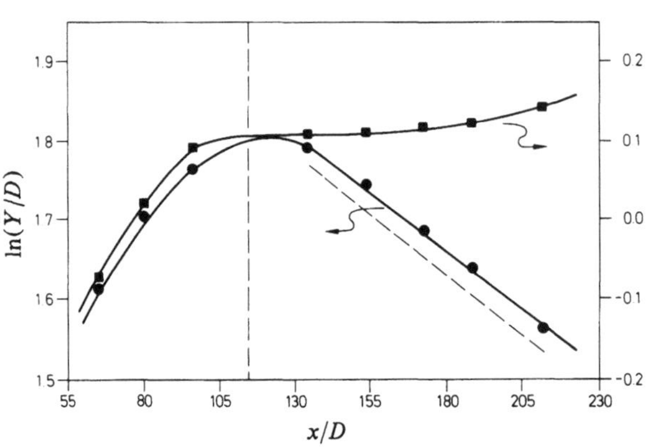

Fig. 11. Length scales based on intermittency factor distributions

In Figs. 12 and 13, we present representative one-point uv coherence functions pertaining to the pre-distorted flow and the distorted flow respectively. One notes that these functions exhibit frequency-centred activity, i.e., more or less broad-band peaks, especially in the highly intermittent flow regions (where γ is much less than 0.5). One point uv phase spectra show that there are definite phase differences between the Fourier components of the u and v signals, for the same frequency ranges within which there is significant coherence; for frequencies beyond which the coherence is essentially zero, these phase differences become random. Typical examples of such spectra pertaining to the distorted wake, are depicted in Fig. 14. A comparison of this latter and Fig. 13 shows that, near the core region of the flow (where $\gamma \cong 0.99$), the coherence between u and v is due to Fourier components that are essentially 180° out of phase; whereas, in the highly intermittent region ($\gamma \cong 0.1$), the coherence is due largely to Fourier components that are 90° out of phase, the implication here being that, in this region, the co-spectrum is virtually zero for all frequencies (so that the Reynolds shear stress, $\overline{uv}$, is zero) but the quadrature spectrum is relatively large (i.e., the 90° out-of-phase Fourier components of u and v are relatively highly correlated) over a certain frequency range. We can infer from these phase and coherence results that coherent structures exist within the flow. These structures are considered to be parcels of (turbulent) fluid with the following basic traits:

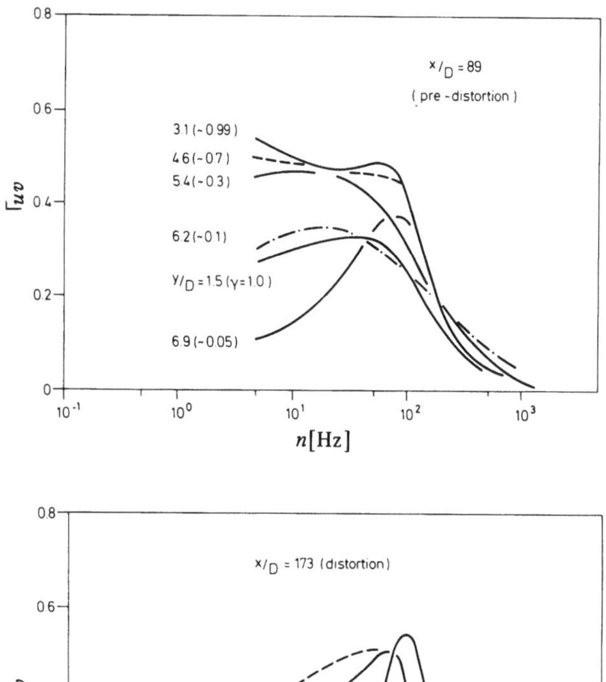

Fig. 12. One-point uv coherence functions

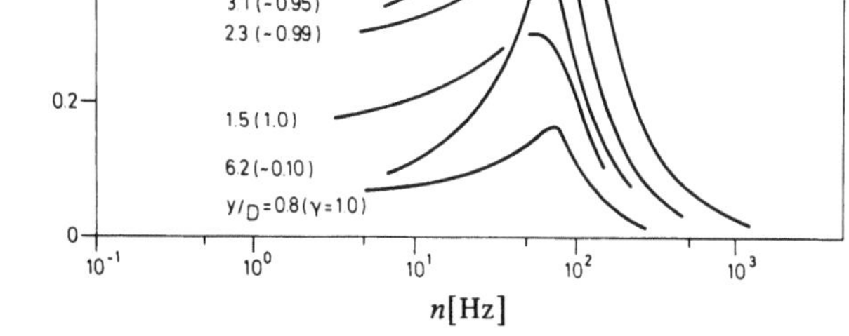

Fig. 13. One-point uv coherence functions

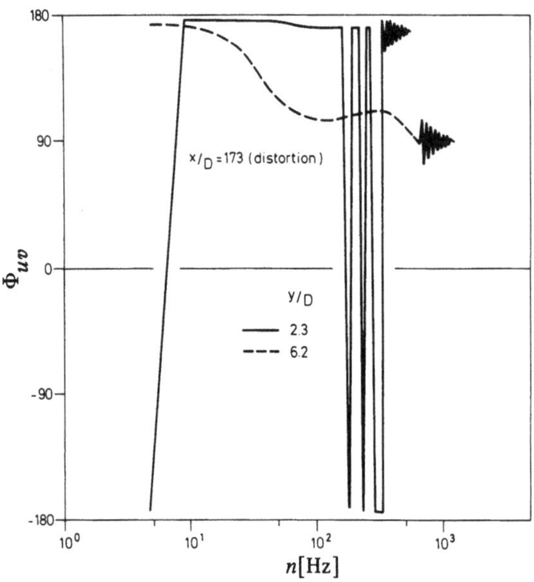

Fig. 14. One-point *uv* phase spectra

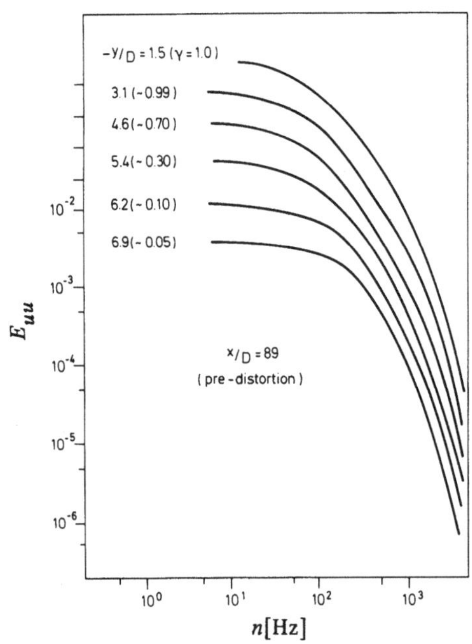

Fig. 15. Streamwise velocity autospectra

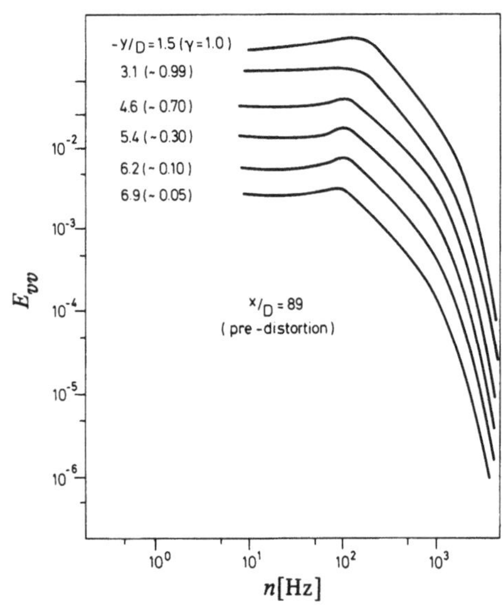

Fig. 16. Lateral velocity autospectra

(a) in general, they occur intermittently; (b) they are all more or less the same size – their streamwise extent determining the nominal frequency at which the frequency-centred activity is located in the coherence functions; and (c) they, individually, possess a definite mean circulation (or vorticity). Notice that, in consequence of (c), with respect to every coherent structure within a flow, there would be a phase difference of 90° between the streamwise and lateral velocity components associated with the mean circulation. (Clearly, this is consistent with the *uv* phase spectrum for $\gamma \cong 0.1$, shown in Fig.

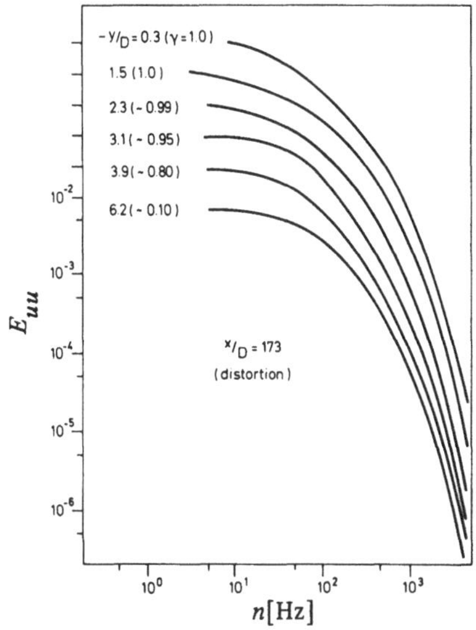

Fig. 17. Streamwise velocity autospectra

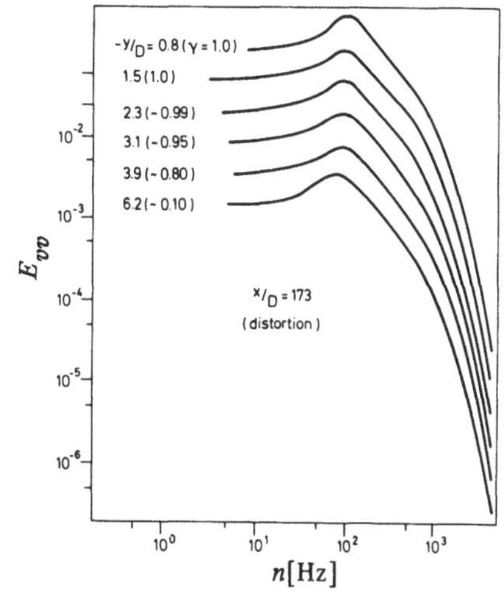

Fig. 18. Lateral velocity autospectra

14). Moreover, it follows from (c) that, during the passage of a structure past a specified point in a flow, the sign of the lateral velocity component associated with the mean circulation would be reversed; whereas, the sign of the corresponding streamwise velocity component would remain unchanged. Accordingly, given that structures having the above characteristics exist within a flow, it is entirely possible that the v (i.e., lateral velocity) autospectra for the flow would display frequency-centred activity at the same nominal frequencies as those at which the corresponding uv coherence functions exhibit this activity; of course, the corresponding u (i.e., streamwise velocity) autospectra would not show any trace of such activity. This is, in fact, the case here, as is apparent from Figs. 15 to 18, which depict representative u and v autospectra pertaining to the pre-distorted and distorted wakes.

In the light of the above findings, we see that the large-scale bulges are organized, non-randomly orientated structures, and, as such, they can be amplified by the imposed strain field, i.e., their vorticity can be increased by the vortex stretching action of the strain field. A comparison of Fig. 12 and Fig. 13 and of Fig. 16 and Fig. 18 show that the bulges become much more organized within the distortion: on the average, the frequency-centred activity becomes more localized, i.e., the coherence functions and the v autospectra become distinctly more "peaky", and the magnitude of the peak in the coherence function increases. This, in conjunction with the behaviour of σ (Fig. 11) leads to the conclusion that the imposed strain field does, indeed, effect an amplification of the bulges, so that they eventually control the development of the flow – as was originally suggested by *Keffer* [2]. And it is for this reason that the uniformly distorted wake fails, in practice, to undergo the predicted self-preserving development. It should be pointed out here that *Keffer* [2] had speculated that the turbulent bulges occurred in a periodic or quasi-periodic fashion. But the results of *Kawall* and *Keffer* [10, 11], concerning the motion of the turbulent/non-turbulent interface, have established that these coherent structures arise randomly within both the undistorted wake and the uniformly distorted wake.

Concluding Remarks

The present investigation of the uniformly strained wake has demonstrated that the mere fact that self-preserving scales for a particular turbulent flow can be derived from the equations of motion does not guarantee that the flow will be characterized by such scales in practice: coherent structures can exist in the flow, and these organized motions are capable of influencing the development of the flow in such a way as to prevent it from attaining the predicted self-preserving state.

Acknowledgement. This work was supported by the Natural Sciences and Engineering Research Council of Canada through Grant A-2746.

Appendix

The Velocity Field

Reynolds [1] and *Keffer* [2] have shown that for the velocity field of the uniformly distorted, plane turbulent wake to be self-preserving, it is necessary that

$$\frac{1}{u_0}\left(\frac{dl_0}{dx} + a\,l_0\right) = \text{const.} = C_1 \tag{A.1}$$

and

$$\beta^{-1}\, l_0\, u_0 = \text{const.} = C_2 \quad , \tag{A.2}$$

where $\beta = \exp\,[-a(x - x_s)]$.
From (A.1, A.2), we obtain

$$\frac{dl_0^2}{dx} + 2a\,l_0^2 = 2C_1\,C_2\,\beta \quad , \tag{A.3}$$

which yields

$$l_0 = C_0\,\beta^{1/2}\,(1 + c\,\beta)^{1/2} \quad , \tag{A.4}$$

where C_0 and c are constants. It then follows from Eq. (A.2) that

$$u_0 = C\beta^{1/2}\,(1 + c\,\beta)^{-1/2} \quad , \tag{A.5}$$

where C is a constant.

It may be remarked that in the case of the undistorted wake $a = 0$ and therefore, from (A.3, A.2), the self-preserving velocity scales are, as shown by *Townsend* [12],

$$l_0 = C_0^*\,(x - x_0)^{1/2} \tag{A.6}$$

and

$$u_0 = C^*\,(x - x_0)^{-1/2} \quad , \tag{A.7}$$

where C_0^*, C^* and x_0 are constants. If the wake is in a self-preserving condition prior to as well as within the uniform distortion, then, $C_0^* = C_0\, a^{1/2}$, $C^* = C/a^{1/2}$ and, in (A.4, A.5), $c = a(x_s - x_0) - 1$.

The Temperature Field

The heat transfer equation governing the mean temperature of the passively-heated, uniformly distorted wake can be approximated as

$$U_1 \left[\frac{\partial \Theta}{\partial x} - ay \frac{\partial \Theta}{\partial y} \right] + \frac{\partial \overline{v\theta}}{\partial y} = 0 \ , \tag{A.8}$$

where $\overline{v\theta}$ is the lateral heat transport term. Since the total heat flux in this flow is conserved (as in the total momentum flux), Θ must also satisfy the integral constraint.

$$\beta^{-1} \int_y \Theta \, dy = \text{const.} \tag{A.9}$$

The following self-preserving relationships are introduced:

$$\Theta = \theta_0 \, g(\eta_\theta) \tag{A.10}$$

and

$$\overline{v\theta} = u_0 \, \theta_0 \, h(\eta_\theta) \ , \tag{A.11}$$

where $\eta_\theta = y/l_\theta$ and $g(\eta_\theta)$ and $h(\eta_\theta)$ are universal functions describing the variations of the mean temperature and the turbulent heat transfer across the wake. Substitution of (A.10, A.11) into (A.8, A.9) gives, after simplification,

$$U_1 \left[\frac{l_\theta}{u_0 \theta_0} \frac{d\theta_0}{dx} \, g - \frac{1}{u_0} \left(\frac{dl_\theta}{dx} + a \, l_\theta \right) \eta_\theta \, g' \right] + h' = 0 \ , \tag{A.12}$$

where the prime denotes differentiation with respect to η_θ, and

$$\beta^{-1} \, l_\theta \, \theta_0 = \text{const.} = K_2 \ . \tag{A.13}$$

For self-preservation of the thermal field to be possible, the coefficients of the terms in Eq. (A.12) must be constant, i.e.,

$$\frac{1}{u_0} \left(\frac{dl_\theta}{dx} + a \, l_\theta \right) = \text{const.} = K_1 \tag{A.14}$$

and

$$\frac{l_\theta}{u_0 \theta_0} \frac{d\theta_0}{dx} = \text{const.} = K_1' \ . \tag{A.15}$$

[It may be mentioned that, on the basis of (A.13), $K_1' = K_1$]. Combining (A.5) and (A.14), we obtain

$$\frac{dl_\theta}{dx} + a\, l_\theta = K_1\, C\, \beta^{1/2}\, (1 + c\, \beta)^{-1/2} \quad , \tag{A.16}$$

from which it follows that

$$l_\theta = K_0\, \beta^{1/2}\, [(1 + c\, \beta)^{1/2} + k\, \beta^{1/2}] \quad , \tag{A.17}$$

where K_0 and k are constants. Equation (A.13) then gives

$$\theta_0 = K\, \beta^{1/2}\, [(1 + c\, \beta)^{1/2} + k\, \beta^{1/2}]^{-1} \quad , \tag{A.18}$$

where K is a constant.

We note that the self-preserving length and intensity scales for the thermal field of the undistorted, passively-heated wake can be obtained by means of (A.13) and (A.14) with $a = 0$, together with (A.7). These scales, which are formally identical to the corresponding velocity scales, (A.6, A.7), are given by:

$$l_\theta = K_0^*\, (x - x_0)^{1/2} \tag{A.19}$$

and

$$\theta_0 = K^*\, (x - x_0)^{-1/2} \tag{A.20}$$

where K_0^* and K^* are constants. If the thermal field is self-preserving before and during application of the uniform distortion, then, $K_0^* = K_0\, a^{1/2}$ and $K^* = K/a^{1/2}$. And if this is also the case for the velocity field, then $k = 0$ in (A.17, A.18).

Now, in the present situation, $a\,(x_s - x_0)$ is about 0.85 and the flow can be assumed to be approximately self-preserving before it enters the distortion (x_s/D being in excess of 100). Accordingly, in (A.4, A.5, A.17, A.18), $0 \leqslant |c|\beta << 1$ for all $\beta < 1$, and k may be taken as zero. It then follows that, once the distorted flow undergoes a self-preserving development, the various scales will be essentially directly proportional to $\beta^{1/2}$.

References

1. Reynolds, A.J.: Observations on distorted turbulent wakes. J. Fluid Mech. **13**, 333–355 (1962)
2. Keffer, J.F.: The uniform distortion of a turbulent wake. J. Fluid Mech. 22, 135–159 (1965)
3. Keffer, J.F.: A note on the expansion of turbulent wakes. J. Fluid Mech. 28, 183–193 (1967)
4. Keffer, J.F., Kawall, J.G., Hunt, J.C.R., Maxey, M.R.: The uniform distortion of thermal and velocity mixing layers. J. Fluid Mech. **86**, 465–490 (1978)
5. Budny, R.S., Kawall, J.G., Keffer, J.F.: "Vortex Street Evolution in the Wake of a Circular Cylinder", 2nd Symp. on Turbulent Shear Flows, Imperial College, London, 1979
6. Keffer, J.F., Kawall, J.G., Giralt, F., Béguier, C.: "Analysis of Turbulent Structures in Complex Shear Flows", 2nd Symp. on Turbulent Shear Flows, Imperial College, London, 1979
7. Keffer, J.F., Kawall, J.G.: "Structural Features of Plane Turbulent Shear Flows", AIAA 13th Fluid and Plasma Dynamics Conference, Snowmass, Colorado, 1980
8. Keffer, J.F., Kawall, J.G., Giralt, F., Béguier, C.: "Characteristics of Coherent Structures in Complex Turbulent Shear Flows", Heat and Mass Transfer and the Structure of Turbulence Joint Symposium, Dubrovnik, Yugoslavia, 1980
9. Kawall, J.G.: "Statistical Properties of a Slightly Heated Uniformly Distorted Plane Turbulent Wake"; Ph.D. Thesis, University of Toronto (1975)

10. Kawall, J.G., Keffer, J.F.: "Uniform Distortion of a Heated Turbulent Wake", in *Structure and Mechanics of Turbulence II*, ed. by H. Fiedler, Lecture Notes in Physics, Vol. 76 (Springer, Berlin, Heidelberg, New York 1978) pp. 85–100
11. Kawall, J.G., Keffer, J.F.: Interface statistics of a uniformly distorted heated turbulent wake. Phys. Fluids, **22**, 31–39 (1979)
12. Townsend, A.A.: *The Structure of Turbulent Shear Flow*, 2nd ed. University Press, Cambridge (1976)

Investigations on a Reaction Model for Turbulent Diffusion Flames

H. Eickhoff
Deutsche Forschungs- und Versuchsanstalt für Luft- und Raumfahrt e.V., Institut für Antriebstechnik, D–5000, Köln 90, Fed. Rep. of Germany

K. Grethe
Universität Karlsruhe (TH), Engler-Bunte-Institut, Bereich Feuerungstechnik, D–7500 Karlsruhe, Fed. Rep. of Germany

Abstract

A study was performed in order to gain new insight into the phenomenology and mathematical modelling of turbulent diffusion flames. The numerical part of the present investigation treats unsteady one-dimensional laminar diffusion flames using an extended kinetic scheme. The result of these predictions yields the instantaneous sub-structure of turbulent diffusion flames. In order to check the validity of the model predictions instantaneous peak temperatures as well as time mean concentrations have been measured in turbulent methane diffusion flames.

Nomenclature

c_A	mixture fraction
$c_{p,i}$	heat capacity of species i at constant pressure
c_p	mixture heat capacity
D_i	binary diffusion coefficient for species i in nitrogen
$F(T^*)$	nondimensional rate of chemical heat release as a function of T
h	mixture enthalpy
h_i	enthalpy of species i
h_i^0	standard heat of formation of species i
NS	number of species
Re	Reynolds-number defined by nozzle diameter and nozzle exit velocity
s	layer thickness
s^0	initial layer thickness
S	stretch rate defined by $-S = (1/s)\cdot(ds/dt)$
t	time coordinate
T	local temperature
T_0	initial
T_{af}	adiabatic flame temperature
T°	reference temperature
T^*	nondimensional temperature defined by $T^* = (T - T_0)/(T_{af} - T_0)$
u_0	normal velocity of convective transport through the flame
$\dot{w}_i$	chemical production/consumption rate of species i
x	coordinate normal to the flame zone
γ	Kolmogorov strain rate (reciprocal value of Kolmogorov time scale)
λ	excess air ratio
ν	viscosity, used in the thermal conduction term of Eq. (1) for Prandtl number unity
ω	nondimensional space coordinate normal to the flame zone
ρ	density
σ_i	mole number of species i in 1 kg mixtures
τ	characteristic chemical time scale, defined so that the peak value of $F(T^*)$ becomes unity
$(\bar{\ })$	time mean values

Introduction

Combustion of nonpremixed reactants has been investigated in turbulent jet flames with attention focussed on hydrogen combustion. Compared with hydrogen combustion, however, turbulent hydrocarbon diffusion flames introduce additional problems due to the more complex combustion reactions.

An essential feature of some combustion models for turbulent diffusion flames is a unique determination of the instantaneous molecular species concentration by a conserved scalar. The assumption of a one-step irreversible reaction leads to the flame sheet model, with the combustion reactions taking place in an infinitesimally thin flame sheet. If the fast chemistry assumption is not restricted to a one-step irreversible reaction it follows that the products are in chemical equilibrium. Due to the shifting equilibrium reactions a large amount of carbon monoxide is predicted by this model in the fuel-rich region of a diffusion flame, more than is found experimentally in laminar or in turbulent diffusion flames. As it has been shown [1] a turbulent natural gas diffusion flame can be predicted rather well by a chemistry model which is based on the assumption that the instantaneous gas composition in a turbulent flame is that corresponding to local chemical equilibrium over a wide range of mixture strength around stoichiometry, and that the shifting equilibrium reactions freeze at a unique fuel-rich mixture fraction. To check the validity and limitations of this simplified instantaneous reaction model, one-dimensional unsteady laminar diffusion flames have been investigated. These flames are assumed to be representative of the diffusive reactive interfaces between fuel and air eddies in a turbulent flame.

The Unsteady Laminar Diffusion Flame

The behaviour of the quasi-laminar substructure of turbulent flames has been discussed qualitatively by *Williams* [2] in terms of the unsteady, one-dimensional temperature equation

$$\frac{\partial T^*}{\partial t} - \gamma x \frac{\partial T^*}{\partial x} + u_0 \frac{\partial T^*}{\partial x} - \nu \frac{\partial^2 T^*}{\partial x^2} = \frac{1}{\tau} F(T^*) \tag{1}$$

This equation is obtained if the coordinate system moves parallel to the reactive layer such that the tangential velocity component becomes zero at the origin of the coordinate system. The second and third term in Eq. (1) represent convective fluxes due to flame stretch and a velocity component normal to the flame front. In the following these convective fluxes will be neglected, but the stretching process is considered with respect to a decrease of the thickness of the reactive layer as proposed by *Spalding* [3]. The resulting balance equations are

Concentration:

$$\frac{\partial \rho \sigma_i}{\partial t} - \frac{1}{s^2} \frac{\partial}{\partial \omega} \left(\rho D_i \frac{\partial \sigma_i}{\partial \omega} \right) = \dot{w}_i \quad ; \quad i = 1, \ldots, \mathrm{NS} \tag{2}$$

Energy:

$$\frac{\partial \rho h}{\partial t} = \frac{1}{s^2} \frac{\partial}{\partial \omega}\left(\frac{\lambda}{c_p} \frac{\partial h}{\partial \omega}\right) + \frac{1}{s^2} \sum_{i=1}^{NS} \frac{\partial}{\partial \omega}\left[h_i \left(\frac{\lambda}{c_p} - \rho D_i\right) \frac{\partial \sigma_i}{\partial \omega}\right] , \tag{3}$$

with the non-dimensionalized space coordinate $\omega = x/s$, $0 < \omega < 1$ and

$$h = \sum_{i=1}^{NS} h_i \sigma_i \;;\; h_i = h_i^0 + \int_{T^\circ}^{T} c_{p,i}\, dT \;;\; c_p = \sum_{i=1}^{NS} c_{p,i}\, \sigma_i \tag{4}$$

The following conditions were used for the calculations:

1. As it is difficult to specify the time dependent thickness of the shear layer, a constant stretch S was assumed, which yields [3]

   $s = s_0 \exp(-St)$

   with s_0 as the initial shear layer thickness.
2. Initial conditions of the computations are characterized by a step change in the concentrations of fuel gas and air with a very thin layer of hot gas at the interface.
3. Dimensions of the fuel gas and air volumes are large compared to the thickness of the diffusion layer.
4. 29 elementary reactions were used to describe the oxidation process of methane adopted as a fuel.
5. Binary diffusion coefficients for the gas components in N_2, Lewis number equal to one.

Details of the numerical computations are given in [4].

In order to get information about the properties of flamelets in turbulent diffusion flames, direct thermocouple measurements of instantaneous temperatures were made in free burning turbulent methane flames at various Reynolds numbers [5]. The different Reynolds numbers were chosen to get flames with very different flow structures, which were visualized by a schlieren technique. They ranged from orderly coherent structures to fully developed three-dimensional small scale turbulence.

In addition, conventional measurements of mean concentration were performed to get further information about the net influence on the reactions.

Results

In Fig. 1 computed rates of CO and O_2 as functions of the excess air ratio are shown. The effects of diffusion are clearly visible in the changing peak values. The limit for the freezing of the reactions assumed in the flame zone model [1] lies within the area in which the reactions are frozen during the successive time steps. The calculations have been performed for a time period of 25 ms wich is beyond the eddy life times of the turbulent flames considered.

The concentrations and temperatures shown in Fig. 2 do not change significantly during the time period considered. This means that even with very large concentration gradients the conditions at the beginning of the lifetime of an eddy are not very different from those at later time steps.

A particularly important result is the fact that the computed peak temperatures of a non-stretched flame zone lie around 1 930 K as did the thermocouple measurements in the turbulent methane diffusion flames. This value is considerably below the adiabatic

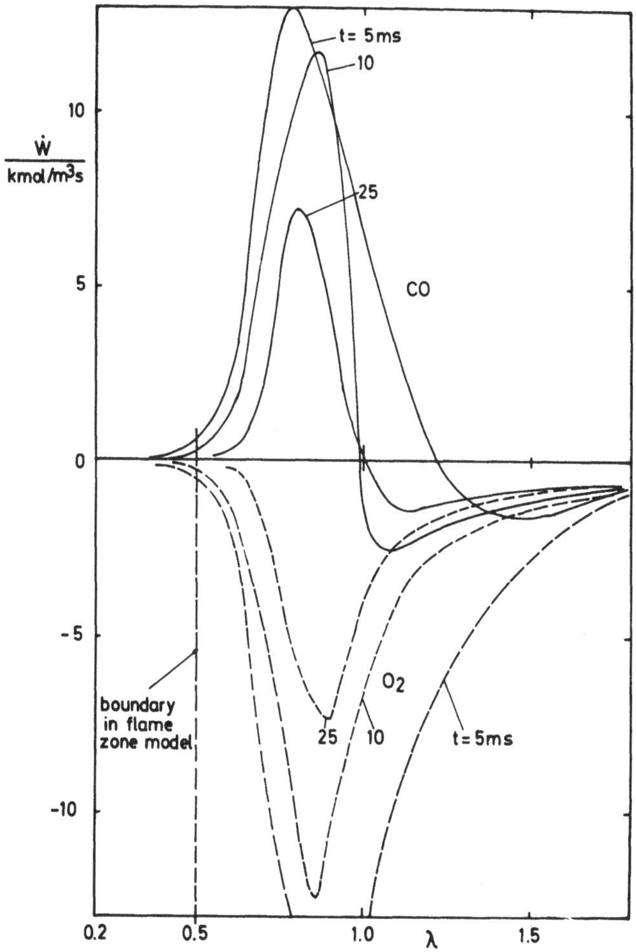

Fig. 1. Production rates of CO and O_2 in a one dimensional unsteady laminar natural gas diffusion flame for different time steps as function of the excess air ratio λ

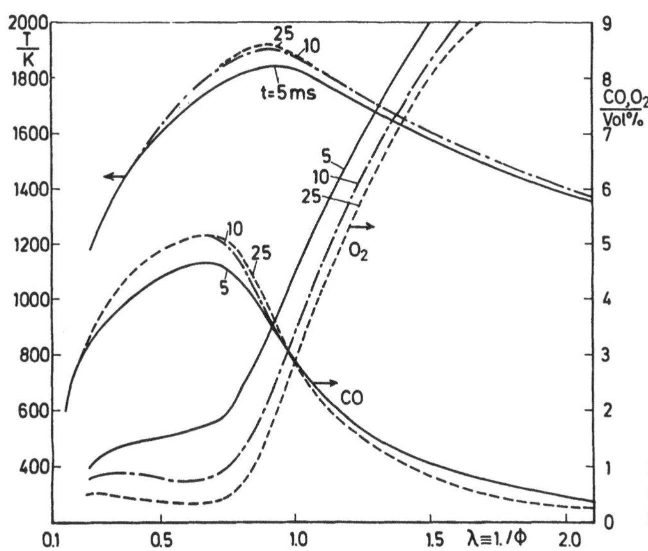

Fig. 2. Concentrations of CO, O_2 and temperature T as functions of the excess air ratio λ

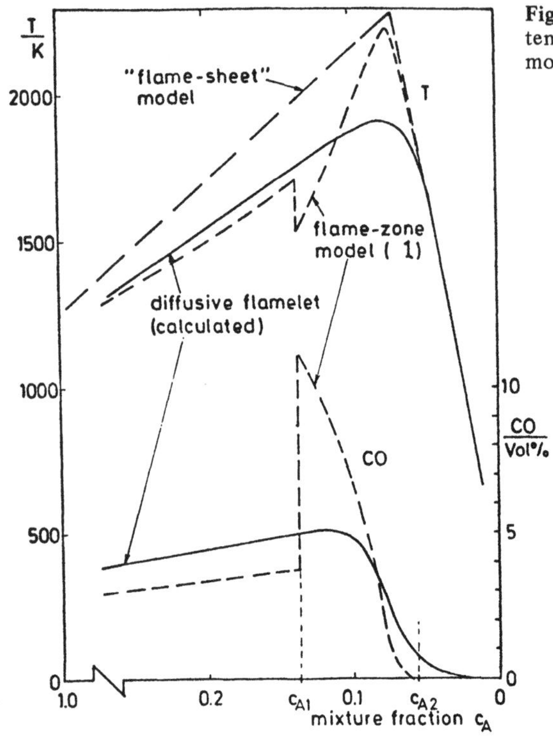

Fig. 3. Comparison of CO-concentration and temperature T calculated by two different flame models and finite kinetics (t = 25 ms)

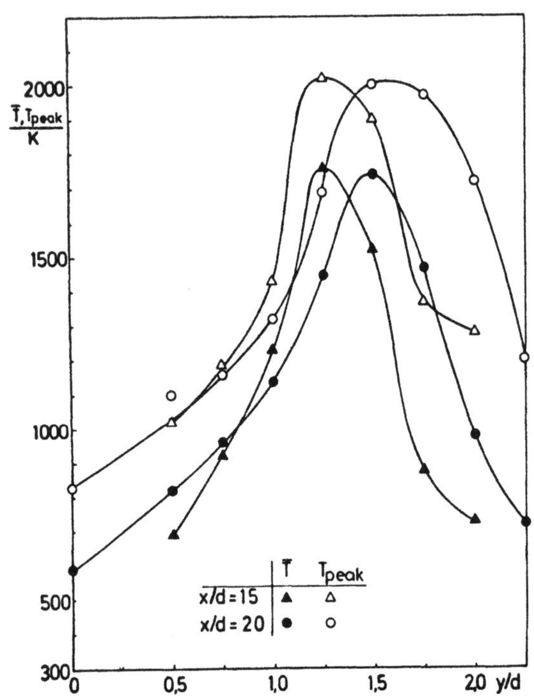

Fig. 4. Radial profiles of measured peak and time mean temperatures in a free burning natural gas flame, Re = 5 500, [5]

equilibrium flame temperature of 2 280 K (Fig. 3). The measured peak temperatures of 1 900–2 000 K (Figs. 4, 5) in all regions of the turbulent flames at different Reynolds numbers [5] may also be compared to the value of about 2 000 K from Laser Raman measurements [6] in a laminar methane diffusion flame and thermocouple measurements in a different laminar diffusion flame [7].

It follows from Figs. 4 and 5 that the maximum peak temperatures in the region of the visible coherent structures ($x/d \leqslant 20$, Re = 5 500) are approximately equal to those of fully developed three-dimensional turbulence ($x/d = 60$, Re = 37 000). It may therefore be concluded that the instantaneous peak temperatures in turbulent diffusion flames are not significantly affected by turbulence. Measurements of mean concentrations also show that in large parts of the flame the combustion takes place in diffusion flamelets.

The influence of the stretch rate S on the behaviour of the diffusive reactive layer was studied parametrically. In Fig. 6 the predicted time dependent temperatures and concentrations of a stretched layer are shown for a constant stretch rate $S = 111\ s^{-1}$ which is a characteristic value for the large turbulent fluid elements in the flame with Re = 37 000. It can be estimated by

$$S = 1/\tau_L$$

where $\tau_L = k/\epsilon$. The decrease of temperature indicates that for higher stretch rates a certain number of flame zones will be extinguished.

Concentration measurements in free burning turbulent natural gas diffusion flames of different Reynolds numbers clearly indicate that even in stably burning flames quenched regions occur near the nozzle, through which unreacted oxygen penetrates to the flame axis, Fig. 7. Due to this cold mixing it is to be expected that premixed combustion occurs to a certain extent further downstream. Figure 8 shows measured axial CO profiles in turbulent flames of different Reynolds numbers. The rapid decrease of down-

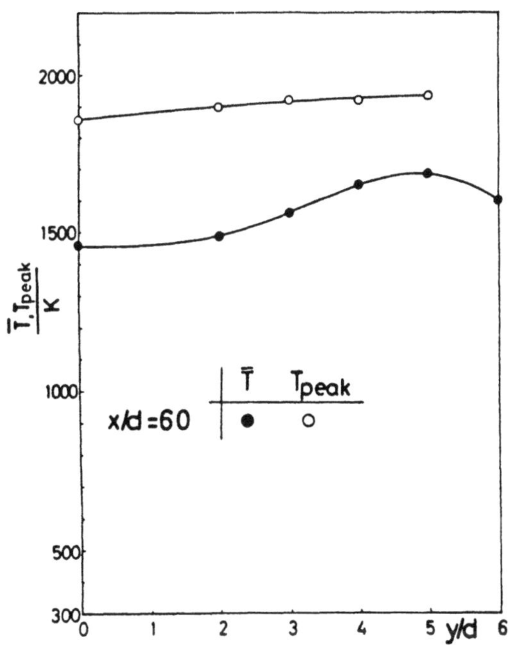

Fig. 5. Radial profiles of measured peak and time mean temperatures in a free burning natural gas flame, Re = 37 000, [5]

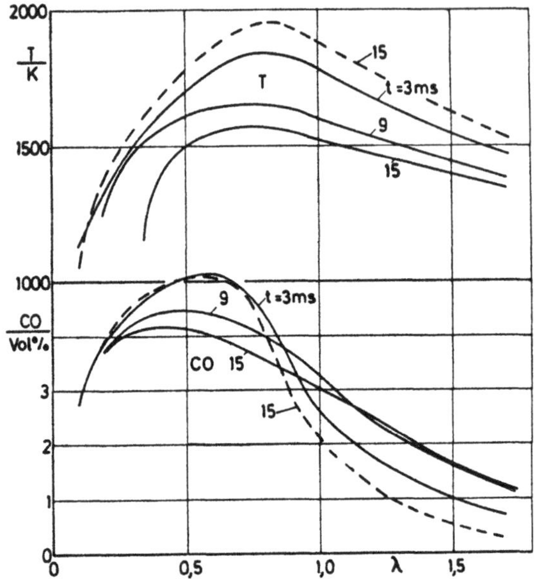

Fig. 6. CO-concentrations and temperatures as function of the excess air ratio λ (——) stretch rate $S = 111\ s^{-1}$; (– – –) without stretch

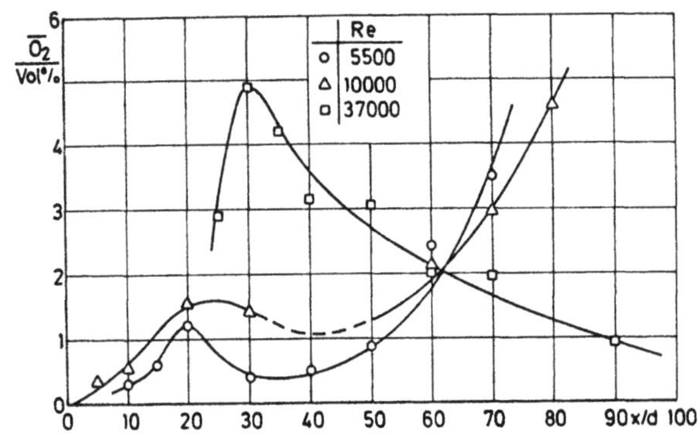

Fig. 7. Measured axial profiles of O_2-concentrations in free burning natural gas diffusion flames at different Reynolds numbers

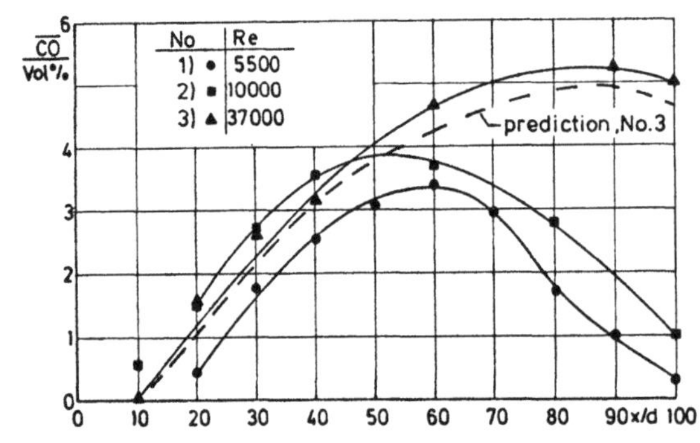

Fig. 8. Measured and predicted axial profiles of CO-concentrations in free burning natural gas diffusion flames at different Reynolds numbers

stream CO-concentration in the lower Reynolds number flames is due to mixing enhanced by buoyancy. The dashed curve which refers to the flame with Re = 37 000 represents computed values. The predictions are based on the instantaneous gas composition and temperature (Fig. 2), a pdf-representation of the mixture fraction and the *k*-ϵmodel for the flow field [8]. Moreover it is assumed that the contribution to the time mean values of concentrations and temperature is much higher for flame zones in large turbulent fluid elements which have small strain rates, than for flamelets in strong sheared small scale elements. Therefore the structure of a laminar diffusion flame zone (Fig. 2) was adopted as a representative flame structure for the spectrum of instantaneous flamelets in a turbulent flame.

Conclusions

1. From comparison of predictions and measurements it can be concluded that turbulent diffusion flames over a large range are composed of laminar diffusion flamelets with instantaneous peak temperatures about 300 K below the adiabatic equilibrium temperature.
   This is an especially useful information with regard to the thermal NO-formation, this being strongly dependent on the peak temperature.
2. Areas of premixed combustion can occur even in flames stably burning at the nozzle exit.
3. The instantaneous temperature and concentration profiles do not change greatly during the most probable life time of an eddy. This result can be compared to some extent to the conclusion of *Bilger* [9] who found by inspection of measurements in laminar diffusion flames that the gas-composition is independent of the position in the flame and therefore independent of the flow structure.

With the diffusive flamelet model which is compared to the other models in Fig. 3, a turbulent diffusion flame based on the k-ϵ-model and a pdf of the mixture fraction can be well predicted as long as premixed combustion does not take place.

References

1. Eickhoff, H.E., Grethe, K.: A flame zone model for turbulent hydrocarbon diffusion flames. Combust. Flame **35**, 267–275 (1979)
2. Williams, F.A.: "Recent Advances in Theoretical Descriptions of Turbulent Flames", in *Turbulent Mixing in Nonreactive and Reactive Flows*, ed. by S.N.B. Murthy (Plenum, New York 1975)
3. Spalding, D.B.: "Chemical Reactions in Turbulent Fluids", in *Physicochemical Hydrodynamics*, Vol. 1, ed. by D.B. Spalding (Advance Publications Ltd., London 1977)
4. Grethe, K.: "Verification of a Flame-Zone Model for Turbulent Diffusionflames", (submitted Doctor thesis), University of Karlsruhe (TH), (1981)
5. Grethe, K., Kohler, W., Leuckel, W.: "Direct Thermocouple Measurements of Instantaneous Temperatures in Turbulent Diffusion Flames", (to be submitted), University of Karlsruhe, Engler-Bunte-Institut (1981)
6. Stricker, W., Just, Th.: "Lokale Temperaturmessung in Flammen mit Laser-Raman-Verfahren", 9. Deutscher Flammentag, Aachen, (1979), VDI-Bericht Nr. 346
7. Yamaoka, I., Tsuji, H.: "An Experimental Study of Flammability Limits Using Counter Flow Flames", 17. Symp. Inter. on Combustion (1978)
8. Eickhoff, H.E., Grethe, K., Thiele, F.: "Numerical Investigations on Turbulent Transport and Reaction Phenomena in Jet Flames", 2nd Symposium Turb. Shear Flow (1979)
9. Bilger, R.W.:Reaction rates in diffusion flames, Combust. Flame, **30**, 277–284 (1977)

Part III

Recirculating Flows

Introductory Remarks

I.P. Castro

University of Surrey
Guildford, Surrey GU2 5XH, England

The emphasis in the corresponding section of the previous volume in this series was on numerical prediction of complex, but not necessarily recirculating flows. By contrast, the majority of the present contributions describe experimental studies and all of them consider recirculating flows. This is an accurate reflection of the scope of the two sessions during the conference itself and is, perhaps, an indication that a two year gap is a rather short time in which to expect significant progress by a significant number of workers in a field as fraught with difficulties as is the prediction of complex flows. After all, thirteen years were to elapse after the first Stanford conference on boundary layer predictions before it was felt appropriate to convene a major conference on Complex Flow Prediction. There were, nonetheless, and despite the proximity of that Stanford Conference, a few papers describing such predictions and these are typified by the two selected for inclusion here. The first, by McGuirk, Taylor and Whitelaw, concerns the familiar problem of numerical diffusion in upwind difference calculations and the second, by Baron, Benque and Coëffe, describes a large eddy simulation of the turbulent flow induced by a jet in a cavity. Some general comments concerning the prediction of recirculating flows and some more particular remarks about these two contributions will be made, before turning to consideration of the remaining four papers, which all describe experimental studies.

In his introduction to the corresponding section of the previous volume McDonald expressed the belief that 'observations and evidence concerning the predictive capabilities of turbulence models for complex flows based on the earlier course mesh results must now be questioned'. He was referring specifically to the numerical errors inherent in such predictions, rather than to the uncertainties in the turbulence models and was, with the present writer, in the minority of those who doubted the accuracy of the standard techniques used to discretize the equations when applied to complex recirculating flows. The respectability of such a view was recently enhanced by a suggestion of the Evaluation Committee of the Stanford Conference, made in the light of predictions presented at the meeting: 'One of the more important steps, which will make (turbulence) model comparisons more meaningful, will result from a more detailed concern for the numerical problems leading to accuracy of solutions. Only in this way is it possible to distinguish the precision of the physical modelling from the inherent numerical errors related to algorithm and grid'. It is encouraging to see that an increasing number of turbulence modellers are not only now recognising the numerical difficulties but also, in some cases, trying to do something about it. Thus a few of the predictions presented at the Stanford Conference of, in particular, the backward facing step flow (fast becoming a 'classic' complex flow despite the lack of real consensus either about its suitability as such or, indeed, the nature of the flow) used more sophisticated numerics than the commonly employed Hybrid Differencing Scheme. The first paper in this section, by McGuirk, Taylor and Whitelaw, is a good example of this encouraging trend. Following the suggestions of, for example, *McGuirk* and *Rodi* (1978) and *Castro* (1979), they perform an 'a posteriori' accuracy check by calculating directly from the predicted flow field, using central differences, each term in each equation solved. The 'out of balance' terms must essentially be the difference between the upwind and central differenced forms of the convective terms (except in regions where the mesh Reynolds numbers are less than two so that the original prediction, utilising a hybrid scheme, would have reverted to central differences). This residual term is therefore simply a direct measure of the numerical diffusion and where it is significant compared with the other terms in the equations grid refinement is clearly desirable. Having improved the grid on that basis, the best predictions of McGuirk et al. for the wake flow behind a confined axisymmetric disc are undoubtedly more grid independent than the earlier calculations of *Pope* and *Whitelaw* (1976), so that proper assessment of the k-ϵ turbulence model used in the work was feasible.

However, the fact that absence of locally large errors at any one section of the flow does not necessarily imply an error free solution (because small errors generated upstream can propagate and accumulate in a way which would maintain *small* local residuals) means that the process of grid refinement is a rather tedious one – strictly, checks should be made over a good proportion of the flow. Indeed, the authors' momentum balances are immediately reminiscent of those of *Castro* (1979), where it was shown from calculations of the backward step flow that errors generated around the separation corner simply propagated downstream, but were not, of course, deducible by a residual check at downstream sections. In fact, the authors recognise that 'only the development of a more accurate discretisation scheme offers long term advantages' so that their procedure is admitted to be only 'a useful interim measure'. It is a relatively easy check to make, however, and would not noticeably increase the cost of running a typical industrial code, so should be seriously considered by those who worry about numerical errors in such calculations and definitely implemented by those who do not.

It is also worth noting that McGuirk, Taylor and Whitelaw chose not to include prediction of the flow upstream of the disc. Although having to supply experimentally determined boundary conditions in the plane of the disc, they therefore avoided one of the fundamental problems in prediction of many complex flows: having to cope simultaneously with regions where the convective terms in the momentum equations are (roughly) balanced by pressure gradients and other regions where the latter are much less significant and convection balances diffusion. It is this dual parabolic/elliptic nature of many complex flows which makes the development of a stable and accurate numerical scheme particularly difficult. This problem was addressed in another paper presented at the conference and since published elsewhere (*Castro* et al. 1982).

The paper by Baron, Benque and Coëffe is at first sight very different. It is novel, in that the authors have attempted to use one of the more sophisticated turbulence modelling approaches, large eddy simulation, to compute a fully three-dimensional recirculating flow which those used to dealing with classic turbulent flows could only describe as horrendous: a jet issuing into a cubical box from a square hole in one wall with flow leaving from another square hole (not aligned with the first) in the opposite wall. The formal filtering procedure was not used, reliance being placed on the inherent filtering effect of their finite differencing scheme. A course grid ($21 \times 21 \times 27$) was used and the authors state that the Leonard stress was found from tests to be insignificant, so it was neglected in the calculation. The only turbulence to come out of the calculation was therefore presumably all of the sub-grid-scales – determined by the subgrid turbulent viscosity used – so it is difficult to see how this procedure can be truly called a large eddy simulation. Indeed, the authors actually made one computation with no turbulent viscosity included at all, in which case the equations solved were just the time-dependent laminar viscous equations. The mean flow results seemed to be quite similar to those obtained from the turbulent calculation and not too dissimilar from their experimental results. Since the flow must be dominated by pressure gradients this is not too surprising. The turbulent calculations showed a rapid decay of the turbulence fluctuations, in contrast to the experimental data, and this led the authors to revert to study of plane Poiseuille flow. Detailed conclusions regarding this part of the paper should perhaps be left to the interested reader, but it is not without significance that rather ad hoc procedures had to be employed to ensure a proper production of turbulent energy and so prevent its gradual decay.

Finally it must be emphasised that although the signs of an increasing 'numerical awareness' amongst turbulence modellers attempting to predict recirculating flows is encouraging, there is evidently much ground still to be covered. One cannot help feeling that there is too little interaction between modellers – usually interested in turbulence per se – and numerical analysts who, over the last decade or so, have become adept at developing sophisticated schemes for particular (laminar) flows. What is still needed is a numerical procedure which, whilst being demonstrably much more accurate and robust than current methods, can be relatively easily implemented in *engineering* codes. It may well be that a modular approach to the numerics like that suggested by *Kline* (1981) for turbulence models in complex flows, will be the best approach to dealing with flows which are described by equations which have different characteristics in different parts of the flow.

Turning now to the remaining four papers in this section, the first three all concentrate on a simple presentation of experimental results, with little attempt being made to discuss the findings. This is not a criticism, but merely reflects the current 'state of the art' in recirculating flow experiments, where the ability of the turbulence community at large to make meaningful measurements of any accuracy at all still appears to lag behind the ability of the predictors to produce plausible results. The situation is therefore in that sense the reverse of that which prevailed in the struggle to understand and predict simpler turbulent flows. Since in principle the accuracy of predictions must depend on the adequacy of the physical input which itself will depend, to some extent at least, on understanding gained from experiments, there are obvious dangers in this situation. It is therefore worth reiterating one of the clear messages to emerge from the Stanford Conferences, that there is a desperate need for further well-documented complex flows, particularly those involving recirculations.

The paper by Dengel, Fernholz and Vagt is really a progress report on their experiment whose aim is to 'complement earlier mean flow data' in an axisymmetric boundary layer with incipient separation) 'by fluctuating flow measurements and to extend measurements into the separation region itself'. A fairly large scale rig allowed the use of pulsed wire anemometry, in addition to more standard instrumentation, and the care with which the authors check and use their instruments is a model to all who aspire to making measurements in *any* flow. They found some interesting behaviour of the longitudinal Reynolds stress as separation is approached which was not, incidentally, observed by *Simpson* et al. (1977) in their separating boundary layer experiment, although the latter's measurements did not extend so near the wall. To aid in a more detailed understanding Dengel et al. intend to measure the other normal stresses and the shear stress (now possible with pulsed wire anemometry, see *Castro* and *Cheun,* 1982).

Okamoto's paper continues his sequence of studies on surface mounted bluff bodies (all with curved surfaces), in which attention is focussed on the surface pressure field and the development of the wake beyond the recirculating region. There are some interesting photographs of the 'vortex' motions occurring behind the bodies, but no attempt is made to compare the results with the many similar studies in the literature of sharp-edged bluff body flows. A comparison of the rate of decay of the mean velocity defect on the wake's centre-line with the theoretical result of *Hunt* (1971) is made, but the author fails to point out that an important ingredient of the latter is the assumed thick upstream boundary layer ($\delta >> h$): in each of the Okamoto experiments the bodies were placed on a smooth flat plate so that $\delta < h$! From the Wind Engineering point of view, experiments of this kind ought really to be motivated, in part at least, by a desire to isolate the dominant flow phenomena and to discover how these are changed in *kind*, not just in strength, by changes in body geometry or upstream flow characteristics. This certainly necessitates more attention being paid to the dynamics of the recirculating zone. If the motivation is simply to provide useful data for engineering codes of practice then proper scaling with a simulation of the atmospheric boundary layer is required of course.

The paper by Farcy, Mercier and Lablanc presents data obtained in the transonic recirculating flow in an air-intake set at an angle of attack in a $M = 0.7$ jet. Laser anemometry was used to obtain mean velocity and longitudinal turbulent intensity in the separated region originating from the sharp leading edge of the intake. Whilst the presence of shocks must considerably complicate matters, time dependent flow visualisation and spectra (measured with a hot wire) seem to indicate that the flow is qualitatively similar to corresponding low speed flows, in that although large structure motions were evident, there was no discernible periodic motion. One interesting result is that a distinct 'double hump' was observed in a velocity probability density distribution measured just beyond reattachment. The authors state that this 'is certainly characteristic of the intermittency of the flow'. It must, in fact, be evidence of a distinct bi-stable behaviour. There is currently considerable discussion concerning the dynamics of separated flows and this is, perhaps, the first evidence of bi-model behaviour of such flows (except in the well known class of diffuser flows), although it must be seen as a rather tentative result – other $p(u)$ measurements presented by Farcy et al. did not contain the same double maxima.

This whole subject of the dynamics of a recirculating region is discussed in much more detail in Eaton and Johnston's paper, the last in this section. They focus attention on the very low frequency unsteadiness apparent in the shear layer separating from a backward facing step. Their findings and also those of Hillier, whose paper[1] should really be read in conjunction with Eaton and Johnston's, are now beyond question. The major point is that these low frequency motions have time scales several times longer than the largest turbulent eddies and contain a significant proportion of the total turbulent energy. The reattachment point seems to move slowly up- and downstream over a distance about twice what might be expected on the basis of a regular passage of the largest eddies in the shear layer. Various suggestions as to the cause of these motions are made but they are all basically of a two-dimensional variety, despite the fact that, as the authors admit, the evidence for the motions themselves being two-dimensional is weak. Their preferred mechanism is based on the simple balance which must exist between the rate of entrainment in the shear layer and the rate of reinjection of fluid into the recirculating region at reattachment, but there is clearly much more work to be done before this conclusion can be regarded as certain.

---

[1] This paper was presented at the meeting and is being published elsewhere. It is currently available in the original conference proceedings.

What is certain is that this low frequency unsteadiness must be taken into account by the turbulence modellers if they are going to use empirical input obtained from flows of this sort to develop more sophisticated models. It is a sign that recirculating flows are not always (and perhaps never) what they may seem at first sight, and a spur to experimentalists to increase the range and effectiveness of their instrumentation to allow much deeper probing of such flows lest the computors are tempted to say that they can do it all.

References

Castro, I.P. "Numerical Difficulties in the Calculation of Complex Turbulent Flows", in *Turbulent Shear Flows I*, ed. by F. Durst, B.E. Launder, F.W. Schmidt, J.H. Whitelaw (Springer, Berlin, Heidelberg, New York 1979)

Castro, I.P., Cheun, B.S. The measurement of Reynolds stresses with a pulsed wire anemometer. J. Fluid Mech. **118**, 41 (1982)

Castro, I.P., Cliffe, K.A., Norgett, M.J. Numerical predictions of the laminar flow over a normal flat plate. Int. J. Num. Meth. Fluids **2**, 61 (1982)

Hunt, J.C.R The effect of single buildings and structures. Philos. Trans. R. Soc. London **A269**, 457 (1971)

Kline, S.J. "Universal or Zonal Modeling – the Road Ahead", Stanford 1981 Conference on Prediction of Complex Turbulent Flows (1981)

McDonald, H. "Introduction. Recirculating and Complex Two- and Three-Dimensional Flows", in *Turbulent Shear Flows II*, ed. by L.J.S. Bradbury, F. Durst, B.E. Launder, F.W. Schmidt, J.H. Whitelaw (Springer, Berlin, Heidelberg, New York 1979)

McGuirk, J.J., Rodi, W. A depth-averaged mathematical model for the near-field of side discharges into open channel flow. J. Fluid Mech. **86**, 761 (1978)

Pope, S.B., Whitelaw, J.H. The calculation of near wake flows. J. Fluid Mech. **73**, 9 (1976)

Simpson, R.L., Strickland, J.H., Barr, P.W. Features of a separating turbulent boundary layer in the vicinity of separation. J. Fluid Mech. **79**, 553 (1977)

Low Frequency Unsteadyness of a Reattaching Turbulent Shear Layer

John K. Eaton and James P. Johnston

Department of Mechanical Engineering, Stanford University,
Stanford CA 94305, USA

Abstract

Low-frequency unsteadiness of a reattaching shear layer was studied in the flow behind a backward-facing step. Spectral data from a hot-wire probe showed that a substantial fraction of the measured turbulence intensity occurs at frequencies significantly lower than the large eddy passing frequency. Several thermal tuft probes were used simultaneously to identify the source of the low-frequency disturbances. It was concluded that the reattaching free shear layer undergoes a non-periodic, roughly two-dimensional, vertical motion. The motion causes the impingement point of the reattaching shear layer to drift slowly up- and downstream over a range of approximately two step heights. The likely cause of the low-frequency motion is an instantaneous imbalance between the entrainment rate from the recirculation zone and the reinjection rate near reattachment.

Nomenclature

- F Frequency [Hz]
- h Step height = 5.08 cm
- U_0 Inlet free stream velocity
- x Streamwise coordinate
- x_R Time mean reattachment location
- γ Percent of time that flow is in downstream direction
- θ Momentum thickness

Introduction

In this paper we shall discuss our observations and measurements of low-frequency motions of a reattaching shear layer in the flow behind a backward-facing step. Low-frequency motions are defined as those motions having time and length scales much longer than the scales of a typical large, turbulent eddy.[1] The time scale for a large eddy in a separated shear layer is approximately equal to the shear layer thickness divided by one-half of the free stream velocity. The length of a separation bubble is typically 5 to 10 times the shear layer thickness. Therefore the low frequency motions have time scales which are greater than or equal to the time it takes the free stream to pass over the separation bubble.

It has long been recognized that low-frequency unsteadiness can occur in separation bubbles bounded by turbulent free shear layers. "Jitter" of the separation point is

1 Here we refer to large turbulent eddies as the largest structures bisible in flow visualization. These typically have a length scale comparable to the shear-layer thickness.

frequently reported in airfoil experiments [1] and transitory stall behavior is a well-known but poorly understood example of very large-scale, low-frequency unsteadiness [2]. In general, low-frequency unsteadiness has been regarded as a nuisance rather than a topic of scientific interest. To date, there have been few studies of long time-scale motions of separation bubbles. However, low-frequency unsteadiness can cause significant difficulties in engineering equipment such as combustors and wind tunnel nozzles. It is also a major source of uncertainty in single-sample measurements with relatively short averaging times.

Perhaps more important is the fact that low-frequency unsteadiness causes significant difficulties in the measurement, understanding and modeling of turbulent separated flows. Unsteadiness has gone unrecognized in many previous experiments and has, we believe, contaminated the available data base. For example, it has often been stated that the structure of a reattaching free shear layer is fundamentally different from a plane-mixing layer. The major evidence for this conclusion is that the measured turbulence intensity is considerably higher in a reattaching free shear layer. It is the authors' contention [3] that this measured difference is caused largely by low-frequency unsteadiness in reattaching flows which has little real effect on the basic structure of the free shear layer.

The contamination of turbulence data by low-frequency unsteadiness could have serious consequences for computational fluid mechanics. In general, turbulence models attempt to represent the effects of eddies up to the scale of the typical "large eddy" described above. However, turbulence data which are used for calibration of models may be strongly biased by low-frequency contributions. It would then be impossible to rationally compare model results to data.

The backward-facing step flow has usually been considered immune from low-frequency unsteadiness of significant magnitude because the separation line is fixed at the sharp corner. It was felt that most unsteadiness of separation bubbles is related to a wandering separation point. *Tani* et al. [4] and *Chandrsuda* [5] both determined that low frequency motions were not present in their back-step flows by examining the fluctuations of manometers attached to static taps in the separated zone. However, during the course of a recent study of the backward-facing step flow [6], we found significant evidence of low-frequency motions which were affecting the time-averaged data. This led us to concentrate in the present study on understanding low-frequency unsteadiness.

Long time scale motions in backward-facing step flows have not been explicitly studied prior to this work. Low-frequency flapping of a plane jet has been studied by *Goldschmidt* and *Bradshaw* [7] and by *de Gortari* [8]. They found that the flapping was quasi-periodic and had an amplitude of approximately 20% of the jet half width. A number of other workers have investigated low frequency oscillations in cavity flows (c.f. *Rockwell* and *Naudascher* [9]), *Knisely* and *Rockwell* [10] concluded that low frequency oscillations of a cavity flow are caused by feedback of disturbances from the impingement point which are then amplified in the free shear layer. *Rockwell* [11] suggested that a similar mechanism may cause low frequency motions in the backward-facing step flow. *McGuinness* [12] observed low-frequency pressure fluctuations in the separation bubble at a pipe entrance. He concluded from spectral data that these fluctuations were caused by eddies moving upstream in the low speed recirculating flow. However, his flow visualization work did not confirm this conclusion.

Objectives

The objective of the present study is to understand the nature and cause of low-frequency motions of separation bubbles. We hope to obtain data to gauge the effects of unsteadiness on time-averaged turbulence data. We also hope to better understand how very low-frequency unsteadiness can be accounted for in turbulence models which currently ignore its presence.

Methods

The experiments were conducted in the same low speed air flow facility used by *Eaton* and *Johnston* [13] (see Fig. 1). After a short entrance duct to develop turbulent boundary layers the flow passes through a two dimensional sudden expansion of area ratio 3:5. The step height was set at 5.04 cm which gave an aspect ratio (channel span/step height of 12). The inlet free stream velocity was approximately 12 m/s for most of the measurements reported here. At that inlet speed the boundary layer at separation was approximately 0.2 step heights thick and turbulent. The momentum thickness Reynolds number of the boundary layer was 950 indicating that the boundary layer was not fully turbulent. Mean velocity and turbulence intensity profiles however did appear quite similar to an ordinary turbulent boundary layer.

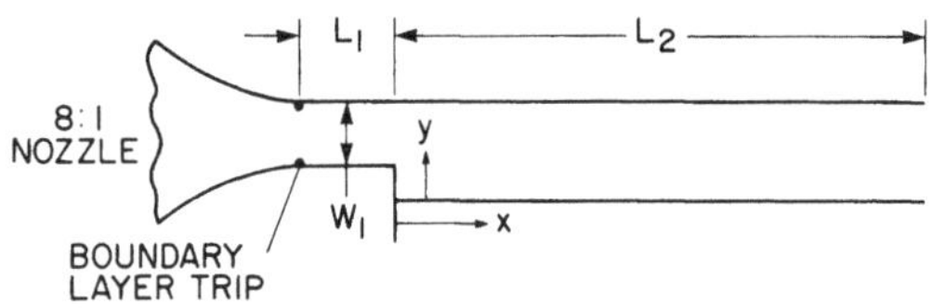

Fig. 1. Sketch of the wind tunnel test section

Mean velocity and turbulence data for this flow were reported in references 6 and 13. The measurements indicated that two-dimensionality in the channel was good; mass and momentum balances for an integral control volume around the entire separation bubble showed balance to within 0.5% and 1.7% of the inlet mass and momentum flux, respectively. However, *Westphal* and *Johnston* [14] in later measurements using the same facility found a narrow region near the channel centerline where the spanwise velocity profile was non-uniform. Although this non-uniformity may have affected some of the quantitative data, it is doubtful that it had a significant effect on the results reported here.

It was desirable to have a laminar boundary layer at separation for some of the experiments. In this case the tunnel was run at an inlet free stream velocity of 3.26 m/s. The boundary layer at separation was then laminar with a thickness of 0.9 cm and a momentum thickness Reynolds number of 240. Transition of the free shear layer to turbulent flow occurred just downstream of separation despite the fact that the upstream boundary layer was laminar.

The primary instrument used in this study was the thermal tuft, a device which indicates the instantaneous flow direction (upstream or downstream) in a thin layer of fluid adjacent to the wall.[2] A sketch of the probe is shown in Fig. 2. The center wire is heated with a DC current of 1.5 A to create a heated wake which moves back and forth as the flow direction reverses. The outer two wires are operated as resistance thermometers and sense the presence of the hot wake. The control electronics compare the temperatures of the two outer wires and give an output signal of 0 V or 5 V depending on which wire

2 Details of thermal tuft construction and operation are described in Ref. [15].

is hotter. It should be noted that the electronics will always indicate an instantaneous flow dierection. If there is no temperature difference between the wires electronic noise will cause the output to toggle between 0 and 5 V at high frequency. In practice, such signal dropout rarely occurs. The signal due to the temperature difference is almost always larger than the electronic noise.

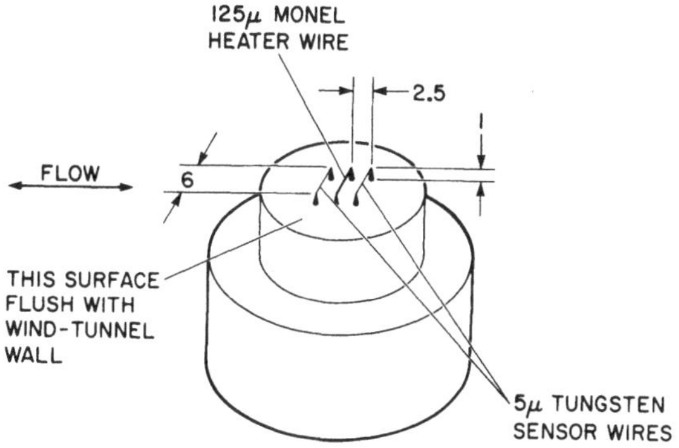

Fig. 2. Sketch of thermal tuft probe (dimensions in mm)

Single hotwires were used for the measurement of mean velocity and turbulence intensity near the outer edge of the shear layer, where the turbulence intensity was less than 30–40%. The probes were operated with a TSI 1050 anemometer and the signal was linearized with a TSI Model 1052 linearizer. A pulsed-wire anemometer [16] was used for mean velocity and turbulence intensity measurements in more highly turbulent regions of the flow including the recirculating region.

Results

The thermal tuft was used to measure the percentage of time that the flow near the wall is in the downstream direction. Plots of this quantity for the two cases (laminar and turbulent boundary layers at separation) are shown in Fig. 3. Also shown are the time mean reattachment points for the two cases.[3]

The curves show a broad region at least 4 step heights long where the near wall flow can be instantaneously moving either upstream or downstream. This result alone suggests that large scale unsteadiness is present in the flow. However, other evidence is far more conclusive. Averaging times of approximately 200 s were needed to obtain converged (± 1%) data. This was true both for thermal tuft data and velocity data. The largest turbulent eddies in the flow were the spanwise (Brown-Roshko) vortices which had a maximum period of approximately 0.06 s or 1/3600th of the averaging time. It was apparent from statistical considerations that unsteadiness was present which had a timescale much greater than the large-eddy period.

Spectral analysis of the turbulence data was used to quantify the effect of low frequency fluctuations of the time-averaged data. Spectra were computed only from the hotwire data because the pulsed-wire sampling rate was too slow to obtain alias-free spectra. Autocorrelations of the hotwire signal were obtained using a Hewlett Packard Model 3721A Digital correlator. The digital autocorrelation data were then Fourier transformed

3 It was shown by *Eaton* et al. [17] that the point of 50% downstream flow is the same as the point of zero mean skin function in reattaching flows.

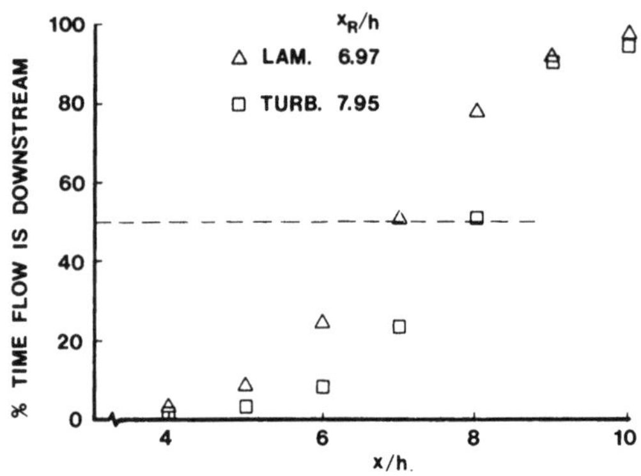

Fig. 3. Time averaged thermal tuft data for laminar and turbulent boundary layers at separation

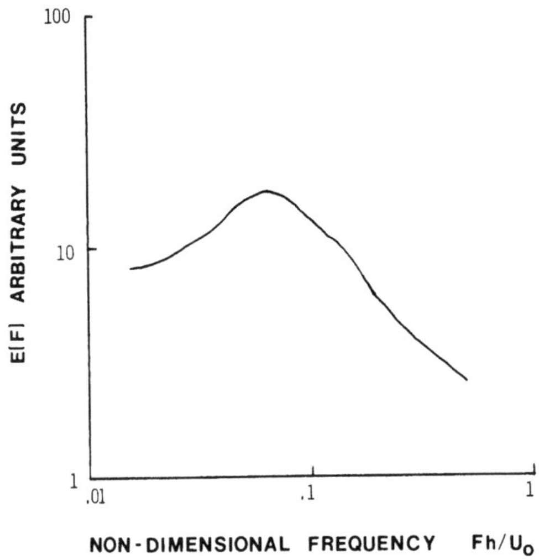

Fig. 4. Power spectrum of streamwise velocity fluctuations near the outer edge of the shear layer at $x/h = 8$

to obtain the power spectrum of the velocity fluctuations. The autocorrelation was formed at 100 equally spaced time delays and approximately 130,000 samples were taken at each time delay.

A typical spectrum of hot-wire data is shown in Fig. 4. This spectrum was measured near the outer edge of the shear layer at $x/h = 8$, very close to reattachment. The broad peak in the spectrum at a non-dimensional frequency of 0.065 is caused by large spanwise vortices in the free shear layer. The spectrum also shows that significant turbulence energy occurs at frequencies much lower than the large eddy frequency. The spectrum in Fig. 4 and others shows that more than 30% of the measured turbulence intensity $(\overline{u'^2})$ occurred at nondimensional frequencies below 0.02, considerably lower than the large-eddy passage frequency. The resolution of the spectra was not sufficient to identify any spectral peaks in the low-frequency range.

The above-described results showed that low-frequency motions occur in the separated flow behind a backward-facing step. These motions have a significant effect on the time-averaged turbulence data. Multiple thermal tufts were used simultaneously in an effort to explore the nature of the low-frequency motions. For this set of experiments the tunnel was run at a somewhat higher speed resulting in a reduction of the reattachment length to 7.6 step heights. Three thermal tuft probes were used at $x/h = 6$, 8 and 9.

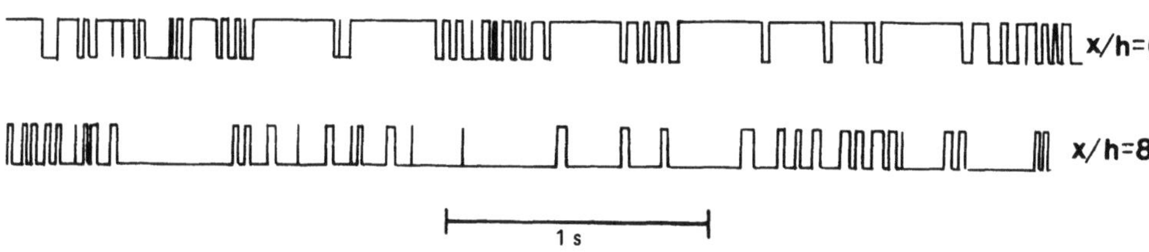

Fig. 5. Simultaneous records of the instantaneous direction of the flow near the wall at two points in the reattachment zone

Thus, all three probes were within the region where flow reversals occurred a significant fraction of the time. Figure 5 is a typical record of output from two of the probes which shows two significant trends. First, at a given x-location flow reversals occur in bursts of rapid activity separated by times when the flow direction is relatively steady. During the bursts of activity, flow reversals occur at roughly twice the passage frequency of the largest turbulent eddies. Secondly, we find that when a burst is occurring at one location the output is relatively quiescent at the other location. These observations taken together suggest that a motion of larger scale than the large tubulent eddies in the flow must be responsible for the wide region of instantaneous flow reversal. A detailed examination of records similar to Fig. 5 shows that for any given short time interval a significant number of flow reversals occur in a region which is only about two step heights long. However, over a long period of time flow reversals occur over a region which is approximately 4 step heights long (see Fig. 3). Our interpretation of the records is that the impingement point of the free shear layer is moving slowly up- and downstream. When the impingement point is centered on a probe, rapid flow reversals caused by the passage of large turbulence structures will be recorded. When the impingement point is well upstream of the probe, the "mean" velocity (short-time average) would be sufficiently large that turbulent fluctuations could not cause flow reversals.

The existence of large-scale, low-frequency motions seem to be irrefutable. Given their existence, two significant questions are: (I) are the very long time scale motions periodic in nature? and (II) what causes the motions? We attempted to resolve the first question using short-time averages of thermal tuft data. The outputs of the three thermal tufts were simultaneously sampled at a rate of approximately 1300 Hz using a microcomputer. Data were recorded for approximately 0.2 s and a number indicating the fraction of samples which indicated downstream flow was recorded for each probe. Sampling recommenced after a very short time delay (< 1 ms). This procedure continued for about 30 s for a given run. Short time-averaged values (0.2 s) of the percent downstream flow time γ were then available at each of three steamwise locations as a function of time. For each time interval a parabola was fitted to the three values of γ and a short-time-averaged reattachment length computed. We assumed that the reattachment point is the same as the point where γ is 50%. This assumption was shown to be valid by *Eaton* et al. [17] for long-time-averaged data. There is considerable statistical scatter in the data because the averaging time of 0.2 s would typically include only about eight flow reversal cycles. The short-time-averaged data varied monotonically for every record examined (several thousand) indicating that statistical uncertainty at least did not cause absurd results. We estimate that the results are within about ± 0.3 step heights of the acutal short-time-averaged reattachment point.

A section of one of the records is shown in Fig. 6. The short-time-averaged reattachment length varies from about 6.5 to 8.5 step heights about the mean reattach-

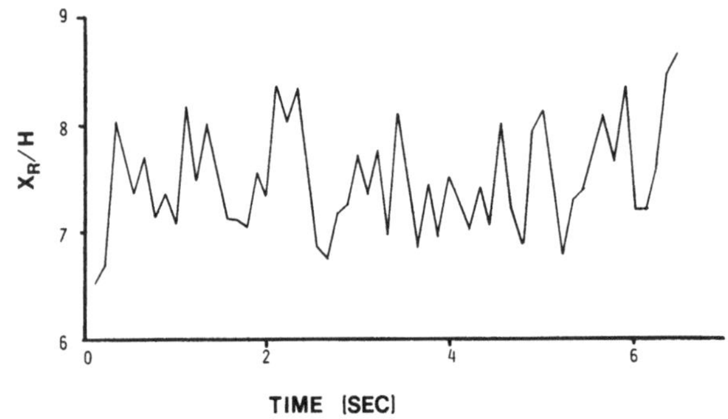

Fig. 6. Variation of the short-time-average (0.2 s) reattachment length

ment length of 7.6 step heigths. No regular periodicity is evident in the record shown or in any of the other records examined. The record also shows that the impingement point may remain nearly stationary for as long as a second before moving to a new location.

Discussion and Conclusions

The results reported above show that low-frequency motions occur in reattaching shear layers. These motions have time scales several times longer than the largest turbulent eddies and make substantial contributions to measured turbulence quantities. Several possible descriptions of the nature and cause of the low-frequency motions are discussed below.

Our own opinion is that the motions are long-time-scale, vertical movements of the separated shear layer. We envision these motions as being roughly two-dimensional, although we have only weak evidence to support this conjecture. The vertical motions of the separated layer cause the shear-layer impingement point to move up- and downstream over a distance of at least two step heights. This type of motion has also been observed computationally by *Mellor* and *Celenligil* [18]. The long time-scale motions are probably caused by an instantaneous imbalance between shear-layer entrainment from the recirculating zone on one hand and reinjection of fluid near reattachment on the other. For example, an unusual event may cause a short-term breakdown of the spanwise vortices in the shear layer. The entrainment rate would be temporarily decreased, while the reinjection rate remained constant. This would cause an increase in the volume of recirculating fluid, thus moving the shear layer away from the wall and increasing the short-time-averaged reattachment length. This conclusion is strongly supported by our thermal tuft data, which show that the impingement point may remain stationary for 10–20 large eddy periods, then move to another location.

Rockwell [11] suggested that the low-frequency motions may be oscillation of the shear layer induced by feedback of disturbances from the impingement point to the separation point. The disturbances would appear as a low-frequency modulation of the shear layer downstream of separation which would be amplified by the free shear layer. This is the mechanism responsible for low-frequency motions in cavity flows. However, the present flow is considerably more complex than a cavity flow. Multiple vortex pairings occur upstream of reattachment, and the shear-layer impingement point is not fixed. Resolution of this question may be difficult because of these complexities.

A third possibility is that a section of the recirculating flow could be "divorced" (or cut off) from the separation bubble by a large structure which reaches the wall well upstream of the mean reattachment point. *McGuinnes* [12] observed that the separation bubble was occasionally cut roughly in two and that the downstream half proceeded downstream as a "divorced eddy". McGuinness's observations were made in the separation bubble at the entrance to a pipe, using smoke visualization. We have never observed similar behavior in several experiments on two-dimensional, backward-facing steps, although admittedly the visualization techniques have been crude. A divorced eddy could cause non-monotonic behavior in the short-time-averaged thermal tuft data. The flow sweeping to the wall would cause forward flow, while the divorced eddy would continue to cause flow reversals downstream. Non-monotonic behavior was never observed in the large number of thermal tuft records we examined.

Intermittent separation on the wall opposite the step could also cause the low-frequency motions. *Armaly* et al. [19] showed that separation could occur on the opposite wall of single-sided sudden expansions at low Reynolds numbers. Opposite wall separation would cause the reattachment length on the step side to shorten. This hypothesis was discounted because thermal tuft results showed that flow reversals never occurred on the opposite wall of the test section.

Finally, the apparent long-time scale could be caused by low-frequency perturbations in the inlet flow. This possibility has been discounted because the wind tunnel has undergone two significant modifications without changing the characteristics of the unsteady motion. These changes included replacing the blower and filter assembly, replacing the inlet diffuser, completely changing the honeycomb and screen arrangement, and lengthening the test section inlet section downstream on the nozzle. None of these changes caused an appreciable change in the length of the region of flow reversals.

Regardless of their cause, the low-frequency motions can cause serious difficulties for both experimenters and modelers. The spectral data showed that a substantial fraction of the measured turbulence energy is due to the low-frequency unsteadiness. Unfortunately, it is not clear where the low-frequency unsteadiness ends and the turbulence begins. Therefore, an experimenter must be very careful to avoid filtering of low-frequency information. In addition, comparisons among different experiments will be difficult if the unsteadiness is wind tunnel dependent.

The low-frequency motions can probably be ignored by modelers if they need only time-averaged results. One would have to assume that energy is not fed from the low-frequency motions back into the turbulence.

The real problem arises in development and testing of a new model. Models are designed to represent a range of scales which typify the turbulence. If empirical input is used which is biased by very-low-frequency content, one cannot expect the model to correctly represent the turbulence. A model tuned for a given flow configuration could not be expected to work well in another configuration when the low-frequency unsteadiness is different. This problem provides substantial impetus to further investigate low-frequency motions of separated shear layers.

In summary, our data show that a separated and reattaching shear layer undergoes a long-time-scale, non-periodic motion of significant amplitude. This motion must be accounted for in future experimental and computational programs if we hope to have a full understanding of reattaching flows.

Acknowledgment. We gratefully acknowledge the support of the Fluid Mechanics Program of the Engineering Division of the National Science Foundation.

References

1. Owen, F.K., Johnson, D.A.: "Separated Skin Friction Measurements – Source of Error: An Assessment and Elimination," AIAA Paper No. 80-1409, 1980
2. Smith, C.R., Jr., Kline, S.J.: An experimental investigation of the transitory stall regime in two-dimensional diffusers, including the effects of periodically disturbed inlet conditions. J. Fluids Eng. **96**, 11–15 (1974)
3. Eaton, J.K., Johnston, J.P.: "A Review of Research on Subsonic Turbulent Flow Reattachment," AIAA paper 80-1438. AIAA J. (1981, to be published)
4. Tani, I., Iuchi, M., Komoda, H.: "Experimental Investigation of Flow Separation Associated with a Step or Groove," Report No. 364, Aero Research Institute, Univ. of Tokyo (1961)
5. Chandrsuda, C.: "A Reattaching Turbulent Shear Layer in Incompressible Flow"; Ph.D. Thesis, Dept. of Aeronautics, Imperial College of Science and Technology (1975)
6. Eaton, J., Johnston, J.P.: "An Evaluation of Data for the Backward-Facing Step Flow," Rpt. Conf. on Complex Turbulent Flows, Stanford Univ. (1980)
7. Goldschmidt, V.W., Bradshaw, P.: Flapping of a plane jet. Phys. Fluids **16**, 354–355 (1973)
8. de Gortari, C.J.: "An Experimental Study of the Flapping Motion of a Turbulent Plane Jet"; Ph.D. Thesis, Purdue University (1978)
9. Rockwell, D., Naudascher, E.: Self-sustained oscillations of impinging free shear layers. Ann. Rev. Fluid Mech. **11**, 67–94 (1979)
10. Rockwell, D., Knisely, C.: Vortex edge interaction mechanisms for generating low-frequency components. Phys. Fluids *23*, 239–240 (1980)
11. Rockwell, D.: Private communication (1980)
12. McGuinness, M.: "Flow with a Separation Bubble – Steady and Unsteady Aspects"; Ph.D. Dissertation, Cambridge University (1978)
13. Eaton, J.K., Johnston, J.P., Jeans, A.H.: "Measurements in a Reattaching Turbulent Shear Layer," Proceedings of the 2nd Symposium on Turbulent Shear Flows, London (1979)
14. Westphal, R.V., Johnston, J.P.: Private communication (1981)
15. Eaton, J.K., Jeans, A.H., Ashjaee, J., Johnston, J.P.: A wall-flow-direction probe for use in separating and reattaching flows. J. Fluids Eng. **101**, 364–366 (1979)
16. Bradbury, L.J.S., Castro, I.P.: A pulsed-wire technique for velocity measurements in highly turbulent flow. J. Fluid Mech. **49**, 657–691 (1971)
17. Eaton, J.K., Westphal, R.V., Johnston, J.P.: "Two New Instruments for Flow Direction and Skin-Friction Measurements in Separated Flows," ISA 2nd Int. Symposium on Flow (April 1981)
18. Mellor, G.L., Celenligil, M.C.: "Summary of Computations of Flows 0331, 0421, and 0431," 1980–81 Stanford Conferences on Complex Turbulent Flows (1981)
19. Armaly, B.F., Durst, F., Schönung, R.: "Measurements and Predictions of Flow Downstream of a Two-Dimensional Single Backward-Facing Step," Report SFB80/ET/172, SFB80 Universität Karlsruhe (1980)

Turbulent Shear Flow Behind Hemisphere-Cylinder Placed on Ground Plane

S. Okamoto

Department of Mechanical Engineering,
Shibaura Institute of Technology, Tokyo, Japan

Abstract

This paper presents an experimental investigation of the flow past a hemisphere-cylinder (a sphere whose lower half is replaced by a circular cylinder). The experiment was carried out in an N.P.L. blow-down type wind tunnel having a working section of 400 x 400 x 2000 mm in size at the Reynolds number 4.74×10^4.

The drag and lift coefficients were determined by integration of the surface pressure. By fitting a skirt to the lower half of a sphere placed on a ground plane, the drag coefficient was reduced by 7.5%. The velocity and pressure distributions and turbulence intensities in the turbulent shear flow behind a hemisphere-cylinder were measured. Since a hemisphere-cylinder is seen to be a hybrid of a sphere and a circular cylinder of aspect ratio one, the results for a hemisphere-cylinder were compared with those for a sphere and a circular cylinder of aspect ratio one.

1. Nomenclature

b	Position of peak velocity defect [mm]
C_D	Drag coefficient
C_L	Lift coefficient
C_p	Pressure coefficient = $(P - P_\infty)/(\rho U_\infty^2/2)$
D	Diameter of cylinder [mm]
H	Height of cylinder [mm]
n	Frequency of vortex shedding [Hz]
p	Static pressure [Pa]
P_∞	Free-stream pressure [Pa]
P_0	Stagnation pressure [Pa]
r	Radial distance from center axis of cylinder [mm]
Re	Reynolds number = $U_\infty D/\nu$
St	Strouhal number = nD/U_∞
U	Time-mean velocity in X-direction [m/s]
U_∞	Free-stream velocity [m/s]
u', v', w'	Velocity fluctuation in X, Y, Z direction [m/s]
X, Y, Z	Coordinate axes [mm]
α	Latitude angle [deg.]
δ	Thickness of shear layer [mm]
θ	Longitude angle [deg.]
ν	Kinematic viscosity of air [m^2/s]
ρ	Density of air [kg/m^3]

2. Introduction

Most of the investigations of the flow past a three-dimensional bluff body placed on a ground plane have concerned themselves with buildings, cooling towers, stacks, and other structures. The bodies hitherto investigated are divided mainly into two groups: the tall bodies of high aspect ratio H/D (for example, tall building, stack) and the short bodies of low aspect ratio (for example, sphere, short circular cylinder). *Hunt* [1] studied the flow past a building, *Peterka* and *Cermak* [2] measured the mean velocity defect and turbulence in the wake behind a building, and *Okamoto* and *Yagita* [3] investigated the flow past a circular cylinder of finite length placed on a ground plane.

On the other hand, *Schlichting* [4] studied surface roughness by using a surface made up of many small spheres, *Klemin* et al. [5] measured the drag of a sphere placed on a ground plane, and *Okamoto* [6] investigated the flow past a sphere placed on a ground plane. Moreover, *Sforza* and *Mons* [7] studied the wall wake behind a rectangular plate placed on the leading edge of a flat plate, and *Okamoto* et al. [8] investigated the flow past a cone placed on a ground plane. Bodies of low aspect ratio (spheres, circular cylinders, and hemisphere-cylinders [spheres whose lower halves are replaced by circular cylinders]) are often seen in fuel-tank yards and at electric power plants.

This paper presents an experimental investigation of the flow past a hemisphere-cylinder. Since a hemisphere-cylinder is seen to be the hybrid of a sphere and a circular cylinder, the results for a hemisphere-cylinder were compared with those for a sphere and a circular cylinder of aspect ratio one.

3. Experimental Apparatus and Procedure of Measurement

The experiment was carried out in an N.P.L. blow-down type wind-tunnel having a 400 × 400 mm working section of 2000 mm length. The ground plate, an aluminum plate of 4 mm thickness, was set with a spacing of 25 mm from the lower wall of the tunnel exit in order to avoid the boundary layer developed on the tunnel wall, as shown in Fig. 1. The model was placed at a downstream distance 500 mm from the leading edge of the ground plate. The thickness of the boundary layer at the model position was about 4 mm.

The origin of coordinates was taken as the bottom center of the model with the X-axis facing downstream, the Y-axis at right angles to the downstream flow and the Z-axis was in the vertical direction.

The hemisphere-cylinder and a circular cylinder used in this experiment are shown in Fig. 2. Two kinds of models, of 70 mm diameter and height and 40 mm diameter and height were used, the former being used for measurements of the surface pressure and the latter used for measurements of velocity, static pressure, and turbulence intensity of the flow. There was no blockage effect even for a model of 70 mm diameter and height, be-

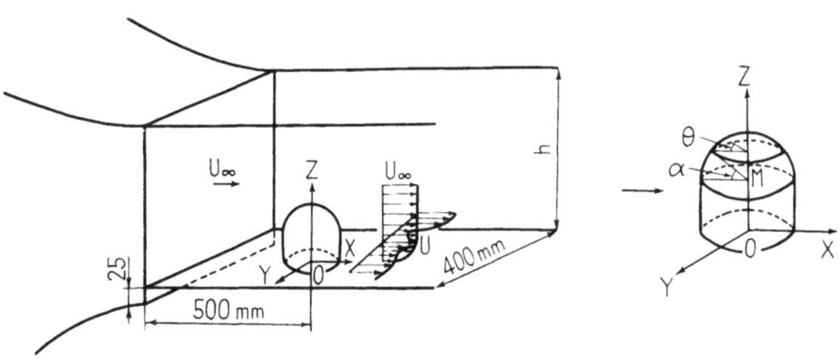

Fig. 1. Schema of apparatus and nomenclature

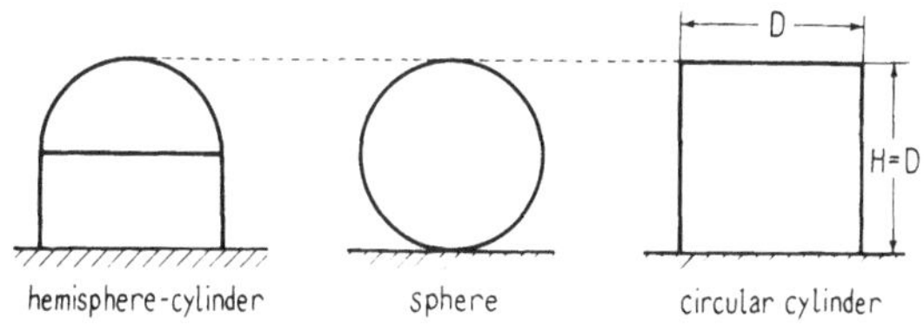

Fig. 2. Models

cause the blockage ratio in this case was 0.026 which is less than the limiting value 0.05 obtained by *Farell* et al. [9].

The pressure holes of 0.5 mm diameter were bored with an angular spacing of $\alpha = 10°$ on the surface of a hemisphere-cylinder in the vertical plane $\theta = 0$. Measurements of the surface pressure for any value of θ were done by rotating the model about the vertical Z-axis. The surface pressures on a circular cylinder were measured through the pressure holes of 0.5 mm diameter bored with a 7 mm vertical spacing on the surface in the vertical plane $\theta = 0$.

The velocity vector was obtained by use of a cylindrical yawmeter of 6 mm diameter. The time-mean velocity and the static pressure were measured by use of a Pitot and static-pressure tube of 1 mm diameter, and turbulence was measured by a hot wire anemometer of constant temperature type. The flow behind a cylinder was measured at various distances $X/D = 1 \sim 14$ at the Reynolds number of 4.74×10^4.

4. Experimental Results

4.1. Surface Pressure

The surface-pressure distribution on the upper hemispherical part and the lower cylindrical part of a hemisphere-cylinder are shown in Figs. 3 and 4 respectively, in which the dotted lines represent the theoretical value $C_p = 1 - (9/4)\sin^2\theta$ of a sphere in the absence of a ground plane and $C_p = 1 - 4\sin^2\theta$ of a two-dimensional circular cylinder obtained by the potential-flow theory. The pressure distributions on the cylindrical

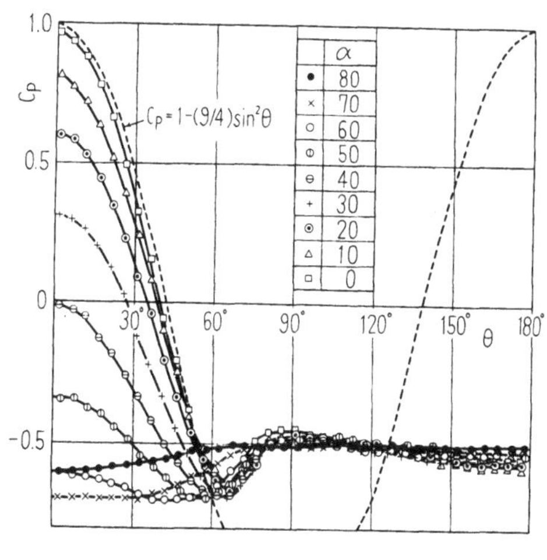

Fig. 3. Surface pressure distribution on upper hemispherical part of hemisphere-cylinder

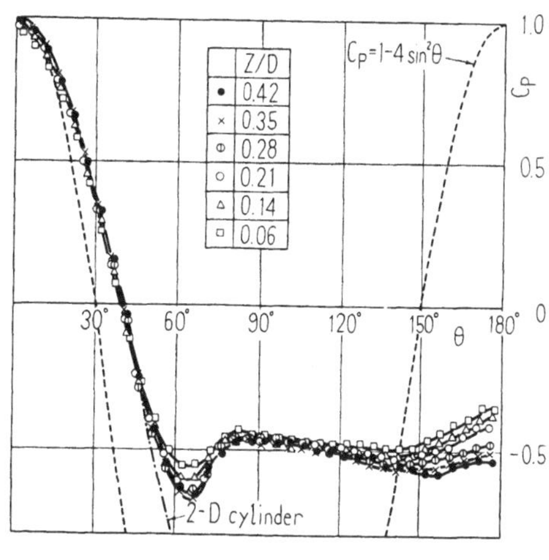

Fig. 4. Surface pressure distribution on lower cylindrical part of hemisphere-cylinder

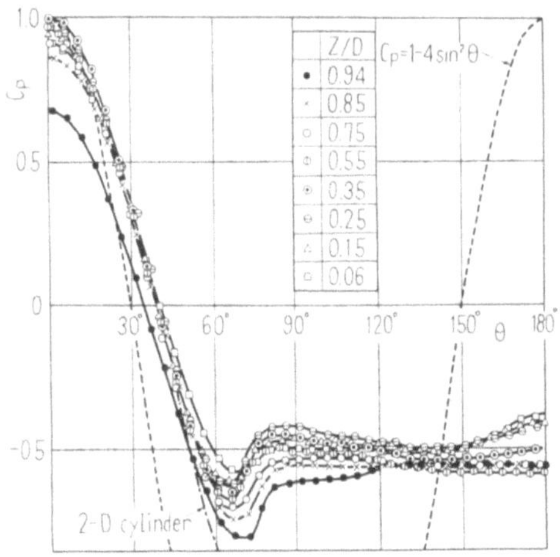

Fig. 5

◁ **Fig. 5.** Pressure distribution on cylindrical surface of circular cylinder

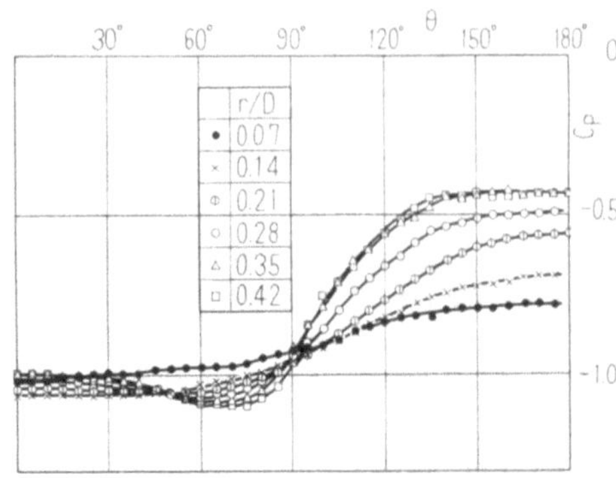

Fig. 6. Pressure distribution on top surface of circular cylinder

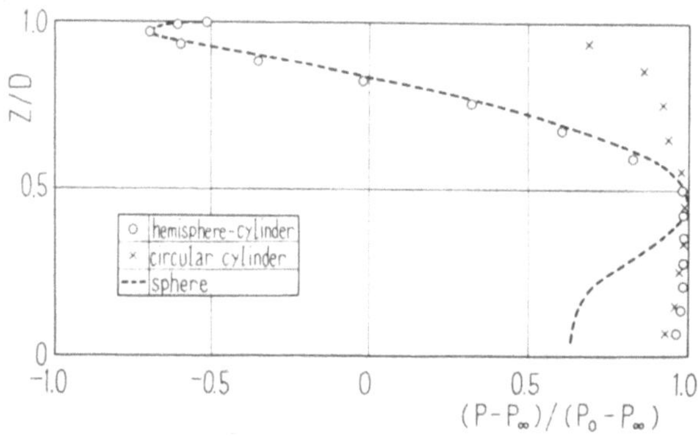

Fig. 7. Pressure at $\theta = 0$

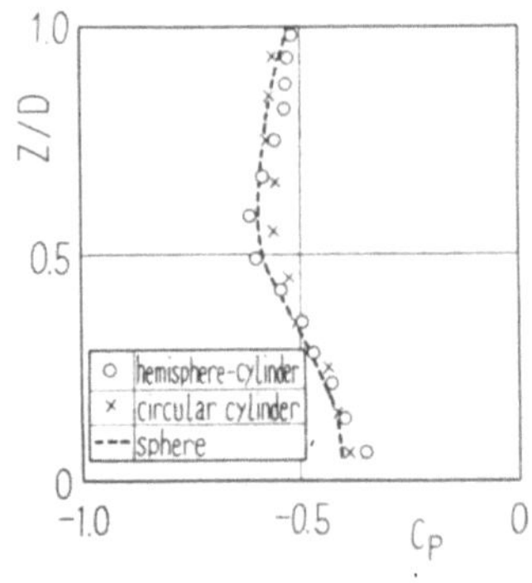

Fig. 8. Pressure at $\theta = 180°$

Fig. 9. Separation line on hemisphere-cylinder

surface and the top surface of a circular cylinder of $H/D = 1$ placed on a ground plane are shown in Figs. 5 and 6 respectively.

The dimensionless surface pressures $(P - P_\infty)/(P_0 - P_\infty)$ at $\theta = 0$ are shown in Fig. 7, from which the stagnation point is seen to be located at $Z/D = 0.45$ for the three bodies.

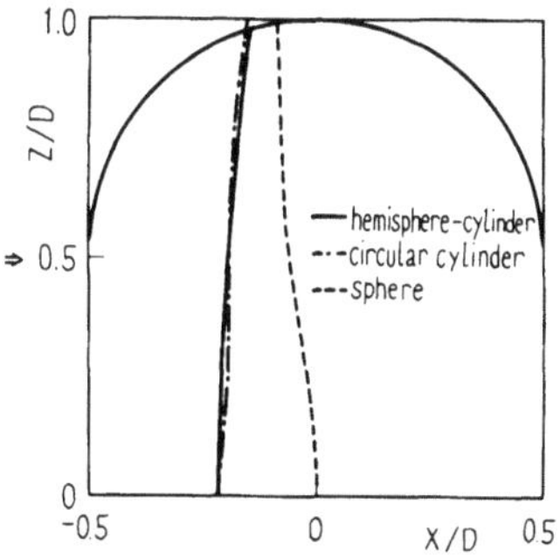

Fig. 10. Separation lines on bluff bodies

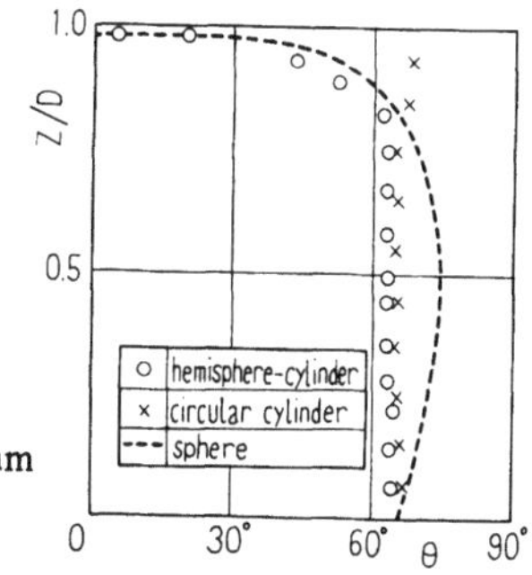

Fig. 11. Angle at minimum pressure on bluff bodies

The location of the stagnation point at $Z/D = 0.45$ is slightly lower than the stagnation point of $Z/D = 0.6$ found by *Good* and *Joubert* [10] for a two-dimensional fence. It should be noted that the pressure on the hemispherical surface of a hemisphere-cylinder at $\theta = 0$ agrees with that of a sphere and the pressure on the lower cylindrical surface is close to that of a circular cylinder of $H/D = 1$, while the pressure at $\theta = 180°$ is nearly equal for the three bodies, as seen in Fig. 8.

Figure 9 shows the separation line obtained by an oil film technique of visualization. It is found that the separation line of a hemisphere-cylinder is close to that of a circular cylinder, but greatly different from that of a sphere, as shown in Fig. 10. This has been confirmed by the variation of the angle at minimum pressure on bluff bodies as shown in Fig. 11.

4.2. Drag and Lift

The drag and lift coefficients of a hemisphere-cylinder are determined from the surface pressure by the following equations.

$$C_D = \frac{2}{\pi} \int_0^{\pi/2} \int_0^{\pi} C_p \cos^2 \alpha \cos\theta \, d\alpha \, d\theta \quad \text{for hemispherical part}$$

$$C_D = \int_0^{1/2} \int_0^{\pi} C_p \cos\theta \, d\theta \, d(Z/D) \quad \text{for cylindrical part}$$

$$C_L = -\frac{1}{\pi} \int_0^{\pi/2} \int_0^{\pi} C_p \sin 2\alpha \, d\alpha \, d\theta \quad \text{for hemispherical part}$$

The above-mentioned drag is, strictly speaking, a pressure drag, but it may be seen to be the total drag, because the frictional drag of a hemisphere-cylinder is negligible compared with the pressure drag. The values of C_D and C_L of a hemisphere-cylinder, sphere, and a circular cylinder of $H/D = 1$ are shown in Table 1. It was found that the drag was smallest for a hemisphere-cylinder and the lift was greatest for a circular cylinder of $H/D = 1$ among the three bodies. It is noteworthy that the drag coefficient of a sphere placed on a ground plane was reduced by 7.5% and the lift coefficient was increased by about 94% by fitting up the lower half of the sphere with a cylindrical skirt.

The drag coefficient of a sphere at the subcritical Reynolds number ($\mathrm{Re} = 4.74 \times 10^4$) was increased 30.6% by the presence of a ground plate. An increase of C_D due to

Table 1. Values of drag and lift coefficients

Bluff bodies	C_D	C_L
Hemisphere-cylinder	0.580	0.470
Circular cylinder	0.698	0.775
Sphere [6]	0.627	0.242

the presence of a ground plane was pointed out by *Klemin* et al. [5], whose experiment of a sphere of 609.6 mm diameter at the supercritical Reynolds number (Re = 1.69×10^6) gave an increase of 186.5% of C_D by the presence of a ground plane and a collar. An increase of C_D of a sphere due to the presence of a ground plane is greater at supercritical Reynolds numbers than at subcritical Reynolds numbers. The drag coefficients $C_D = 0.698$ of a circular cylinder of $H/D = 1$ are compatible with the drag coefficients $C_D = 0.70$ found by *Okamoto* and *Yagita* ([3]).

4.3. Vortices

It has been determined from experiments carried out using a cone [8] and a sphere [6] placed on a ground plane that there exist vortices of two kinds:

1) Horse shoe vortices, generated at the corner between the frontal face of the obstacle and the ground plane.
2) Arch vortices, generated behind the separation line on the surface of the obstacle.

The separation, which induces the horse shoe vortices, occurs on the ground plate, as shown in Fig. 12. It was visually found by coating the ground plate in a solution of tooth powder. Figure 13 shows the separation lines on the ground plate for the three obstacles used. The separation line for a hemisphere-cylinder is close to that of a circular cylinder, but different from that of a sphere.

The arch vortices are shed downstream and are gradually inclined by a strong downwash behind the obstacle until they almost touch the ground plate. The movement of the arch vortices can be surmised from the pressure distribution on the ground plane shown in Fig. 14. There are two areas in which the pressure is lowest. It is within these two areas that the roots of the arch vortices are attached to the ground plane. The position of these roots is shown in Table 2. It is hypothesized that the arch vortex rotates about these attachment points and is then gradually inclined.

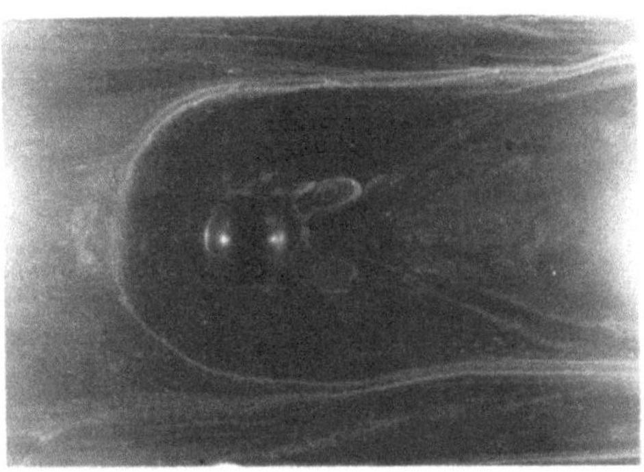

Fig. 12. Separation line on ground plate for a hemisphere-cylinder

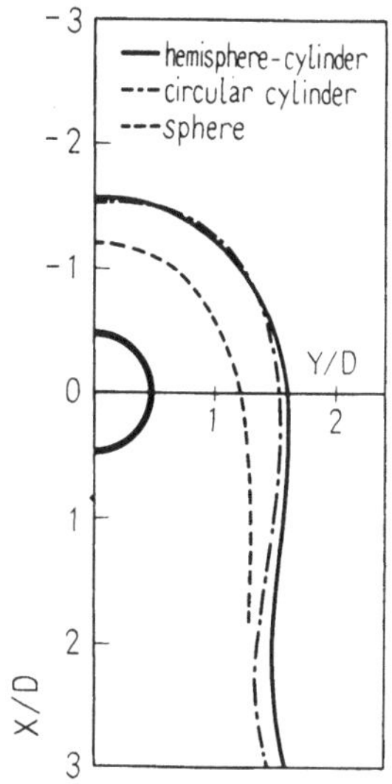

Fig. 13. Separation lines on ground plate

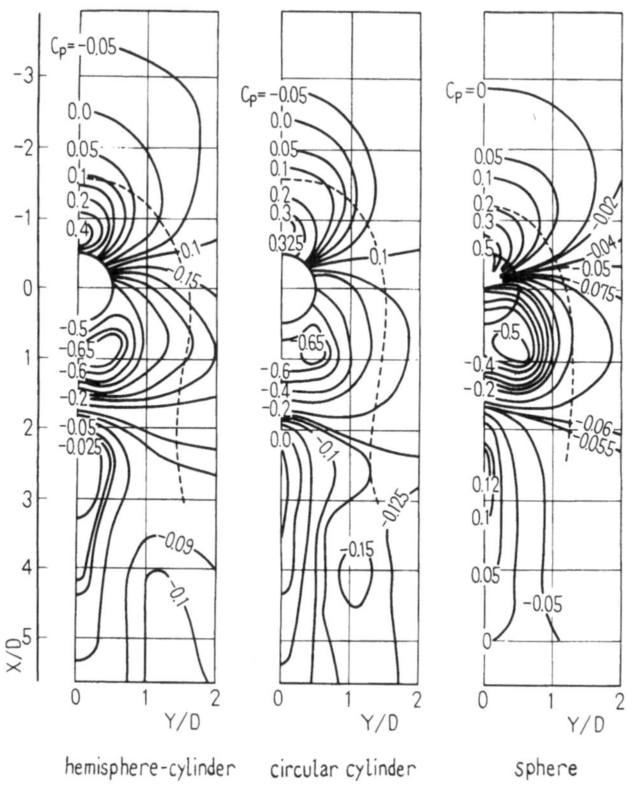

Fig. 14. Static pressure distribution on ground plate

Ink shedding in a water tank at Reynolds number Re = 1.06×10^3 was the method used to visualize vortex generation. Horse shoe and arch vortices which were generated while using a hemisphere-cylinder and a circular cylinder are shown in Figs. 15–18. Figure 17 illustrates the arch vortices inclination. Figure 19 is a rear view photograph of the hemisphere-cylinder taken at a slight angle.

Table 2. Position of roots of arch vortices

Bluff bodies	X/D	Y/D
Hemisphere-cylinder	0.95	0.30
Circular cylinder	0.90	0.50
Sphere [6]	0.86	0.35

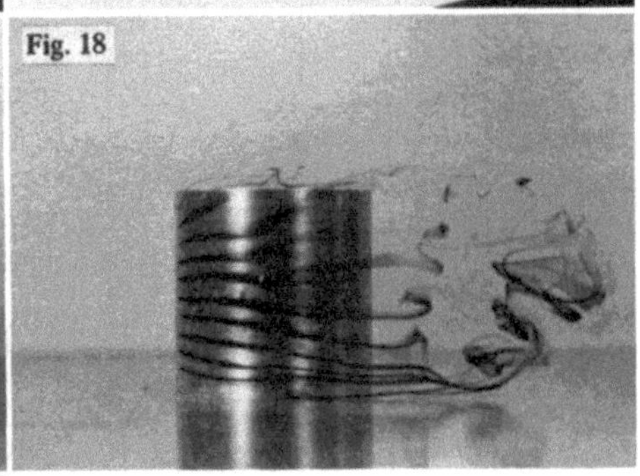

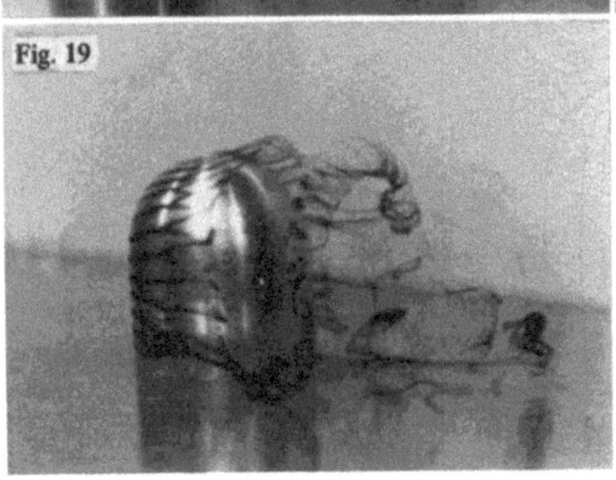

Fig. 15. Horse-shoe vortices for a hemisphere-cylinder

Fig. 16. Horse-shoe vortices for a circular cylinder

Fig. 17. Arch vortices for a hemisphere-cylinder

Fig. 18. Arch vortices for a circular cylinder

Fig. 19. Arch vortices for a hemisphere-cylinder

The Strouhal number of the arch vortex shedding, measured by a hot-wire anemometer, is shown in Table 3. It was found that the Strouhal number for a hemisphere-cylinder was smaller than that for a circular cylinder and a sphere.

A recirculation region is formed behind an obstacle. Because of a strong downwash behind a hemisphere-cylinder, the recirculation region is remarkably reduced compared with a two-dimensional fence. The position of highest pressure on the ground plane

Table 3. Strouhal number of arch vortex shedding

Bluff bodies	$St = nD/U_\infty$
Hemisphere-cylinder	0.152
Circular cylinder	0.225
Sphere [6]	0.222

Table 4. Reattachment point

Bluff bodies	X/D
Hemisphere-cylinder	2.40
Circular cylinder	2.92
Sphere [6]	2.50

is seen to be the point of reattachment of the outer streamline enclosing the recirculation region. The reattachment is in the rear end of the recirculation region, which is shown in Table 4. It was found that the recirculation region for a hemisphere-cylinder was smaller than that for both a sphere and a circular cylinder. The rear end of the recirculation region is located at $X/D = 2.40$ for a hemisphere-cylinder, while it is located at $X/D = 13.2$ for a two-dimensional fence according to the experiment by *Good* and *Joubert* [10].

4.4. Shear Layer Behind Bluff Body

Figure 20 shows the distribution of velocity vectors past a hemisphere-cylinder in the center section ($Y = 0$). It can be seen from this figure that the recirculation region ends at $X/D = 2.40$ as surmised from the pressure distribution on the ground plane, and the turbulent shear layer is formed along the ground plane at $X/D > 4$.

Figure 21 shows the time-mean velocity profiles at $X/D = 5$–28. The thickness of the shear layer is nearly equal to $0.9\,D$ at $X/D = 5$, and it decreases with increasing downstream distance. Figures 22–24 show the X-component, Y-component and Z-component of the turbulence intensity. The turbulence intensities of the free stream are 0.005, and they increase remarkably in the shear layer. It follows that the profiles of the turbulence intensities in the center section for a hemisphere-cylinder are nearly equal to these for a circular cylinder.

Figure 25 shows the thickness of the shear layer in the center section ($Y = 0$), which is expressed as

$$\delta/D = 1.63(X/D)^{-0.40}$$

for the three bluff bodies. It follows that the wall wake behind a three-dimensional obstacle becomes thin with an increase in the downstream distance.

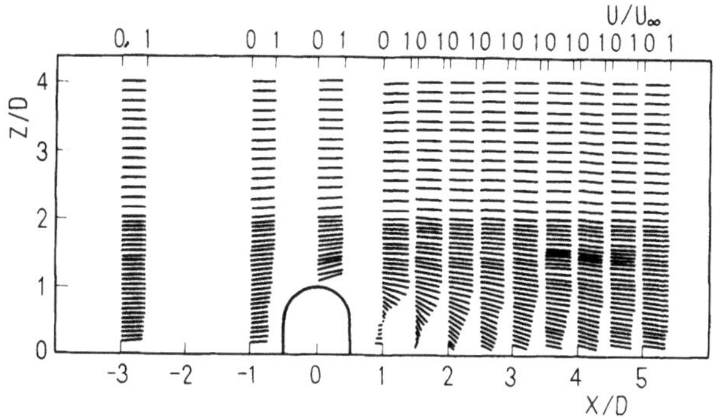

Fig. 20. Distribution of velocity vectors in the center section ($Y = 0$)

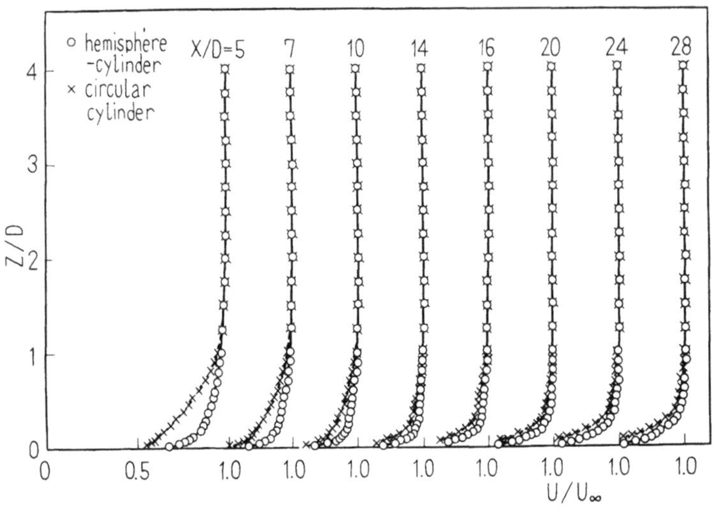

Fig. 21. Profiles of mean velocity in the center section ($Y = 0$)

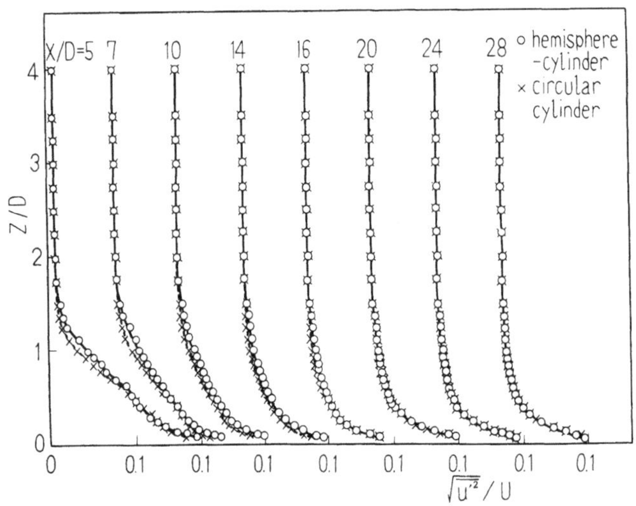

Fig. 22. Profiles of turbulence intensity in the X-direction in the center section ($Y = 0$)

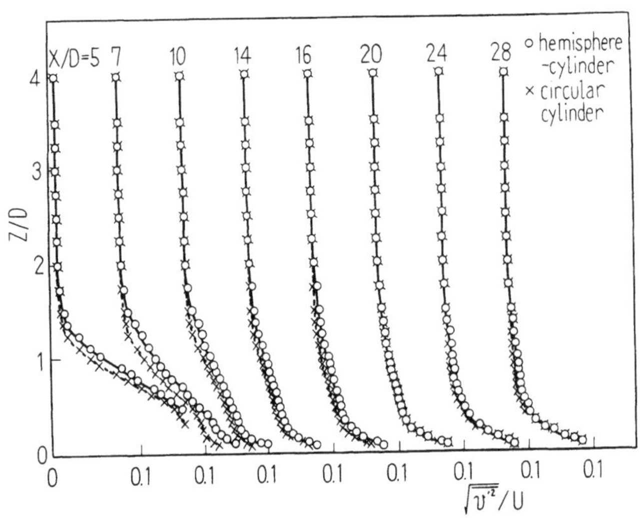

Fig. 23. Profiles of turbulence intensity in the Y-direction in the center section ($Y = 0$)

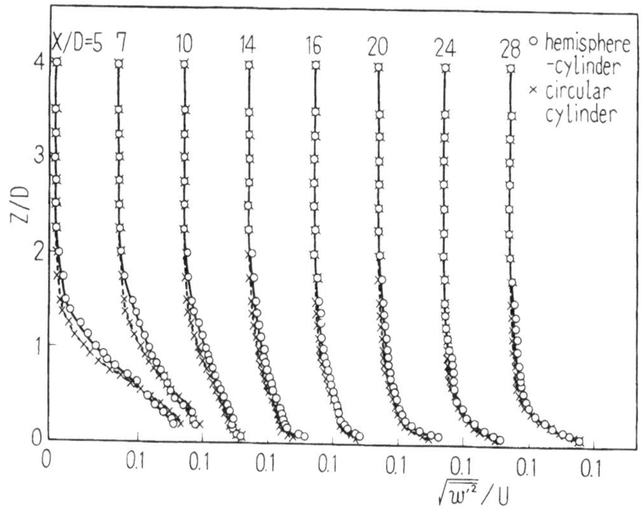

Fig. 24. Profiles of turbulence intensity in the Z-direction in the center section ($Y = 0$)

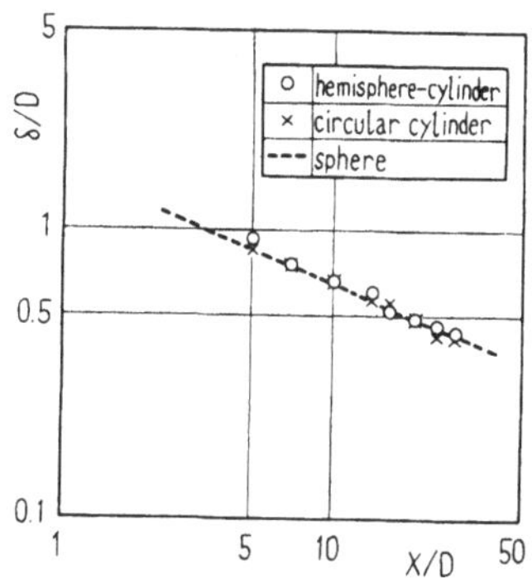

Fig. 25. Thickness of the shear layer in the center section ($Y = 0$)

4.5. Velocity Defect in Horizontal Center Section ($Z/D = 0.5$)

Figure 26 shows the distribution of mean velocity defect in the horizontal center section ($Z/D = 0.5$). The position of the peak velocity defect is shifted in the Y-direction with an increase in the downstream distance. Figure 27 shows the velocity defect in the center section ($Y = 0$), which is expressed as

$$(U_\infty - U)/U_\infty = 0.95(X/D)^{-1.62}$$

at $X/D > 3.0$ for the three bluff bodies. This is close to Hunt's result: $(U_\infty - U)/U_\infty \propto X^{-1.5}$ for the wake behind a building [1].

Figure 28 shows the position ($Y = b$) of the peak velocity defect, which is expressed as

$$b/D = 0.21(X/D)^{0.84}$$

for a hemisphere-cylinder and a circular cylinder, while

$$b/D = 0.26(X/D)^{0.69}$$

for a sphere. It is found from Figs. 25 and 28 that the wall wake behind a three-dimensional obstacle becomes thin and spreads transversely with increasing the downstream distance. This change in the wall wake can be clearly seen from the lines of $(U_\infty - U)/U_\infty = 0.10$ shown in Fig. 29.

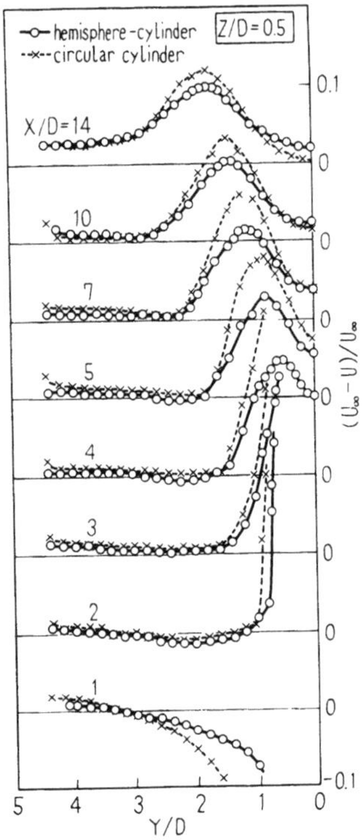

Fig. 26. Profiles of the velocity defect in the horizontal center section ($Z/D = 0.5$)

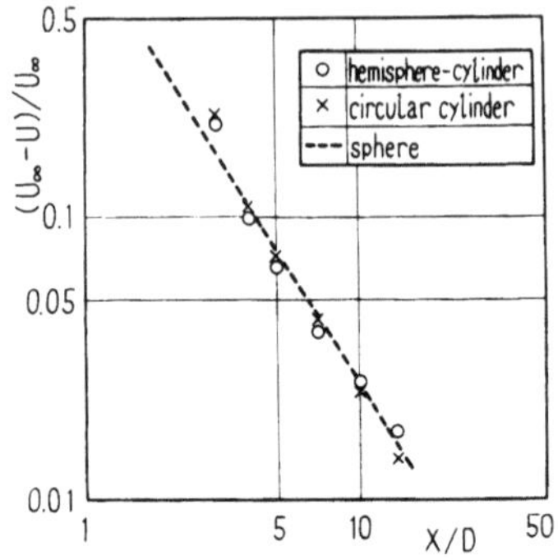

Fig. 27. Velocity defect in the center section ($Y = 0$, $Z/D = 0.5$)

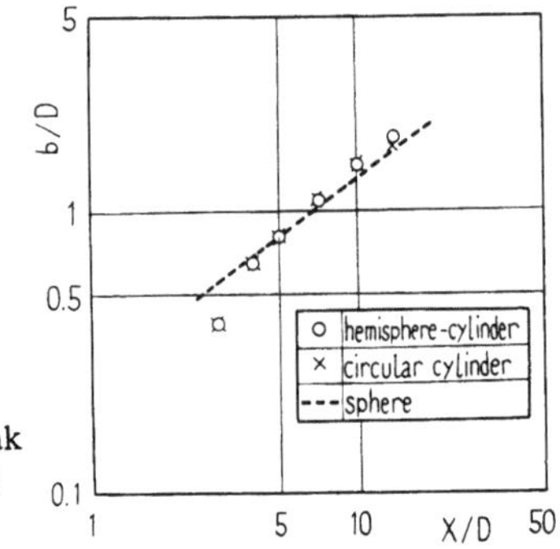

Fig. 28. Position of the peak velocity defect ($Z/D = 0.5$)

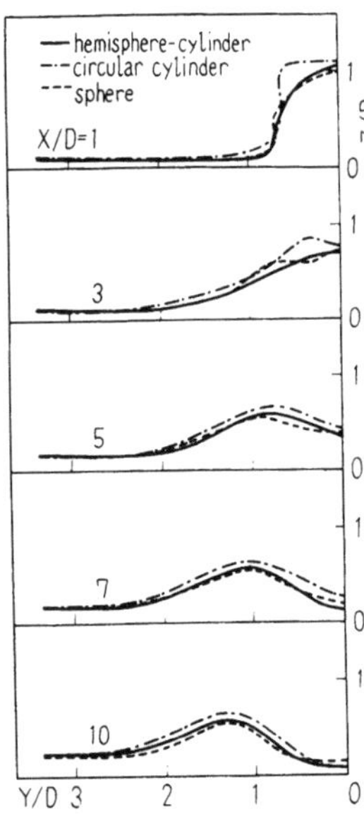

Fig. 29. Lines of $(U_\infty - U)/U_\infty = 0.10$

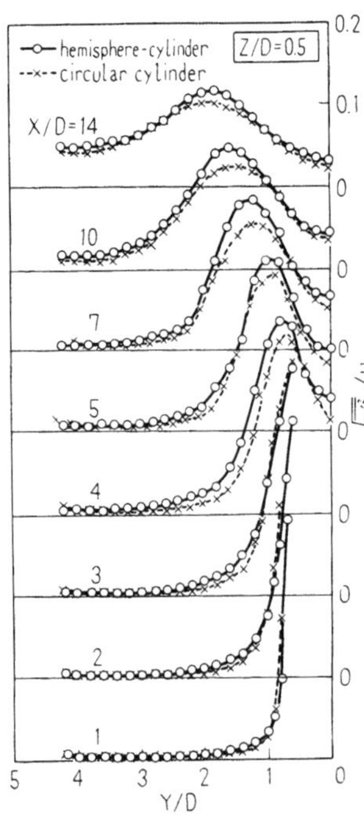

Fig. 30. Turbulence intensity in the X-direction ($Z/D = 0.5$)

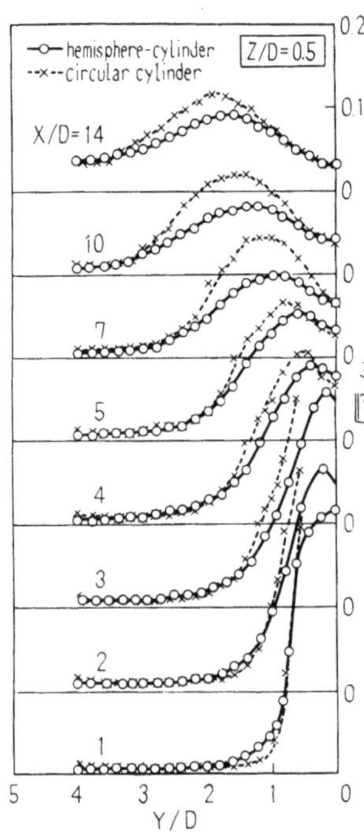

Fig. 31. Turbulence intensity in the Y-direction ($Z/D = 0.5$)

4.6. Turbulence Intensities in Horizontal Center Section (Z/D = 0.5)

Figures 30–32 show the turbulence intensities in the X-, Y- and Z-directions in the horizontal center section (Z/D = 0.5). The position of the peak turbulence intensity is shifted in the Y-direction with increasing the downstream distance. The peak turbulence intensities in the X-, Y- and Z-directions occur at Y/D = 1.00, 0.80 and 1.00 for X/D = 5, and at Y/D = 1.58, 1.50 and 1,45 for X/D = 10.

Figure 33 shows the decay of the X-component of the turbulence intensity in the center section (Y = 0), which is expressed as

$$\sqrt{\overline{u'^2}}/U - \left(\sqrt{\overline{u'^2}}/U\right)_\infty = A(X/D)^{-1.40} \quad ,$$

where A = 0.85 for a hemisphere-cylinder, and A = 0.69 for a circular cylinder and a sphere. Figure 34 shows the decay of the Y-component of turbulence intensity, which is expressed as

$$\sqrt{\overline{v'^2}}/U - \left(\sqrt{\overline{v'^2}}/U\right)_\infty = B(X/D)^{-1.37} \quad ,$$

where B = 1.05 for a hemisphere-cylinder and a circular cylinder, and B = 0.57 for a sphere. Figure 35 shows the decay of the Z-component of turbulence intensity, which is expressed as

$$\sqrt{\overline{w'^2}}/U - \left(\sqrt{\overline{w'^2}}/U\right)_\infty = 0.55\,(X/D)^{-1.09} \quad ,$$

for the three bodies.

It follows that the decay of the X- and Z-components of turbulence intensity in the section (Z/D = 0.5) for a hemisphere-cylinder is nearly equal to that of a circular cylinder and a sphere, and the decay of the Y-component of turbulence intensity of a hemisphere-cylinder is nearly equal to that of a circular cylinder, but different from that of a sphere.

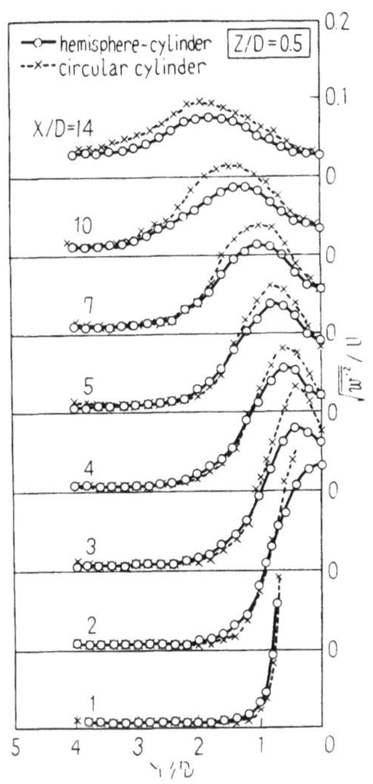

Fig. 32. Turbulence intensity in the Z-direction (Z/D = 0.5)

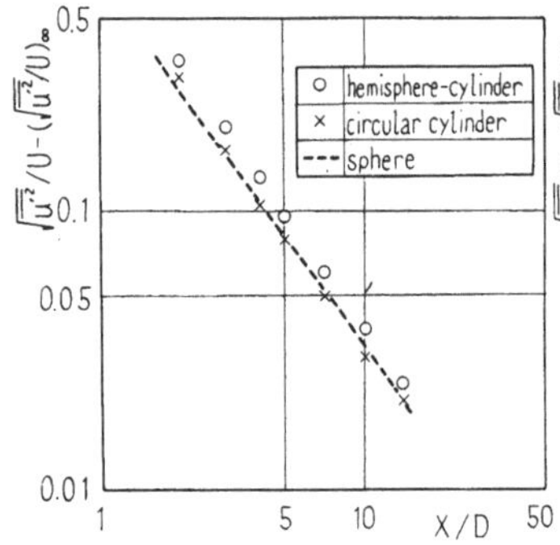

Fig. 33

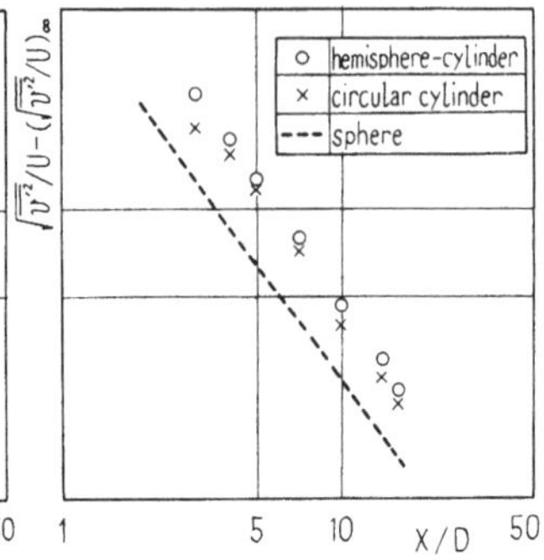

Fig. 34

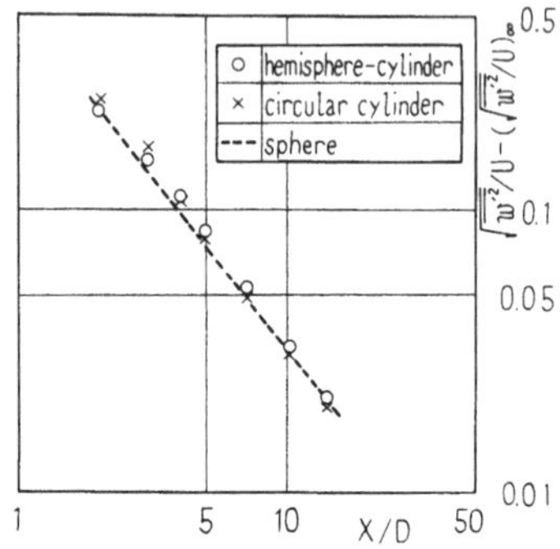

Fig. 35

Fig. 33. Decay of the X-component of turbulence intensity ($Y = 0, Z/D = 0.5$)
Fig. 34. Decay of the Y-component of turbulence intensity ($Y = 0, Z/D = 0.5$)
Fig. 35. Decay of the Z-component of turbulence intensity ($Y = 0, Z/D = 0.5$)

5. Conclusion

The experimental results are summarized as follows:

1) The drag is smallest for a hemisphere-cylinder and the lift is largest for a circular cylinder among the three bluff bodies.
2) The Strouhal number of arch vortex shedding is 0.152 for a hemisphere-cylinder, while it is 0.225 for a circular cylinder and 0.222 for a sphere.
3) The thickness of the shear layer behind a hemisphere-cylinder in the center section ($Y = 0$) is equal to that of a circular cylinder and a sphere.
4) The velocity defect in the center section ($Y = 0, Z/D = 0.5$) of a hemisphere-cylinder is nearly equal to that of a circular cylinder and a sphere.
5) The decay of the X- and Z-components of turbulence intensity in the section ($Z/D = 0.5$) of a hemisphere-cylinder is nearly equal to that of a circular cylinder and a sphere. And the decay of the Y-component of turbulence intensity is nearly equal to that of a circular cylinder, but different from that of a sphere.

Acknowledgement. The author expresses his thanks to Dr. Tetsushi Okamoto, professor emeritus of Tokyo Institute of Technology, for his cordial guidance.

References

1. Hunt, J.C.R.: The effect of single buildings and structures. Philos. Trans. R. Soc. London **A269**, 457–467 (1971)
2. Peterka, J.A., Cermak, J.E.: "Turbulence in Building Wakes", Proc. 4th Int. Conf. on Wind Effects on Building and Structures, 447–463 (1977)
3. Okamoto, T., Yagita, M.: The experimental investigation on the flow past a circular cylinder of finite length. Bull. Jpn. Soc. Mech. Eng. **16**, 805–814 (1973)
4. Schlichting, H.: Experimentelle Untersuchungen zum Rauhigkeitsproblem. Ing. Arch. **7**, 1–34 (1936)
5. Klemin, A., Schaefer, E.B., Beerer, J.G.: Aerodynamics of the perisphere and trylon at world's fair. Trans. Am. Soc. Civ. Eng. **2042**, 1449–1472 (1939)

6. Okamoto, S.: "Turbulent Shear Flow Behind a Sphere Placed on a Plane Boundary", in: Turbulent Shear Flows 2, ed. by L.J.S. Bradbury, F. Durst, B.E. Launder, F.W. Schmidt, J.H. Whitelaw (Springer, Berlin Heidelberg New York 1980) 246–256
7. Sforza, P.M., Mons, R.F.: Wall-wake: Flow behind a leading edge obstacle. AIAA J. **8**, 2162–2167 (1970)
8. Okamoto, T., Yagita, M., Kataoka, S.: Flow past cone placed on flat plate. Bull. Jpn. Soc. Mech. Eng. **20**, 329–336 (1977)
9. Farell, C., Carrasquel, S., Güven, O., Patel, V.C.: Effect of wind tunnel walls on the flow past circular cylinders and cooling tower models. Trans. ASME, Series I **99**, 470–479 (1977)
10. Good, M.C., Joubert, P.N.: The form drag of two-dimensional bluff-plates immersed in turbulent boundary layers. J. Fluid Mech. **31**, 547–582 (1968)

Experimental Investigations in Transonic Highly Separated, Turbulent Flow

A. Farcy, V. Mercier, and R. Leblanc

Laboratoire d'Aérodynamique, LA 191, Centre d'Etudes Aérodynamiques et Thermiques,
43 Rue de l'Aérodrome, F–86000 Poitiers, France

Abstract

Experimental results in transonic, highly separated turbulent flow are presented. The model is an air intake of $20 \times 80\ mm^2$ section set with an angle of attack from 20 to 40° inside a $M = 0.7$ free jet flow. The Mach numbers and mass flow rates are deduced from pressure measurements. Flow visualization, laser-Doppler and hot wire anemometers are used to analyse the mean velocity field, unsteadiness and turbulent flow characteristics.

Introduction

Important difficulties occur in the measurement of the turbulent characteristics in transonic flow. The case of the flow inside an air intake set with a large incidence in transonic speed is particularly difficult (Fig. 1).

This type of configuration is complex due to the combination of compressibility, large separation regions and unsteady flow effects. A separation bubble occurs at the lower leading edge which involves a turbulent wake with high velocity distortion and fluctuations harmful for the engine. Then it is of major interest to know the mean velocity field, the unsteady frequencies, the turbulent intensities and the scale of the fluctuations.

This paper describes some results of experimental investigations inside a model of air intake set with an angle of attack α from 20 to 40° in a $M = 0.7$ stream. Because important difficulties arise in the total and static pressure measurements due to the transonic nature of the flow, laser-Doppler and hot wire anemometers were used in combination with high speed visualizations.

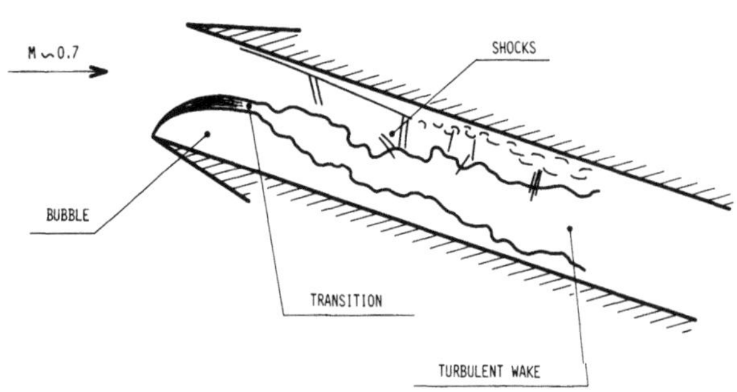

Fig. 1. Transonic separated flow in air intake

Experimental Set Up

A 100 x 40 mm² transonic nozzle produces a free jet at $M = 0.7$ max and $\mathrm{Re} = 4 \times 10^5 \cdot$ cm for a total temperature maintained approximately to the ambiant conditions. The 20 x 80 mm² air intake model is mounted on a support which can be adjusted in incidence with respect to the axis of the jet (Fig. 2).

The side walls of the model exceed the lateral section of 20 mm in order to limit the 3 D effects. The leading edge of the walls is sharp (10°). Windows allow visualizations and laser-Doppler measurements. The total angle of the diffuser is 7° and the variable $-h$ throat permits adjustment of the mass flow rate σ inside the air intake.

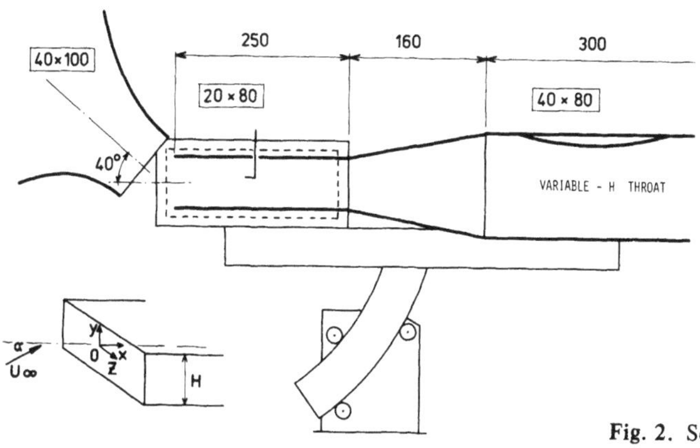

Fig. 2. Schematic view of the test set up

As reported above, the flow configuration is analysed for various values of incidence between 20° and 40°. Preliminary tests [1] having shown the effect of mass flow rate, $\sigma = \rho u/\rho_\infty u_\infty$. It was then decided to establish the curves $\sigma(\overline{M})$ as shown in Fig. 3a. σ and $\overline{M}$ are deduced from pitot and static pressure measurements at the section $x/H = 8$.

For $h = 16$ and 20 mm the mass flow rate increases with increasing $\overline{M}$ and decreasing h. A reverse trend appears at the lower value of h (8 and 12 mm) and for $\alpha = 30°$. Note, as a possible explanation, the proximity of the critical condition ($M^* = 1$) for these throat openings. The flow visualizations conducted simultaneously reveal an increasing separation bubble with increasing incidence, without significant effects on h. Most LDA and HW anemometer measurement were then conducted at $\alpha = 40°$ (large separation bubble) for the contractual fixed point P ($\sigma \cong 0.57, \overline{M} \cong 0.39$).

Visualizations of the Flow

It is well known that compressibility allows shadowgraph or strioscopy visualizations. These techniques are used extensively and the pictures are fixed other by video recording, photography or movies.

The magnetoscope is very useful to detect the low frequency instabilities as a function of incidence and mass flow rate. The following graph is a synthesis of the observations.

The photographs were taken on 9 x 10 plates of 400 ASA with an exposure time of 1/500 s (continuous Hg vapor light) or 1 µs (spark light). The movie of 25 pictures of 15 x 24 mm² is made with a LCA-CI4 high speed camera between 8×10^4 and 10^6 fr/s.

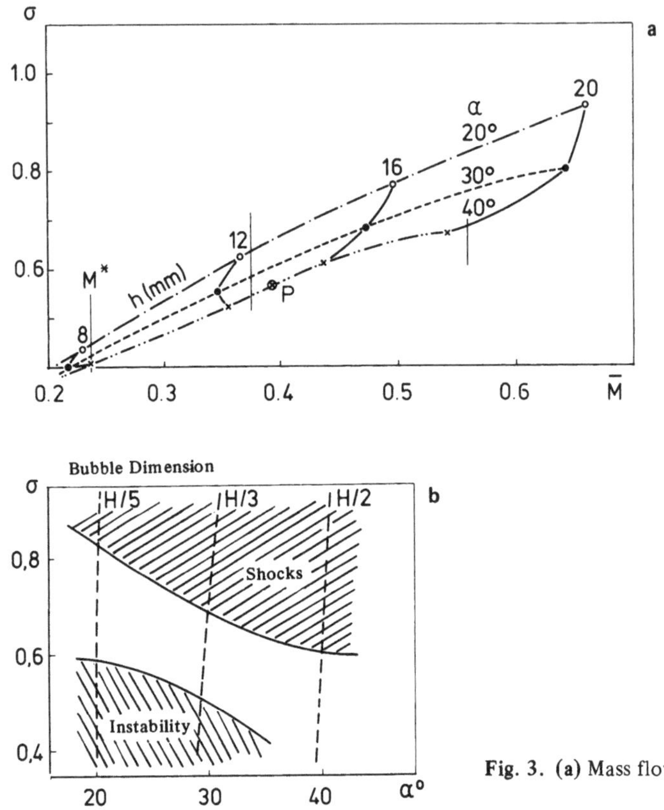

Fig. 3. **(a)** Mass flow rate. **(b)** Flow configuration

An adjustable flash tube is then used as a light source. The exposure time is equal to 314 ns for 10^6 fr/s and 1.25 μs for 250,000 fr/s. Some examples of these different technics are shown on Fig. 4. The description of the flow is first issued from the observations of the photographic plates, Fig. 4a.

For the incidence of 20° there is a small bubble which produces a sonic throat. A λ-shock lies at the rear part of the bubble and produces an interaction with the upper wall laminar boundary layer. Turbulent flow seems to become uniform after $x = 3$ to $4H$. The most important feature at $\alpha = 40°$ is the well established bubble of $H/2$ in height and more than $2H$ in length. Its forward limit is clearly laminar and becomes promptly turbulent at $x = y \sim H/2$. This shear layer between recirculating and highly accelerated flow spreads across the air intake as a turbulent "wake" for the bubble. Downstream from $x = 3H$ inclined large turbulent structures appear. The strioscopies reveal clearly these large movements whereas small structures can be seen on the shadowgraph pictures. Normal shocks, in the upper transonic flow, vary in position as also indicated on the shadowgraphs. More information about the evolution of the turbulent flow is then obtained from the pictures movies analysis, Fig. 4b.

Ten successive strioscopic pictures show the evolution of the turbulent shear layer for $\alpha = 40°$. One see clearly a characteristic "ovoid" structure above the probe (picture 1) with a core which appears on picture 4 and disappears on picture 8. Its persistence is then equal to 20 μs which is a relatively slow variation. Likewise the normal shock position is constant along the sequence which indicate that the frequence of the mean limit oscillations of the shear layer is smaller than 20 kHz.

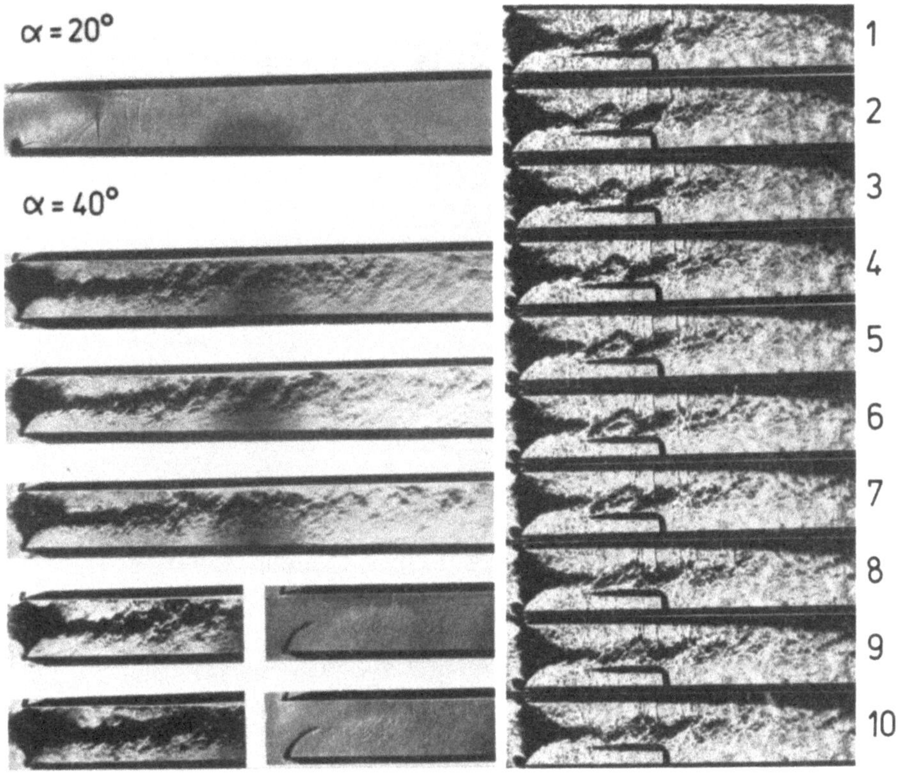

Fig. 4. **(a)** 9 x 12 plates $t = \mu s$. **(b)** High speed movies (250 000 fr/s, $t = 1.25\ \mu s$, $\Delta t = 3.88\ \mu s$)

From a colour "equidensity" exploitation of the movie picture, using the special Agfacontour process, it is concluded, for a sequence at 100,000 fr/s, that the limit of the bubble oscillates at 10–15 kHz with turbulent structure entrainment as a result. It is also clear that the position of the transition point varies some mm ($\sim H/10$) about its mean value $x = H/2$. Further, the mean stability of the bubble (i.e. no pulsation) was confirmed using a HICAM camera at 8000 fr/s.

Finally it is important to note the limit of the visualizations to show the reattachment point. Further information with detailed velocity and turbulent intensity measurements are obtained with the LDA measurements.

Laser Anemometry Measurements

A Laser-Doppler Anemometer (LDA) was developed in 1974 by *Ardonceau* [3]. It is made up of a 5W Coherent Radiation laser, a 40 MHz TSI-Bragg cell and a RTC, XP 1110 photomultiplier. The signal is analysed by a home-made digital counter for frequences between 10 kHz and 200 MHz with a 4 ns resolution. A carriage supports the LDA. The air intake is traversed while the data acquisition and reduction is being performed in the IBM 1800 computer. The diameter of the measurement volume varies between 0.3 and 0.6 mm. Small oil particles (ϕ 0.7 to 1.3 μm) are injected in the settling chamber of the transonic jet. Detailed measurements are conducted inside the air intake only for $\alpha = 40°$.

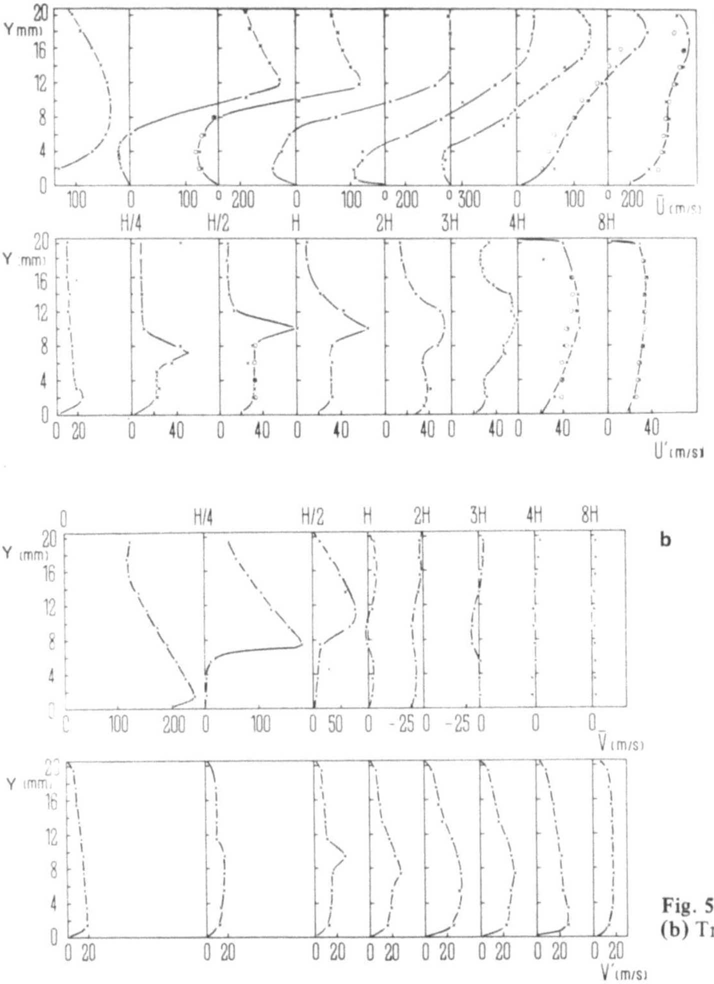

Fig. 5. (a) Longitudinal velocity (b) Transversal velocity

Longitudinal and transversal mean velocities $\overline{u}$ and $\overline{v}$ and corresponding RMS fluctuation values u' and v' are shown on Fig. 5a,b as measured in the eight sections $x = H/4$, $H/2, H, 2H, 4H$ and $8H$. The bubble appears clearly with recirculating flow extending beyond $x = 3H$. The maximum revers velocity, larger than 50 m/s, is found at $x = 2H$. The highly ac celerated shear layer above the bubble leads to a maximum velocity of 280 m/s at $x = H/4$ and H. Considering the high level of the velocity fluctuations (40 m/s for $x = H$ and $Y = 12$ mm) it is then possible to explain the existence of the shock waves shown in the flow visualizations. The instability of the shock would be due to the combination of mean and fluctuating velocities with a probable influence on the displacement of the transition point. The u' peak observed at $x = H/4, H/2$ and H, which spreds across the channel after $x = H$, can be born of other the burst of the highly accelerated shear layer limiting the bubble into a turbulent wake or the bubble frontier oscillations. With this later hypothesis such a peak must also appear for the v' component in the three above mentioned sections, what does not occur in Fig. 5b. Except for this peak, u' is constant across the bubble and equal approximately to three times its value in the upper "potential" flow. The turbulent diffusion across the air intake appears just after the reattachment (between $3H$ and $4H$) and a uniformity of both u and u' is seen at $x = 8H$. At this position, the intensity of the turbulence is two times higher in the x-direction (40%) than in the y-direction (20%).

Three u-velocity histograms are shown in Fig. 6 at $x = 4H$ for $y = H/4$, $H/2$ and $3H/4$. The double maxima of the probability density function is certainly characteristic of the intermittency of the flow.

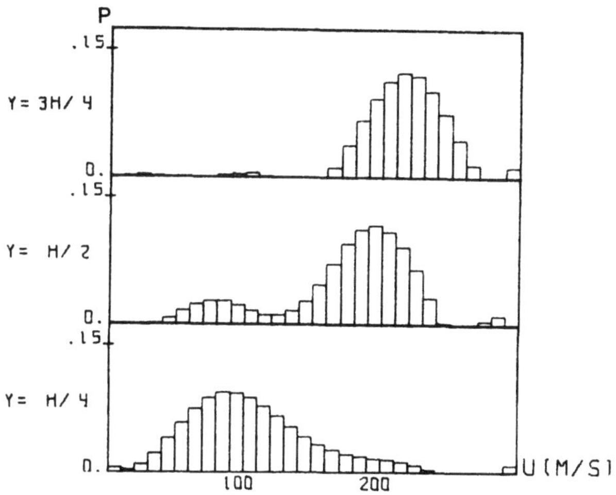

Fig. 6. Histogram LDA

Hot Wire Anemometry Results

It is generally admitted that the velocity and specific mass sensitivity coefficients of the hot wire are equal in supersonic flow ($M \geqslant 1{,}2$). The hot wire response is then dependent upon two parameters only ρu and T_t and the u component is deduced by using a separation procedure as diagram of fluctuation for example. These properties disappear in transonic flows, and some cautionary measures are necessary, that are:

- to limit the T_t' fluctuations maintaining the total temperatures T_t constant
- to measure with a high overheat ratio (0.8)
- to avoid the pure transonic regions ($M \sim 1$) and the shock areas.

A subminiature DISA 55A53 probe is used with a DISA 55M main unit and a CTA 55M10 bridge. The bandwidth is higher than 100 kHz. A FM Bell & Howell RD-378/U tape recorder stores the output signals with a bandwidth of 500 kHz for a nominal signal/noise of 32 dB at a 120 in/s recording speed. The data are then directly treated with a ND 100 mini-computer through a CAMAC data acquisition system.

The evolution of the power spectra for six longitudinal positions of the hot wire on the line $y = H/2$, $z = 0$ is shown on Fig. 7. The frequency f is normalized using H and the velocity of the free jet u_∞. The downstream energy damping of the large structures is clearly seen, approximately equal to 8 dB for example in the sections $x = 1.35H$ and $8H$, (for structures of characteristic dimension about $H/2$).

The spectrum analysis and the application of the Taylor hypothesis allows estimation of an integral scale Λx which is represented in Fig. 8, at the height of the six previous sections. From the section $x = 4H$ the value of Λx vary smoothly both in the x- and y-directions. The mean value of Λx is about 0.3.

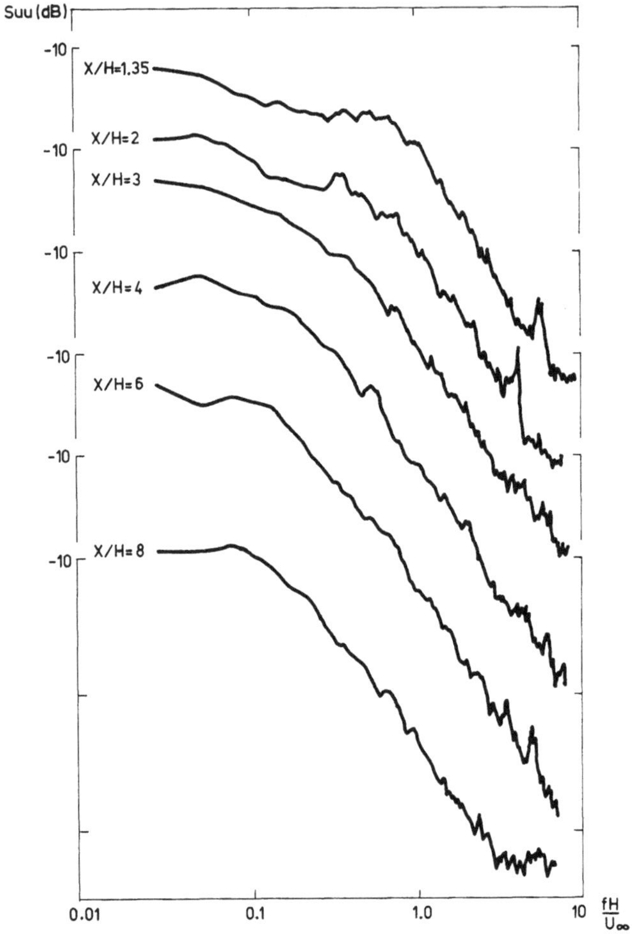

Fig. 7. Power spectra ($y = H/2$)

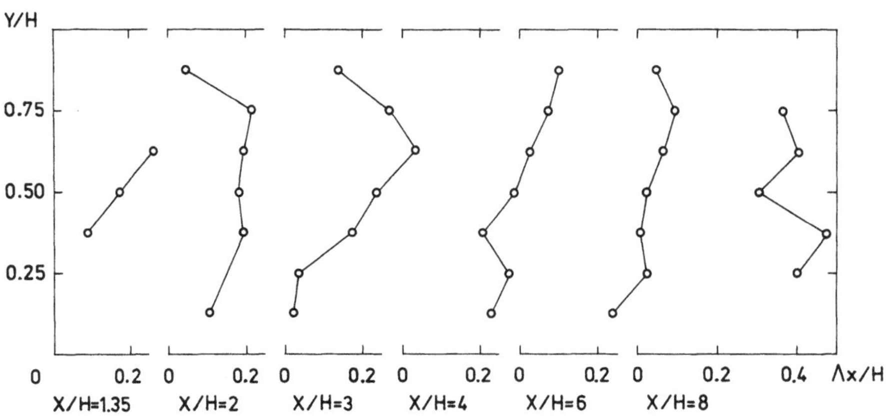

Fig. 8. Integral scale

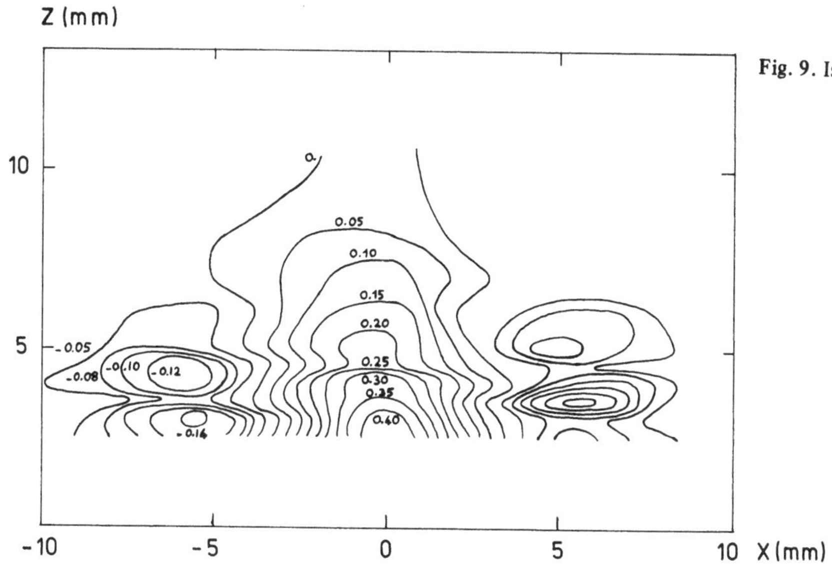

Fig. 9. Isocorrelations

Some correlations between two hot wire measurements are shown on Fig. 9. The first probe coordinates are $x = H, y = H/2, z = 0$ ($z = 0$ corresponds to the vertical plane of symmetry) whereas the second is moving on the half span from the same point.

An interesting feature is the existence of two negative correlation cores in the x-direction which reveal the emission of large vortex structures having a x-scale of about $H/2$ in accordance with the flow visualizations. The integral scale ratio $\Lambda z/\Lambda x$ is approximately equal to 3 which reinforce the two-dimensionality hypothesis. More detailed conclusions would be hazardous because of the non-linearity of the hot-wire response at transonic velocities and the application of the Taylor hypothesis in a such region with high velocity gradients and inhomogeneity.

Conclusion

The combination of visualizations, pressure measurements, laser-doppler and hot wire anemometers allows a detailed desription of a typical turbulent and highly separated transonic flow inside an air intake set at an angle of attack from 20° to 40° in a $M = 0.7$ jet flow.

The most important features of the mean flow configuration are: (1) the dimension of the separation bubble; (2) the existence of shocks and their stability; (3) the mass flow rate.

The turbulent nature of the flow is dependent upon the three above features. The transition of the bubble limiting shear layer occures at an unsteady position and induces a turbulent intensity peak which spreads across the height of the model. Large inclined structures are observed which promptly evolve. The turbulence intensity is constant inside the separation region, outside the shear layer and far downstream.

As a conclusion, it seems of large interest to complete the informations on the high lateral correlation and quasi-periodicity of the turbulent structures in the transition region and to evaluate the distorsions of the unsteady velocity field in the downstream part of the air intake.

Acknowledgements. This work was supported by the Grant DRET 79/069 from the French Defense Ministry.

References

1. Leblanc, R., Thiebault, P., Ardonceau, P., Goethals, R.: "Etude expérimentale des entrées d'air transoniques aux grandes incidences", Final Report, Contrat DRET 77/214 (1978)
2. Thiebault, P., Ardonceau, P., Leblanc, R., Goethals, R.: First measurements in highly separated transonic flow. Mech. Res. Commun. **6**, 113–114 (1979)
3. Ardonceau, P.: "Application de l'anémométrie laser à l'aérodynamique supersonique", Thèse de 3ième Cycle, University of Poitiers (1974)

Turbulent Flow Induced by a Jet in a Cavity-Measurements and 3D Numerical Simulation

F.Baron, J.P. Benqué, Y. Coëffé

Electricité de France, Laboratoire National d'Hydraulique, 6 quai Watier, F–78400 Chatou, France

Abstract

Velocity measurements using a laser anemometer of the turbulent flow induced by a jet in a cubic cavity are presented.

A numerical code solving the non-stationary Navier-Stokes equations in three dimensions for the large-scale field is described. Subgrid scale turbulence can be modelled by a turbulent viscosity.

The results of two cases of large-eddy simulation made with this code on the flow induced by a jet in the cubic cavity are then compared with measurements.

The same numerical code is applied to large eddy simulation of plane Poiseuille flow with periodic boundary conditions. Some results on mean velocities and Reynolds stresses are presented and compared with measurements.

Introduction

The flow induced by a jet in a cavity is studied as a typical case of industrial flows requiring 3-D numerical prediction with turbulence simulation.

A similar computation was performed on plane Poiseuille flow in order to better understand the turbulence production mechanism in a simpler case of wall-bounded flow.

Measurements in the Cubic Box

A cubic box (side a = 50 cm) is included in a loop of cold water. The flow enters the box through the window F_1 (see Fig. 1) and exists the box through the window F_2. F_1 and F_2 are square windows (side b = 15 cm) and are such that there is no symmetry plane for the flow. The flow consists of a jet entering the box and impinging the opposite wall, generating on each side of the jet a recirculating flow which goes down and progressively moves towards the exit area near F_2.

The flow rate is kept as constant as possible by a free overflow control. The entrance mean velocity is U_0 = 24 cm/s, and the entrance Reynolds number is Re = 36,000.

Several flow visualizations were made by taking pictures of particles illuminated when passing through a plane light beam. These visualizations show that except for the planes cutting the jet along its main axis, no big and stable structures are present but mainly middle sized (5 to 10 cm) eddies of relatively short life.

Laser doppler anemometry measurements are performed to obtain simultaneously two components of the velocity in one point of the box. The signals are then treated to get an average value or a filtered value, but also the standard deviation and the spectrum of the fluctuations.

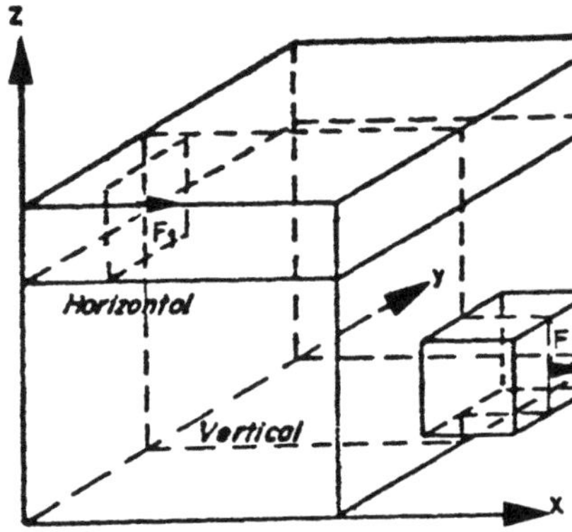

Fig. 1. Cubic box geometry and location of the planes of comparison

Numerical Simulation: "TRIDI" Code

The non-stationary Navier-Stokes equations for the large-scale 3D flow of an incompressible fluid are discretized in a time-advancing finite-difference scheme on two staggered grids.

Large-Scale Field Computation [7, 8]

The velocity field $U_i(\boldsymbol{x}, t)$ is separated into a large-scale component $\overline{U}_i(\boldsymbol{x}, t)$ and a subgrid-scale component $U_i'(\boldsymbol{x}, t)$ by a filtering operation:

$$\overline{U}_i(\boldsymbol{x}, t) = \int G(\boldsymbol{x} - \boldsymbol{x}')\, U_i(\boldsymbol{x}', t)\, d\boldsymbol{x}'$$

$$U_i'(\boldsymbol{x}, t) = U_i - \overline{U}_i$$

The filtered Navier-Stokes equations are then:

$$\frac{\partial \overline{U}_i}{\partial t} + \frac{\partial}{\partial x_j}(\overline{U}_i \overline{U}_j) = -\frac{1}{\rho}\frac{\partial \overline{P}}{\partial x_i} - \frac{\partial}{\partial x_j}\left(L_{ij} + S_{ij} - \nu \frac{\partial \overline{U}_i}{\partial x_j}\right) \tag{1}$$

$$\frac{\partial \overline{U}_L}{\partial x_L} = 0 \;. \tag{2}$$

where

$$L_{ij} = \overline{\overline{U}_i \overline{U}_j} - \overline{U}_i \overline{U}_j \tag{3}$$

is the "Leonard Stress", and

$$S_{ij} = \overline{\overline{U}_i U_j'} + \overline{U_i' \overline{U}_j} + \overline{U_i' U_j'} \tag{4}$$

is the "subgrid stress".

In a finite difference scheme, the characteristic length λ of the filter has to be several times bigger than the mesh size of the scheme, to make sure that all filtered wavelengths are really computed.

In our wall-bounded flows calculations, this would imply a very large number of grid points if λ is in the inertial subrange, as it has to be.

These are the reasons why we did not use a formal filtering operation, but the inherent filtering effect of the finite difference scheme (*Moin* et al. [9] did the same in the inhomogenous direction of channel flow, although their number of grid points was quite important).

With this kind of filtering, tests showed that the Leonard stress was within the round-off errors of the finite difference scheme. We have thus neglected these terms.

The subgrid scale stress tensor is modelled with a turbulent viscosity:

$$S_{ij} = -2\nu_T \bar{D}_{ij} + \frac{1}{3} S_{LL} \delta_{ij} \quad , \quad \text{where} \tag{5}$$

$$\bar{D}_{ij} = \frac{1}{2}\left(\frac{\partial \bar{U}_i}{\partial x_j} + \frac{\partial \bar{U}_j}{\partial x_i}\right) \tag{6}$$

is the large-scale strain-rate tensor. We use the *Smagorinsky*'s [11] subgrid model:

$$\nu_T = C_s^{\,2} \lambda^2 (\bar{D}_{ij} \bar{D}_{ij})^{1/2} \quad , \tag{7}$$

where C_s is a constant and λ a characteristic length of the mesh size.

This model is the most classical one and usually gives rather good results, although it does not reproduce the backscatter transfer from subgrid scales to large scales [8] as was recently pointed out by a comparison with direct simulations [4].

Note that the spherical part of the subgrid stress tensor is incorporated with the pressure:

$$P = \bar{p} + \frac{1}{3}\rho S_{LL} \quad . \tag{8}$$

Discretization of the Equations

The $(n+1)^{\text{th}}$ time step is resolved in three steps (The method follows the pattern proposed by *Chorin* [2]):

Step I: Advection of Momentum (split into 3 directions)

$$\frac{\tilde{\tilde{U}}_i - U_i^n}{\Delta t} + \left(U_j \frac{\partial U_i}{\partial x_j}\right)^n = 0 \quad . \tag{9}$$

This equation is solved by a method of characteristics with third-order interpolation:

$$\frac{\partial U_i}{\partial t} + U_j \frac{\partial U_i}{\partial x_j} = 0 \quad .$$

U_i is constant on the characteristics:

$$dx = U_j \, dt \quad .$$

The intersections of one of the axes:

$$(t = t^n) \text{ or } (x = x_n)$$

with the characteristics containing the points

$(x_i, t^n + \Delta t)$, $i = 1, ..., N$ are computed.

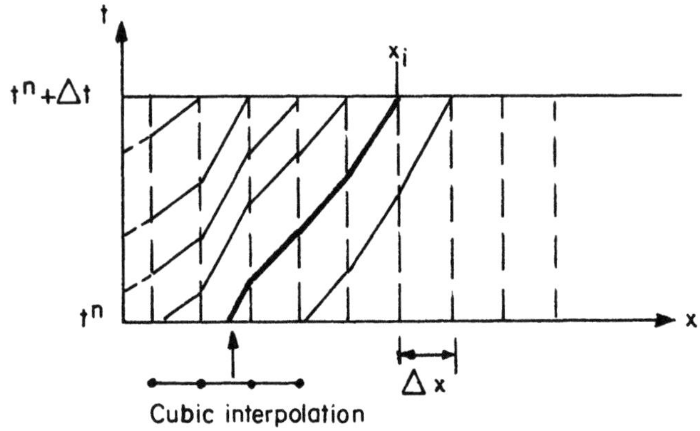

Then $\widetilde{U}_i$ is computed with third-order interpolation at those intersections.

This method is unconditionally stable in each direction, but, due to the splitting in three directions, the Courant criterion has to be observed to insure accuracy. On the other hand this scheme is not energy-conserving, but the third-order interpolation prevents any significant numerical diffusion.

Step II: Diffusion of Momentum (split into 3 directions)

$$\frac{\widetilde{U}_i - \widetilde{\widetilde{U}}_i}{\Delta t} = \frac{\partial}{\partial x_j} (2(\nu + \nu_T) \widetilde{D}_{ij}) \quad , \tag{10}$$

This equation is solved by a double-sweep implicit scheme.

Step III: Continuity and Pressure

$$\begin{cases} \dfrac{U_i^{n+1} - \widetilde{U}_i}{\Delta t} = -\dfrac{1}{\rho} \dfrac{\partial P^{n+1}}{\partial x_i} & (11) \\[2ex] \dfrac{\partial U_L^{n+1}}{\partial x_L} = 0 & (12) \end{cases}$$

$$\Longleftrightarrow \begin{cases} \dfrac{\partial \widetilde{U}_L}{\partial x_L} = \dfrac{\Delta t}{\rho} \dfrac{\partial^2 P^{n+1}}{\partial x_i \partial x_i} & (13) \\[2ex] U_i^{n+1} = \widetilde{U}_i - \dfrac{\Delta t}{\rho} \dfrac{\partial P^{n+1}}{\partial x_i} & (14) \end{cases}$$

Equation (13): The Poisson equation is solved by a Gauss-Seidel iterative method with over-relaxation and Neumann boundary condition:

$$\frac{\partial p^{n+1}}{\partial n} = 0 \quad \text{(see below)}$$

Pressure is defined on a staggered grid ("pressure points" are in the center of "velocity cubes").

Boundary Conditions

Entrance: $\boldsymbol{U}^{n+1} = U_0$ given

Exit: no condition: $\boldsymbol{U}^{n+1}$ is computed assuming that advection is dominating

Walls: The boundary condition should be $\boldsymbol{U}^{n+1} = 0$. But, since the number of grid points we use is not sufficient to allow such a condition, we set: $\boldsymbol{U}^{n+1} \cdot \boldsymbol{n} = 0$ for the component normal to the wall and we assume a relationship between total shear stress and instantaneous tangential velocity U_τ^{n+1}.

This relationship, based on the logarithmic law for mean velocity near a rough wall, is quite a crude assumption for instantaneous velocities. *Schumann* [10] had rather good results with similar boundary conditions in channel flow.

In the calculation, we choose arbitrarily as a boundary condition for step II (diffusion) the same condition as for the whole time step:

$$\tilde{U}_n = U_n^{n+1} = U_{0n}$$

then Eq. (11) yields at boundaries:

$$\frac{\partial p^{n+1}}{\partial n} = -\frac{\rho}{\Delta t}\left(U_n^{n+1} - \tilde{U}_n\right) = 0 \tag{15}$$

(In fact, tests showed that the results are the same with any normal velocity boundary condition for step II).

And then, boundary condition for step III (Poisson equation) is justified:

$$\frac{\partial p^{n+1}}{\partial n} = 0$$

Equation (11) yields also at boundaries:

$$U_\tau^{n+1} - \tilde{U}_\tau = -\frac{\partial p^{n+1}}{\partial \tau} \tag{16}$$

$$\frac{\partial U_\tau^{n+1}}{\partial n} - \frac{\partial \tilde{U}_\tau}{\partial n} = \frac{\Delta t}{\rho}\,\frac{\partial^2 p^{n+1}}{\partial n\,\partial \tau} = 0 \tag{17}$$

So the boundary condition on shear gradient is also the same for step II and whole time step.

Numerical Computation of the Flow Induced by a Jet in a Cubic Cavity and Comparison with Measurements

Figure 2 gives a sample of comparison between measurements and numerical mean flow results in the case of an algebraic turbulent subgrid scale viscosity model. Figure 3 gives

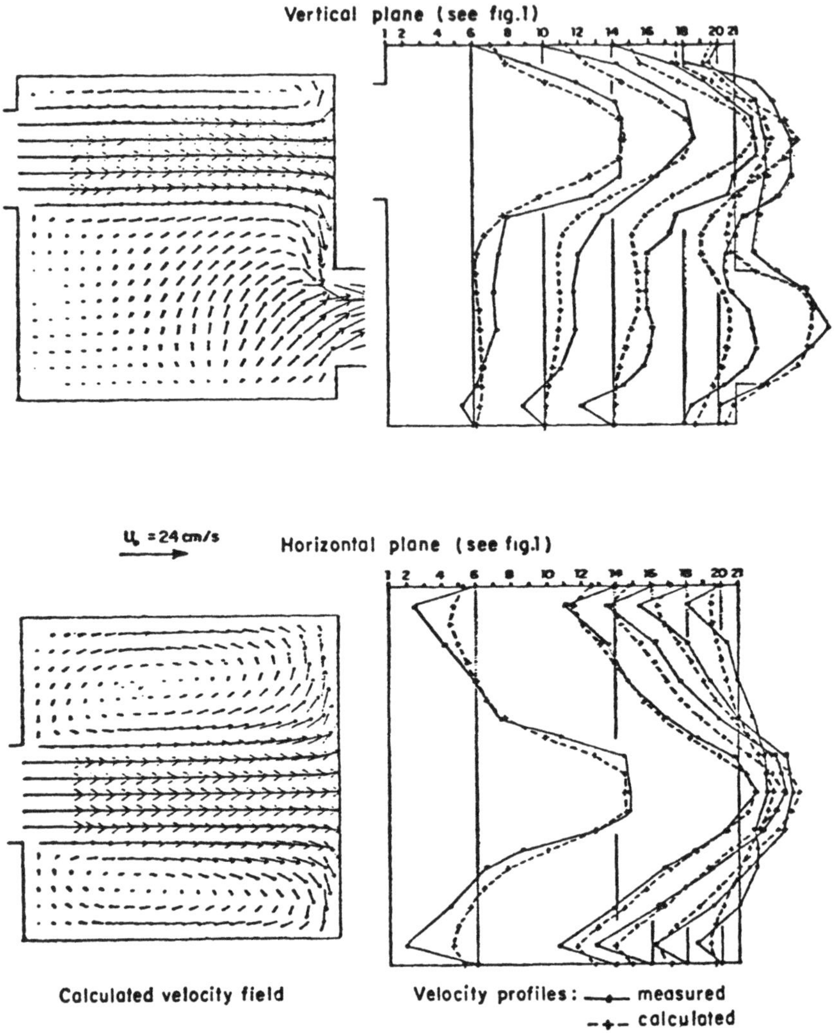

Fig. 2. Flow in the cubic box (grid 21 × 21 × 27). With turbulent viscosity

the same comparison for a case with no such turbulent viscosity. The full comparison between measurements and numerical results shows that the main aspects of the flow are present even in the case with no turbulent viscosity; the turbulent viscosity makes the comparison better in the jet region but some discrepancies remain in the recirculating area under the jet.

These differences between the two cases confirm that the numerical diffusion, inherent to the advection scheme, is not too high: it does not exceed the turbulent diffusion.

But an important feature observed in experiments – the increase of turbulent energy from the entrance to the inside of the box – is not reproduced in this simulation, even with fluctuating boundary conditions at the entrance. This result led us to perform the same simulation scheme on plane Poisseuille flow in order to test its turbulent production mechanism.

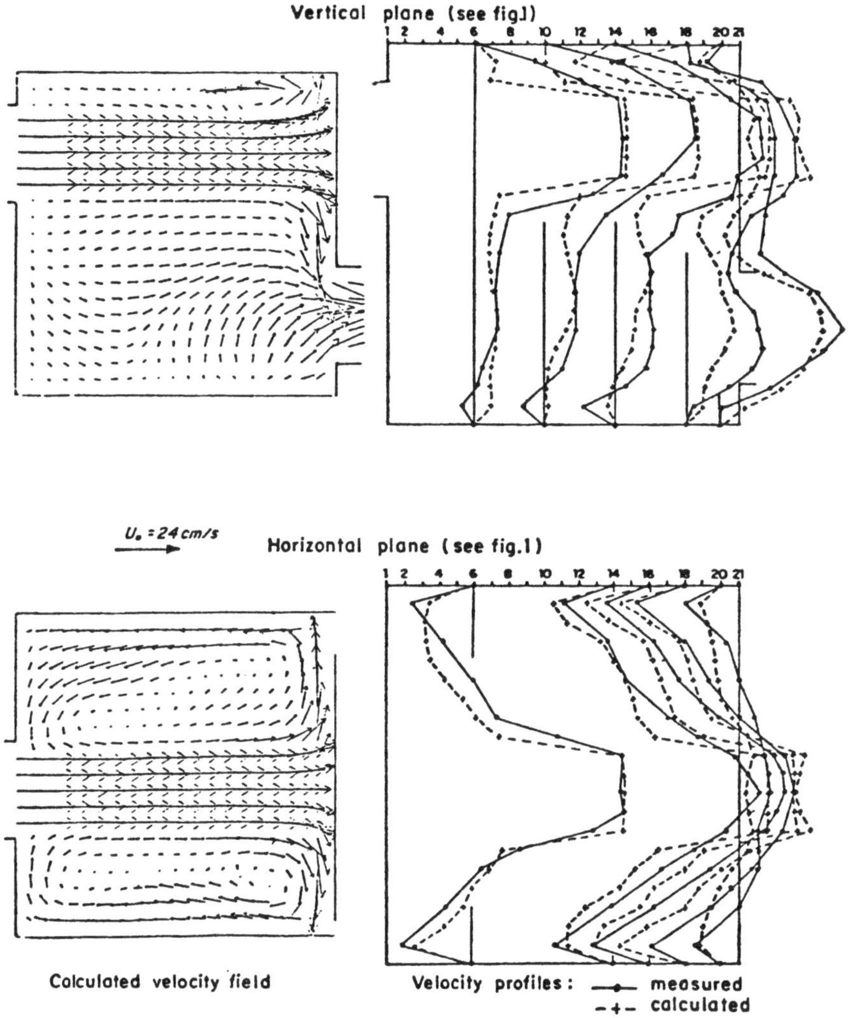

Fig. 3. Flow in the cubic box (grid 21 × 21 × 27). Without turbulent viscosity

Large Eddy Simulation of Plane Turbulent Poiseuille Flow

This calculation aims to improve our LES scheme and its turbulence production mechanism on this simple case of wall-bounded shear flow about which a large number of measurements – made by *Comte-Bellot* [3] – are available.

Computational Specifications for this Flow

The computational domain has the following dimensions

L_x = 8 D in the streamwise direction x
L_y = 4 D in the spanwise direction y
L_z = 2 D in the vertical direction z (perpendicular to the walls)

Boundary conditions in the horizontal directions. According to measurements [3], the double space correlations R_{11} $(L_x/2, 0, 0)$ and R_{ii} $(0, L_y/2, 0)$ between the boundaries in these directions are negligible. So the flow between infinite plates can be represented with spatial periodicity for both velocity and pressure fluctuations:

velocity: in the convective operator, a characteristic leaving the domain at one face of the parallelepiped in the X or Y direction re-enters it at the opposite face in the same direction.

pressure fluctuations: $p' = p - P_0$ [P_0 being a given constant gradient pressure field $P_0 = -\rho U_*^2 (x/D)$].

The Poisson equation is solved in a domain which is "closed on itself" in two directions.

The number of grid points is

$n_x = 65$ in the streamwise direction with uniform spacing
$n_y = 33$ in the spanwise direction with uniform spacing
$n_z = 33$ in the vertical direction with variable grid spacing

Boundary conditions on the walls

- vertical velocity component: $W/_{\text{wall}} = 0$
- for the horizontal velocity components, since the above number of grid points is not sufficient to describe a boundary layer with no-slip boundary conditions, we assume a relationship between total shear stress and instantaneous tangential velocity, as described above.

Statistical operator. A statistical operator is needed to compare results with theoretical and experimental data.

We choose to define it – at each time step of the computation – as the mean value over a plane parallel to the walls:

$$<Q>(t,z) = \frac{1}{L_x L_y} \int Q(t^n, x, y, z)\, dx\, dy \quad \text{and} \quad q = Q - <Q>$$

(An actual time averaging at one point can be used to study large wavelength phenomena, such as pressure-strain correlations...).

Initial condition. The spatial periodicity implies some equivalence between an entrance boundary condition ($x = 0, 0 < t < T$) and an initial condition function of x $(0 < x < L_x,$ $t = 0)$.

We thus defined an initial fluctuating velocity field as a series of periodic functions of x with a $(-5/3)$ power energy spectrum.

$$\bar{u}_L = \sum_{i=1}^{n} a_i^L \cos[ik_x x + \phi_i^L(y,z)]$$

$$n = \frac{n_x + 1}{2} \quad ; \quad k = \frac{-2\pi}{L_x}$$

where: $(a_i^L)^2 = C_1(z) i^{-5/3}$ and $\phi_i^L(y,z)$ are random functions [pressure field and incompressibility are obtained by resolution of Poisson equation (13), as in time step].

With no more initial correlation between the $\bar{u}_i$, the turbulent energy level had a rapid decrease during the first time steps.

In order to insure a high-level initial turbulent production:

$$P = -<\overline{u}_i\,\overline{u}_j> \frac{\partial <\overline{u}_i>}{\partial x_j}$$

we add to the initial fluctuating velocities a sheared mean velocity profile $<U>$ and we choose the differences:

$$(\phi_i^2 - \phi_i^1) \text{ and } (\phi_i^3 - \phi_i^1)$$

such as the turbulent shear stresses have the theoretical values:

$$<\overline{u}\,\overline{w}> = u_*^2\,(1 - Z/D) \text{ [except near the walls]}$$

$$<\overline{u}\,\overline{v}> \cong 0 \text{ and } <\overline{v}\,\overline{w}> \cong 0$$

With these initial conditions, the computed energy decrease is much slower.

Computed Turbulent Kinetic Energy

$$(k_c = 1/2 <\overline{u}_i \overline{u}_i>)$$

Despite the improvement on initial conditions, the decrease of k_c still remained, due to the advection scheme.

As a first approximation to face this problem, we maintain at a fixed (experimental) level K_c the *integral* value I_c of k_c over the whole computational domain, at each time step:

$$I_c(t^n) = \frac{1}{V} \int_D k_c(\mathbf{x}, t^n)\,dV$$

(Practically, we multiply the three components of the fluctuating velocity at each point by the same ratio $\sqrt{k_c/I_c}$).

A few remarks about this operation: 1) the relative change of I_c decreases in time, and, after a few time steps, it becomes:

$$\frac{|K_c - I_c|}{K_c} < 5 \cdot 10^{-3}$$

It seems that the first time steps can be interpreted as a time scale for the non-linear process.

2) If we stop maintaining I_c, its integral value stays over k_c during 6 crossings of the domain and then decreases very slowly.

Results

Compared with *Comte-Bellot*'s [3] measurements.

The computed fluctuations (Fig. 4) are rather close to measurements in streamwise (u) and spanwise (v) direction. In the vertical (w) direction, there is certainly an incorrect redistribution by the velocity-pressure correlations.

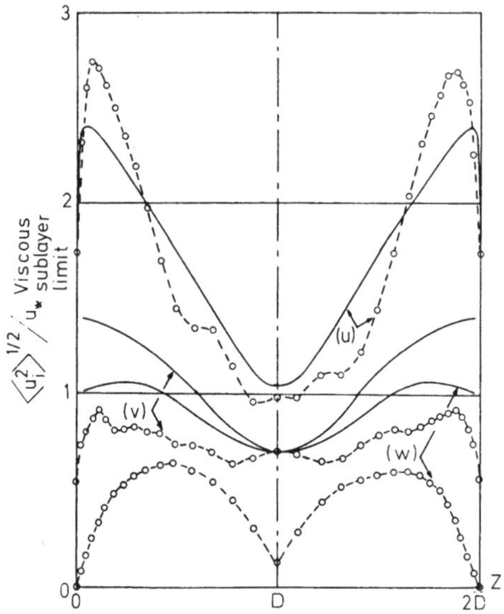

◁ **Fig. 4.** Turbulence intensities in streamwise (u), spanwise (v) and vertical (w) directions — measurements (3); – o – calculations (resolvable part only)

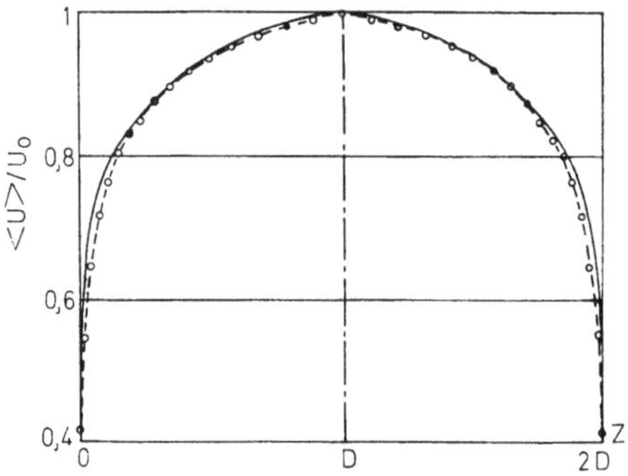

Fig. 5. Mean velocity; — measurements; – o – calculation

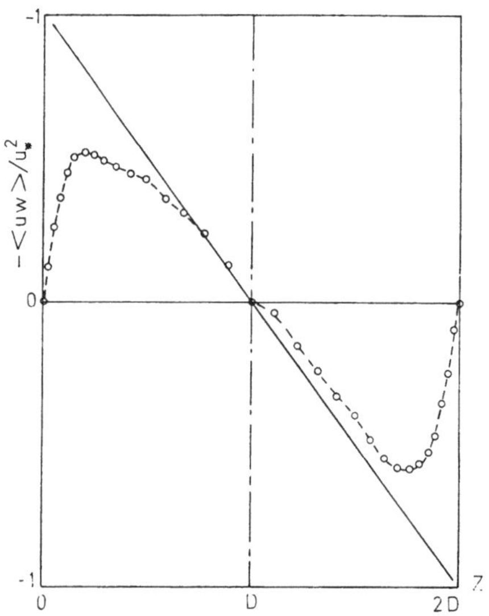

Fig. 6. Turbulent shear stress; — theoretical value; – o – computation (resolvable part only)

The mean velocity profile $\langle U \rangle$ (Fig. 5) is also near the measurements, despite a lack of accuracy on the shear gradient value on the boundary.

The computed part of turbulent shear stress $\langle \overline{u}\,\overline{w} \rangle$ is about what can be expected (Fig. 6).

Conclusion

The lack of computed fluctuations observed in the flow induced by a jet in a cubic cavity led us to study the turbulent production in plane Poisseuille flow, and to define on this flow an integral turbulent energy control procedure.

An analogous procedure could be applied to the jet in a cavity (with time-mean statistical operators) in order to obtain, with an improvement on subgrid scale modelling, a more accurate large eddy simulation of that flow with our stable numerical scheme.

References

1. Benqué, J.: "Modélisation d'écoulements turbulents isothermes", Report EDF-LNH No. HE041/ 78.12 (1978)
2. Chorin, A.J.: The numerical solution of the Navier-Stokes equations for an imcompressible fluid. Bull. Am. Math. Soc. **73**, 928–931 (1967)
3. Comte-Bellot, G.: "*Ecoulement Turbulent Entre Deux Parois Parallèles*" (Publications scientifiques et techniques du Ministère de l'Air, Paris 1965)
4. Ferziger, J.H.: "Higher-Level Simulations of Turbulent Flows", Von Karman Institute Lecture Series 1981–5
5. Grötzbach, G., Schumann, U.: "Direct Numerical Simulation of Turbulent Velocity-, Pressure- and Temperature Fields in Channel Flows", *Symp. on Turbulent Shear Flows*, Pennsylvania State Univ. (1977)
6. Kim, J., Moin, P.: "Large Eddy Simulation of Turbulent Channel Flow", AGARD Symp. on Turbulent Boundary Layers, The Hague (1979)
7. Leonard, A.: Energy Cascade in large-eddy simulation of turbulent fluid flows. *Adv Geophys.* **A18**, 237 (1974)
8. Leslie, D.C., Quarini, G.L.: The application of turbulence theory to the formulation of subgrid modelling procedures. J. Fluid Mech. **91**, 65–91 (1979)
9. Moin, P., Reynolds, W.C., Ferziger, J.H.: "Large Eddy Simulation of Incompressible Turbulent Channel Flow", Report TF-12, Dept. of Mech. Engineering, Stanford University (1978)
10. Schumann, U.: Subgrid scale modelling for finite difference simulations of turbulent flows in plane channels and annuli. J. Comp. Phys. **18**, 376 (1975)
11. Smagorinsky, J.: General circulation experiments with the primitive equations, I. the basic experiments. Mon. Weather Rev. **91**, 99 (1963)

The Assessment of Numerical Diffusion in Upwind Difference Calculations of Turbulent Recirculating Flows

J.J. McGuirk, A.M.K.P. Taylor and J.H. Whitelaw

Department of Mechanical Engineering, Imperial College of Science and Technology, London, SW7 2BX, England

Abstract

The problem of numerical diffusion in calculations of turbulent recirculating flow which use the upwind difference approximation for convection is addressed in the paper. A technique is described for estimating the local numerical diffusion error using the difference between upwind and central difference expressions for the convection terms, and this error is made small by selective grid refinement. It is demonstrated that a 55 × 50 mesh produces a grid-independent solution for the case of flow over a confined axisymmetric disc and this enables the true performance of the two-equation k-ϵ model in this recirculating flow to be examined. Application of the technique to the turbulence kinetic energy equation shows that although the source and sink terms are large, turbulent transport is still important over the whole flow and hence numerical diffusion effects can still intrude; the turbulence energy solutions are, if anything, more sensitive to grid refinement than the velocity profiles. The predicted recirculation length is, in the grid independent results, about 8% larger than measured. The rate of recovery from the near-wake region is, however, substantially too slow, and the calculated levels of turbulent kinetic energy are much smaller than those measured.

Nomenclature

$C_{\epsilon_1,\epsilon_2,\mu}$	Constants in turbulence model
D/Dt	Substantial derivative
d	Diameter of disc
k	Turbulence kinetic energy
l	Turbulence length scale
P	Production of turbulence kinetic energy
p	Static pressure
Pe	Computational Péclet number
R	Radius of confining pipe
r	Radial co-ordinate
T	Out-of-balance residual term
U	Mean axial velocity
U_0	Bulk velocity at disc trailing edge
u	Fluctuating axial velocity
V	Mean radial velocity
v	Fluctuating radial velocity
w	Fluctuating azimuthal velocity
x	Axial co-ordinate

Greek Symbols

δx	Grid spacing
ϵ	Rate of dissipation of k
μ_t	Turbulent viscosity
ν	Kinematic viscosity
ρ	Fluid density
$\sigma_{k,\epsilon}$	Turbulent Prandtl numbers for k, ϵ
Φ^*	Converged numerical solution for ϕ
ϕ	Surrogate dependent variable

Subscripts

E, W, P	Grid-node notation
i,j	Tensor notation

Operators

$< >$	Time-averaged quantity

Introduction

In the past decade many solutions have been presented for two-dimensional turbulent recirculating flows in which first-order accurate upwind differencing was used to approximate the convection terms in the governing differential equations and finite-difference meshes of order 20 x 20 were employed. Marginal grid refinement has sometimes been claimed to demonstrate grid independence of the solutions, the implication being that the solutions were untainted by the numerical diffusion errors which are known to accompany first-order upwind-difference methods (see, e.g. *Roache* [1], *Raithby* [2]). More recently it has been realised (see, e.g., *Castro* [3]) that in those calculations of these complex flows where the errors inherent in the use of a particular turbulence model are under investigation, the possible presence of numerical errors in the predictions must be examined much more closely than has previously been the case. Indeed, the presence of numerical diffusion in the important thin shear layer regions embedded in the recirculating flow can render meaningless the comparison of, say, a two-equation and Reynolds stress turbulence model on grids of moderate density. *Leonard* et al. [4] have demonstrated that on a 25 x 25 grid in the flow they were considering, only a more accurate finite-difference scheme responded to alterations in the turbulence model, the upwind difference solutions on the same grid being insensitive to changes in the turbulent viscosity. An obvious remedy for this problem would be the use of much finer meshes (100 x 100 say); however, this is certainly not the most cost-effective approach to adopt and in three dimensions this level of grid-refinement is probably not feasible, even if it could be guaranteed that it were sufficient. In addition, limitations on available computer storage and time together with problems commonly experienced obtaining converged solutions using presently available fluid flow solution algorithm on fine meshes have led to the search for an alternative strategy.

The use of a higher-order accurate scheme is one approach which has been adopted by some workers. The obvious difficulty here is that all these higher-order methods (as is well known for second-order central differencing) generate non-physical oscillations (wiggles) in the vicinity of steep gradients when the convection terms in the equations dominate. These under- and over-shoots represent a fundamental drawback of higher-order schemes since not only may they make converged solutions difficult, or impossible, to achieve if iterative techniques are used to solve the discretised equations, but also they violate the boundedness properties of, for example, positive definite quantities such as species mass fractions or turbulence kinetic energy, and their use in the differential equations for these quantities may be problematic. In addition, it has often been demonstrated (see, for example, *Cheng* and *Shubin* [5]) that the formal order of accuracy of a scheme (as derived via Taylor series truncation analysis) in terms of the smallest neglected power of the local finite-difference mesh size (Δx say) does not reflect the accuracy of the computed results, especially for finite and moderate Δx. A clear explanation why reducing the local truncation error does not guarantee a reduction in the solution error has been given recently by *Stubley* et al. [6]who demonstrate that it is not just the error introduced in the approximation of local function values or gradients by a particular discretisation scheme which is important (this is always reduced by going to higher-order methods), but also the nature of the way this local error is propagated by the discretised version of the convection and diffusion terms in the differential equation. Hence ill-chosen higher-order schemes reduce local interpolation errors, but particularly at high mesh Reynolds numbers propagate this lower error in such a way (which does not reflect the transportive properties of the differential equation) that the overall errors in the whole solution domain are increased.

It is interesting to note that the conventional Galerkin finite-element technique, being essentially also a second-order accurate method suffers from the same oscillation problem mentioned earlier; this has led some workers in the field to re-discover upwind methods and apply them to finite-element solution algorithms in order to suppress the wiggles, a practice viewed with great suspicion by many members of the finite-element community (see *Gresho* and *Lee* [7]); with some justification for the future calculation of turbulent flows with these methods for the reasons given above. Methods for filtering the oscillations or computational noise from higher-order finite-difference solutions have been suggested by, for example, *Forester* [8], but more promising techniques for the development of more accurate finite-difference schemes are those of [6, 9, 10] which attempt to derive discretisation schemes which reflect basic physical properties of the differential equations without resorting to crude interpolation formulae.

The above discussion indicates that only the development of a more accurate discretisation scheme offers long-term advantages in terms of calculations of turbulent elliptic flows on reasonable meshes which are free of numerical error. Although some work has been done on the construction and evaluation of such methods, these schemes are still under development and no single approach has yet demonstrated clear superiority for sets of non-linear equations including, for example, dependent variables for positive-definite turbulence quantities. It is also likely to be some time before a generally acceptable scheme emerges. An interim, alternative approach is examined in the present paper. It is proposed that if upwind difference methods continue to be used in turbulence model assessment calculations of two-dimensional elliptic flows that a more rational way of distributing grid nodes must be undertaken so that grid-independent solutions may still be obtained using moderate mesh densities (up to about 50×50 say). The problem is how to decide in which regions of the flow grid refinement is most needed since ad hoc refinement would undoubtedly require much finer meshes. As has been pointed out by *Gresho* et al. [7], it might even be looked upon as an advantage of the presence of oscillation in a solution that at least these help to identify the location in the flow where the mesh cannot resolve the steep gradients and hence the locations where judicious mesh refinement is necessary to eliminate the wiggles – numerically diffusive solutions give no such clear indication of the critical or problem regions in the flow. *McGuirk* and *Rodi* [11], however, have shown how the problem regions in the flow can be identified a posteriori by evaluating the amount of local numerical diffusion present in a solution obtained with a given grid via an estimate of the magnitude of the local truncation error in the converged solution of the discretised equations. This identifies the local regions where the mesh Reynolds (or Péclet) number is high and *at the same time* the spatial curvature of the solution in the corresponding co-ordinate direction is sufficiently large to make the product of the numerical diffusion coefficient and the second gradient of the dependent variable a significant contribution in the balance of the local discretised transport equation. Regions where this error term is large in comparison with other terms in the equation are thus regions where grid refinement is necessary. A related technique has been used to increase the order of accuracy of the solution for a developing boundary layer by *Krause* and *Kordulla* [12] and a similar approach was used by *Castro* [3] to demonstrate that upwind difference generated numerical diffusion can sometimes cause large local errors in boundary layer flows without seriously affecting mean velocity profiles but that in complex flows such errors could propagate and degrade the whole solution.

The purpose of this paper is then to demonstrate the selective grid refinement approach for a typical two-dimensional turbulent recirculating flow, namely the case of axisymmetric flow over a confined disc. Further details of the flow considered are given in the following section and the mean-flow and turbulence model equations are presented. A description is given of the method used to evaluate the truncation error terms, hereafter called out-of-balance residuals, and in the following section the reduction of these residuals to a level where a grid-independent solution may be claimed is illustrated. The relative magnitude of the various terms in the U-momentum equation are then examined as an aid to understanding the development of the flow. A comparison of the predictions with the experimental results of *Taylor* [13] is used to assess the performance of the two-equation k-ϵ model in this flow, and finally the truncation error analysis is applied to the transport equation for turbulence kinetic energy k; this last investigation is of importance to examine whether numerical diffusion errors are important in the k balance since, for the reasons given above, higher-order methods may have problems with a positive definite variable such as k.

Flow Configuration

The flow which is to be calculated is that immediately downstream of a confined, axisymmetric baffle, as shown in Fig. 1. Measurements were made in a water tunnel using a directionally-sensitive, forward-scatter laser-Doppler velocimeter at a Reynolds number ($U_0 d/\nu$) of 34,700. Various blockage ratios and forebody shapes were investigated in [13], but for the purposes of this study a sharp-edged disc of 25% area blockage ratio was chosen because of the steep trajectory of the separation streamline near the baffle tip; this provides a stringent test of a numerical scheme. Most importantly, experimental measurements of the axial and radial velocity components were available in the plane of the disc's trailing edge; this removes any uncertainty in the initial conditions for the velocities to be used in the calculations.

Calculation Method

Mean Flow Equations and Turbulence Model

The following equations of motion are to be solved and are written in the Reynolds-averaged form for cylindrical polar co-ordinates, assuming steady axisymmetric flow and constant density:

$$\frac{\partial U}{\partial x} + \frac{\partial V}{\partial r} + \frac{V}{r} = 0 \tag{1}$$

$$\rho \frac{\partial}{\partial x}(U^2) + \frac{\rho}{r}\frac{\partial}{\partial r}(rUV) = -\frac{\partial p}{\partial x} - \frac{\partial}{\partial x}\rho \langle u^2\rangle - \frac{1}{r}\frac{\partial}{\partial r}(r\rho\langle uv\rangle) \tag{2}$$

$$\rho \frac{\partial}{\partial x}(UV) + \frac{\rho}{r}\frac{\partial}{\partial r}(rV^2) = -\frac{\partial p}{\partial r} - \frac{\partial}{\partial x}\rho\langle uv\rangle - \frac{1}{r}\frac{\partial}{\partial r}(r\rho\langle v^2\rangle) + \rho\frac{\langle w^2\rangle}{r} \tag{3}$$

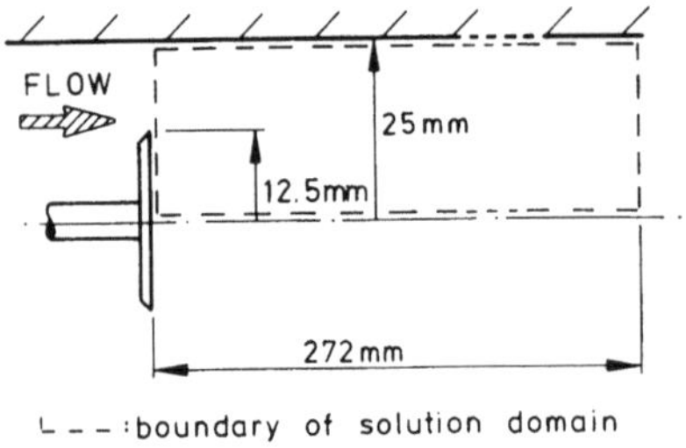

Fig. 1. Flow configuration and solution domain

(The Reynolds number is assumed sufficiently high for laminar transport to be negligible)

The Reynolds stresses are represented by the two equation ("k-ϵ") turbulence model of *Jones* and *Launder* [14]which incorporates a scalar turbulent viscosity hypothesis:

$$\rho <u^2> = \frac{2}{3}\rho k - 2\mu_t \frac{\partial U}{\partial x} \tag{4a}$$

$$\rho <v^2> = \frac{2}{3}\rho k - 2\mu_t \frac{\partial V}{\partial r} \tag{4b}$$

$$\rho <w^2> = \frac{2}{3}\rho k - 2\mu_t \frac{V}{r} \qquad \text{(no mean azimuthal velocity)} \tag{4c}$$

$$\rho <uv> = -\mu_t \left[\frac{\partial U}{\partial r} + \frac{\partial V}{\partial x}\right] \quad \text{where} \tag{4d}$$

$$k \equiv \frac{<u^2> + <v^2> + <w^2>}{2} \tag{5}$$

The turbulent viscosity is given by:

$$\mu_t = C_\mu \rho \frac{k^2}{\epsilon} \tag{6}$$

where C_μ is a constant (comp. below) and ϵ is the rate of isotropic dissipation of k:

$$\epsilon = \nu \left\langle \frac{\partial u_i}{\partial x_j} \frac{\partial u_i}{\partial x_j} + \frac{v^2}{r^2} + \frac{w^2}{r^2} \right\rangle \tag{7}$$

The values of k and ϵ are obtained from the solution of the (modelled) transport equations:

$$\frac{D}{Dt}\rho k = \frac{\partial}{\partial x}\left[\frac{\mu_t}{\sigma_k}\frac{\partial k}{\partial x}\right] + \frac{1}{r}\frac{\partial}{\partial r}\left[r\frac{\mu_t}{\sigma_k}\frac{\partial k}{\partial r}\right] + P - \rho\epsilon \tag{8}$$

$$\frac{D}{Dt}\rho\epsilon = \frac{\partial}{\partial x}\left[\frac{\mu_t}{\sigma_\epsilon}\frac{\partial \epsilon}{\partial x}\right] + \frac{1}{r}\frac{\partial}{\partial r}\left[r\frac{\mu_t}{\sigma_\epsilon}\frac{\partial \epsilon}{\partial r}\right] + C_{\epsilon_1} P\frac{\epsilon}{k} - C_{\epsilon_2}\rho\frac{\epsilon^2}{k} \tag{9}$$

where

$$P = -\rho <uv> \left[\frac{\partial U}{\partial r} + \frac{\partial V}{\partial x}\right] - \rho <u^2> \frac{\partial U}{\partial x} - \rho <v^2> \frac{\partial V}{\partial r} - \rho <w^2> \frac{V}{r} \qquad (10)$$

The turbulence model includes five empirical constants whose values are given below; this set of constants is found to give good agreement with experiment in a wide range of two-dimensional shear flows (see *Launder* and *Spalding* [15]):

$$C_\mu = 0.09 \;\; ; \;\; C_{\epsilon_1} = 1.44 \;\; ; \;\; C_{\epsilon_2} = 1.92 \;\; ; \;\; \sigma_k = 1.0 \;\; ; \;\; \sigma_\epsilon = 1.22$$

Computational Procedure and Boundary Conditions

The numerical integration of Eqs. (1–3), together with the turbulence model equations, is achieved using a version of the finite difference procedure of *Gosman* and *Pun* [16] which incorporates the 'SIMPLE' algorithm of *Patankar* and *Spalding* [17] to deal with the pressure-velocity coupling. The implicit, linearised finite-difference equations are solved iteratively and retain the transportive property of the original partial differential equations [1]. The differencing scheme for the convection terms is of the "hybrid" type [18] which is described below.

Boundary conditions for U, V, k and ϵ must be prescribed on the closed boundary surrounding the area of integration, shown in Fig. 1. The following headings present the conditions which were applied along each side.

Boundary at $x = 0$. Measured radial profiles of axial and radial velocity in the annular gap were used. A uniform profile of $k = 0.01\ \text{m}^2\text{s}^{-2}$ was assumed, derived from a typical (measured) value of k in the region $d/2 < r < R$. The value of ϵ was then set at:

$$\epsilon = \frac{k^{3/2}}{l} \quad \text{where} \qquad (11)$$

$$l = R - \frac{d}{2} \qquad (12)$$

Computational tests (see *Taylor* [13]) showed that the solutions were insensitive to these estimated values of k and ϵ over the inflow plane.

Solid Boundaries. Standard wall functions were used to provide boundary conditions for grid nodes adjacent to the baffle surface and to the confining pipe walls and thereby obviated the need for a fine grid to resolve the distributions in the low turbulence Reynolds number near-wall regions. The prescriptions are based on the logarithmic law of the wall and a near wall flow which is in equilibrium. Details are to be found in [15].

Axis of Symmetry. The following conditions are imposed by virtue of symmetry at $r = 0$:

$$\frac{\partial U}{\partial r} = \frac{\partial k}{\partial r} = \frac{\partial \epsilon}{\partial r} = V = 0 \qquad (13)$$

Downstream Boundary. At the downstream boundary, axial gradients of the dependent variables are set to zero. This boundary is sufficiently far downstream of the baffle (about seven recirculation bubble lengths) to have negligible influence on the flow in the region of interest.

The Assessment of Numerical Diffusion

Provided that iteration convergence is attainable, and that computer round-off errors are negligible, the numerical accuracy of the solution is limited only by the approximation of the partial derivatives by their finite-difference analogues. The approximation arises because these analogues represent truncated Taylor series expansions and it can be shown that the solution of the finite-difference equations in fact corresponds to the exact solution of a partial differential equation which is made up of the original partial differential equation together with additional terms due to the truncated expansion. The accuracy of the finite-difference solution is thus determined by the relative importance of the added terms arising from the truncation expansion to the other (exact) terms in the differential equation.

For reasons to be explained below, it is the truncation error of the "hybrid" difference approximation of convection which gives most cause for concern. Considering the convective transport of a surrogate variable ϕ and assuming for simplicity that U and μ_t are constant, hybrid differencing gives the following at $x = x_p$ (referring to Fig. 2 for nomenclature):

for $|\mathrm{Pe}| < 2$, central differencing:

$$\rho U \frac{d\phi}{dx} \cong \rho \frac{U}{2\delta x}(\phi_E - \phi_W) \tag{14}$$

for $|\mathrm{Pe}| > 2$, upwind differencing:

$$\rho U \frac{d\phi}{dx} \cong \rho \frac{U}{\delta x}[\phi(x) - \phi(x - \delta x)] \quad \text{where} \tag{15}$$

$$\mathrm{Pe} = \rho \frac{U\delta x}{\mu_t} \quad \text{and } x = x_P \quad \text{if Pe} > 2$$
$$x = x_E \quad \text{if Pe} < 2$$

The reason for changing the representation with Pe, the computational Péclet number, is relevant to the maintenance of numerical stability of the iterative solution since it guarantees that the matrix of coefficients is always diagonally dominant.

Taylor series expansions of the variable ϕ about $x = x_P$ allows the following expressions to be derived for the right hand sides of Eqs. (14, 15) (for well-behaved functions):

$$\rho \frac{U}{2\delta x}(\phi_E - \phi_W) = \rho U \left[\frac{d\phi}{dx} + \underline{\frac{1}{3!}\frac{d^3\phi}{dx^3}\delta x^2 + \ldots \frac{1}{\gamma!}\frac{d^\gamma\phi}{dx^\gamma}\delta x^{(\gamma-1)} + \ldots}\right] \tag{16}$$

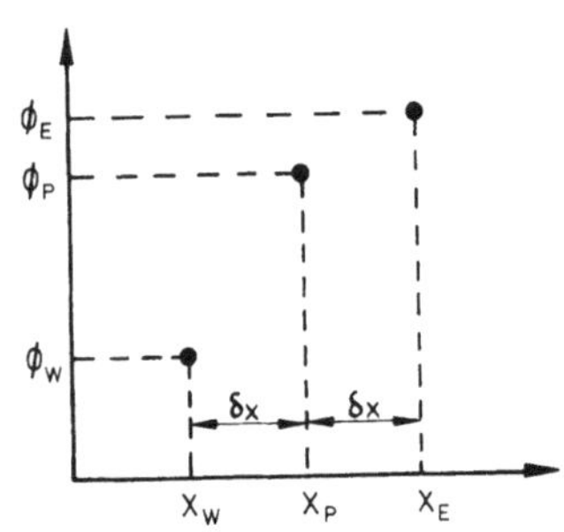

Fig. 2. Definition sketch for variation of variable ϕ at three consecutive gird nodes

(where $\gamma = 2n + 1$; n integer)

$$\rho \frac{U}{\delta x} (\phi_P - \phi_W) = \rho U \left[\frac{d\phi}{dx} - \underline{\frac{1}{2!} \frac{d^2\phi}{dx^2} + \dots \frac{(-1)^n}{n!} \frac{d^n\phi}{dx^n} \delta x^{n-1} + \dots} \right] \tag{17}$$

(assuming Pe > 2)

Both expansions have been given for $\phi = \phi(x)$ only and uniform grid spacing, for simplicity. In the case of non-uniform spacing, additional truncation terms are introduced which depend on the mesh expansion between one grid node and the next: for small expansion factors, as were used in the calculations of this work, the errors are of secondary importance. The terms which are truncated in discretising the differential equation are underlined. Comparing Eq. (17) with (16), it is apparent that upwind differencing gives rise to a leading truncation term which resembles a diffusion term and is thus referred to as numerical diffusion.

A method for assessing the local magnitude of this numerical diffusion has been presented by *McGuirk* and *Rodi* [11] and its basis is as follows. Let $\Phi^*(x,r)$ represent an iteratively converged solution of the finite difference equations for a given grid and consider a location at which upwind differencing has been used in the x-direction, because Pe > 2. Then $\Phi^*(x,r)$ satisfies exactly the finite-difference analogue of the relevant differential equation with x-direction convection represented as:

$$\rho U \frac{\partial \phi}{\partial x} = \rho \frac{U}{\delta x} (\Phi_P^* - \Phi_W^*) \tag{18}$$

If the left hand side of Eq. (18) were to be replaced by a central difference approximation:

$$\rho U \frac{\partial \phi}{\partial x} = \rho \frac{U}{2\delta x} (\Phi_E^* - \Phi_W^*) \tag{19}$$

then, in general, the terms in the finite-difference analogue of the whole differential equation when summed will no longer satisfy an equality but will give rise to an out-of-balance residual term, T, which is arithmetically the difference between the right hand sides of Eqs. (18) and (19). Formally, T is the difference in the truncation series for upwind and central differencing which is, from (16) and (17):

$$T \equiv \rho U \left[\frac{1}{2!} \frac{d^2\Phi^*}{dx^2} \delta x + \dots \frac{1}{\beta!} \frac{d^\beta \Phi^*}{dx^\beta} \delta x^{\beta-1} + \dots \right] \tag{20}$$

(where $\beta = 2n$; n integer) and the leading term in this series is the numerical diffusion term. Whether numerical diffusion is important or not is examined point-by-point by comparing the magnitude of the out-of-balance residual with the original terms (convection, diffusion, source/sink) in the transport equation for ϕ, these terms also being evaluated from the numerical solution. The case of T being small is a necessary, but not sufficient, condition for a solution to be grid independent because the truncation terms of order higher than δx are not properly evaluated [compare coefficients of the higher-order terms in (17) and (20)]. It is noted, however, that δx is small compared to a typical dimension of the solution domain (0.1–0.01) and that for smooth distributions of the function Φ^* its higher derivatives will undoubtedly tend to zero. It is, therefore, plausible to assume that the higher powers of δx are small multiplicative coefficients in the series and that the numerical diffusion will be the dominant contributor to the magnitude of T and this term (numerical diffusion) is properly evaluated. A final condition on the validity of our assumption (that if T is small numerical diffusive errors are small) is that the equations must be well-posed in the sense that a small term will only have a small effect in determining the solution. Whilst this seems a plausible assumption to make, we know of no formal proof for the type of partial differential equations under consideration here.

The preceding analysis can be simply extended to include two-dimensional flow for linear (i.e. convection of k and ϵ) and non-linear (i.e. convection of momenta) terms. This method has been used in the present study to examine both the axial momentum and turbulence kinetic energy equations.

In summary, the method is applied as follows. Converged solutions to the finite-difference equations are obtained for a given grid. The values of convection, diffusion and source/sink terms (pressure gradient or turbulence generation) are evaluated using central difference approximations, a posteriori, together with the out-of-balance residual, for each grid node. Radial profiles of these quantities are plotted at selected axial stations and the magnitude of T is examined to see whether it constitutes an important term in the local balance of the relevant transport equation at any grid node. If this is the case, then the grid line spacing is reduced in this locality in the direction(s) in which upwind differencing is (are) being invoked. A new solution is then generated on the refined grid and again checked for large values of T; the method is repeated until a grid with acceptably small values of T is attained.

Discussion of Results

The method described in the previous section has been used to guide selective grid refinement for the confined disc problem shown in Fig. 1. The various stages of grid refinement and the identification of locations where numerical diffusion was significant are illustrated in Fig. 3 where the balance of terms in the axial momentum equation is plotted in the form of radial profiles at selected axial stations. The terms have been normalised by $\rho U_0^2/R$. In each plot four lines appear, corresponding to the contribution from nett convection [i.e. $\rho\partial\,(U^2)/\partial x + (\rho/r)\,\partial(rUV)/\partial r$], nett turbulent diffusion, axial pressure gradient and the out-of-balance residual, T. All these terms have been brought to the left hand side of the equality so that a positive value corresponds to a local gain of axial momentum. Emphasis has been placed on axial locations within the first quarter of the recirculation length ($\cong$ 10 mm) as even calculations with a coarse grid (20 × 20) indicated that the worst errors were occurring in this region. The magnitude of these errors on such a coarse grid compared to the other terms in the equation is shown in Fig. 3a where it is clear that the development of the shear layer emanating from the tip of the disc is subject to large local errors, since the residual term, representing numerical diffusion, is of the same magnitude as the maximum value of the term representing true, physical, diffusion at both x = 5 and 10 mm stations. At the 5 mm station the outer region, which essentially represents the core of the annulus flow accelerating over the recirculation bubble maximum radial height, is also contaminated by numerical diffusion. As has been pointed out by *Castro* et al. [19], this leads to the numerical solution failing to conserve total pressure in this region, as would undoubtedly be indicated by the measurements. It was clear then that the axial grid required refinement in the region x = 0 to about 15 mm and the radial grid was too coarse in the region of the shear layer centred about a radius of 15 mm (this is slightly higher than the disc tip as the radial velocities

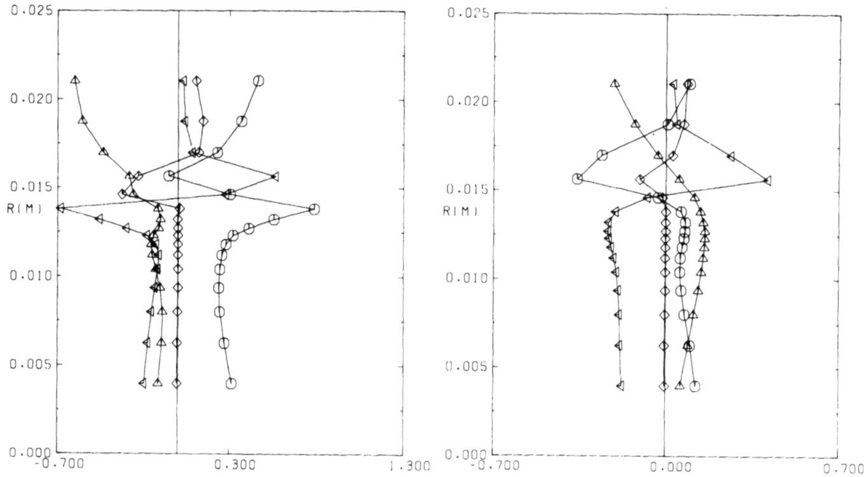
0.025
0.020
0.015
R(M)
0.010
0.005
0.000
-0.700
0.300
1.300
0.025
0.020
0.015
R(M)
0.010
0.005
0.000
-0.700
0.000
0.700

a

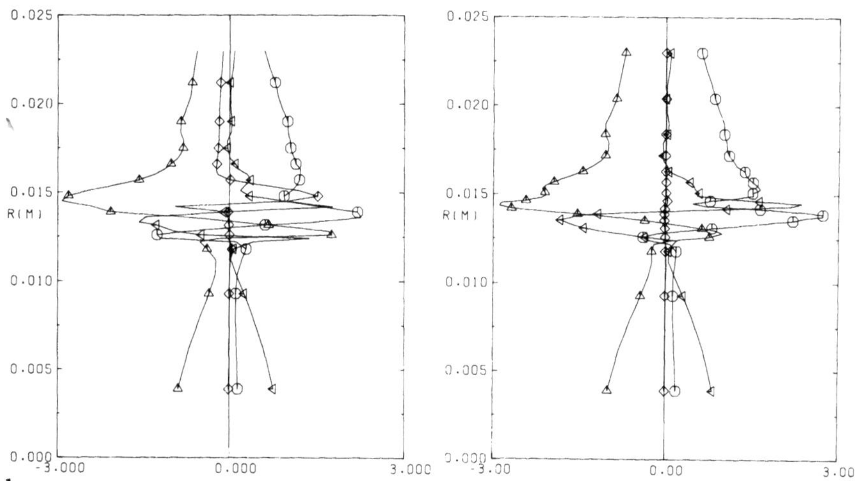
0.025
0.020
0.015
R(M)
0.010
0.005
0.000
-3.000
0.000
3.000
0.025
0.020
0.015
R(M)
0.010
0.005
0.000
-3.00
0.00
3.00

b

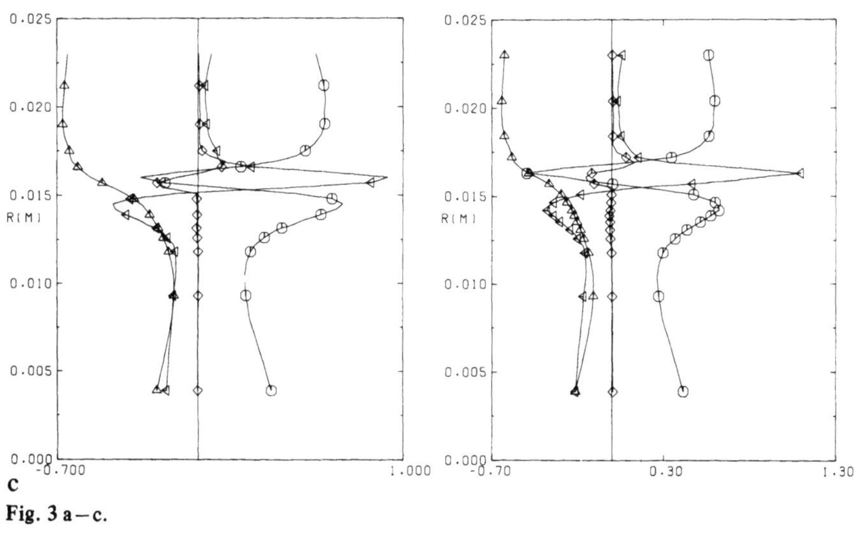
0.025
0.020
0.015
R(M)
0.010
0.005
0.000
-0.700
1.000
0.025
0.020
0.015
R(M)
0.010
0.005
0.000
-0.70
0.30
1.30

c

Fig. 3 a–c.

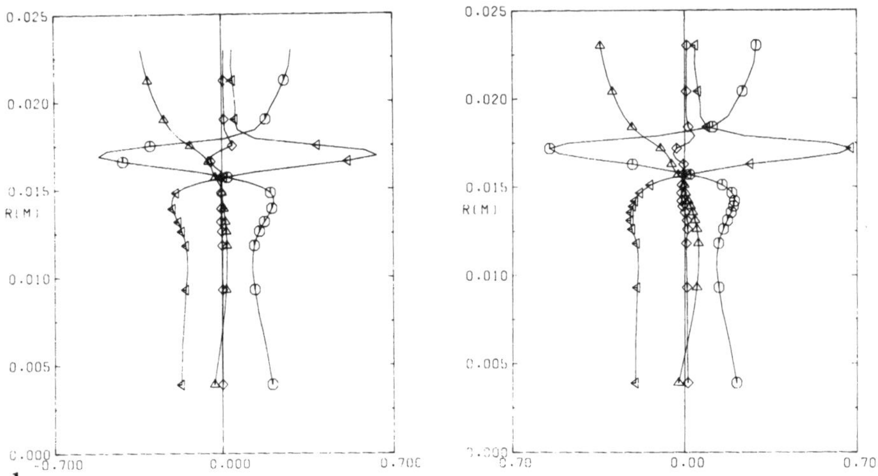

d

Nett convection (by central differencing) — Pressure gradient

Nett diffusion — Out-of-balance residual

Fig. 3 a–d. Radial profiles of terms in the axial momentum equation for three grid sizes. Abscissa in units of $\rho U_0^2/R$. **(a) left:** x = 5 mm (20 × 20 grid); **right:** x = 10 mm (20 × 20 grid); **(b) left:** x = 0.8 mm (40 x 40 grid); **right:** x = 0.8 mm (55 × 50 grid); **(c) left:** x = 5 mm (40 × 40 grid); **right:** x = 5 mm (55 × 50 grid); **(d) left:** x = 10 mm (40 × 40 grid); **right:** x = 10 mm (55 × 50 grid)

at the trailing edge cause the separation streamline to rise before curving down towards the rear stagnation point). A grid of 40 × 40 was used next and, as can be seen from Fig. 3c,d, the residuals at the 5 and 10 mm stations were substantially reduced. An examination of the balance at a station approximately 1 mm from the disc indicated, however, the continued presence of large local errors (see Fig. 3b) so a further grid refinement up to 55 × 50 points was undertaken with, again, additional points being distributed where the residual terms were largest. A comparison between the 40 × 40 and 55 × 50 predictions of the balance at this first axial station, Fig. 3b, shows that in the finest grid calculation the residuals are everywhere small; the large peaks in the profiles of the pressure gradient, convection and turbulent diffusion terms are resolved properly and the out-of-balance residual is reduced to negligible size in the pressure gradient/convection balance region at large radii. The 55 × 50 calculation also shows a slight improvement over the 40 × 40 grid at the 5 mm and 10 mm stations, as can be seen from the sharpening of the peaks in the turbulent diffusion term. The 5 mm station represents perhaps the worst condition in the whole solution domain and even here the residual contributes less than 10% to the local balance and is only of this order of magnitude at two or three isolated grid points which mainly lie in a region where the turbulent diffusion term is very large. A detailed examination at other axial stations indicated no large residual contributions and hence it is believed that the 55 × 50 solution represents a grid-independent solution where the turbulence model is solely responsible for the diffusion of momentum and that the numerical diffusion errors are negligible.

The momentum balances obtained during grid refinement tests are also useful in understanding the development of the flow. Figure 4 presents the balance plots of the finest grid solution at four axial stations covering the recirculation bubble length. At x = 1 mm, shown in Fig. 3b, the pressure gradient is a very large term over the whole

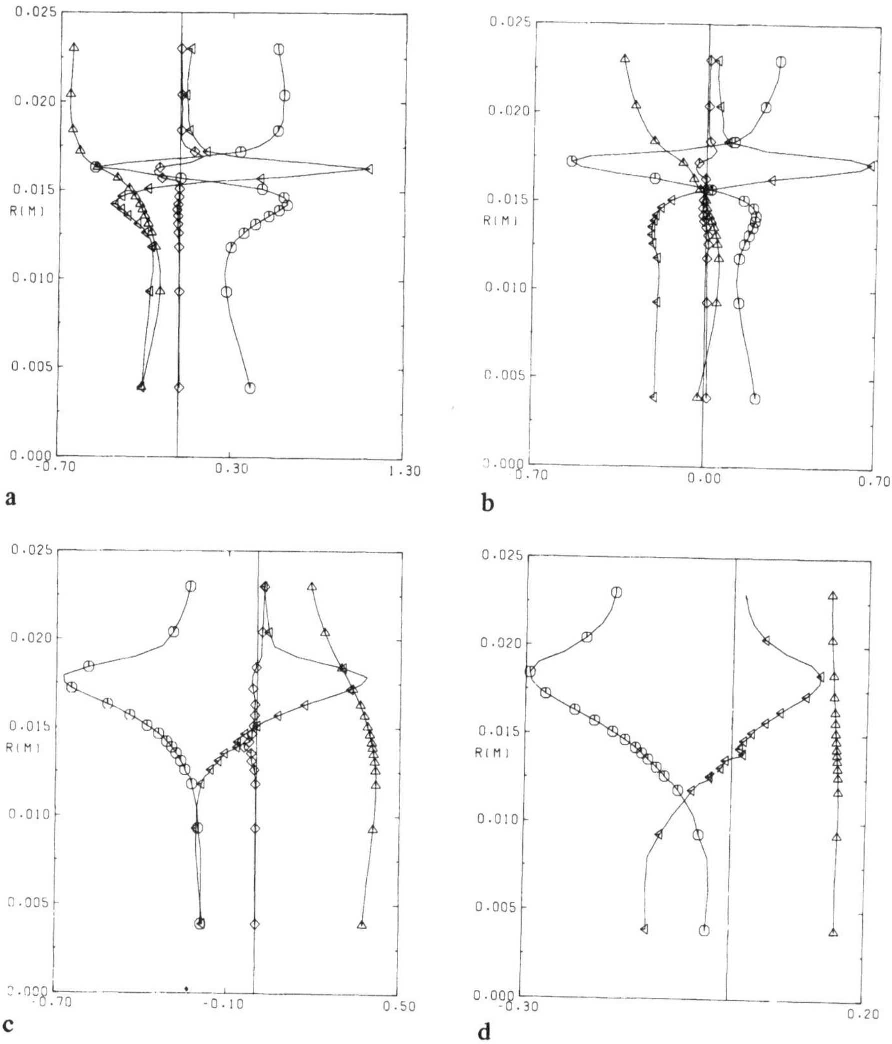

○ Nett convection (by central differencing)

◁ Nett diffusion

△ Pressure gradient

◇ Out-of-balance residual

Fig. 4 a–d. Radial profiles of terms in the axial momentum equation for 55 × 50 grid. Abscissa in units of $\rho U_0^2/R$. **(a)** $x = 5$ mm; **(b)** $x = 10$ mm; **(c)** $x = 20$ mm; **(d)** $x = 40$ mm

pipe radius and is responsible, as stated above, for the acceleration of the flow around the recirculation bubble and for balancing the turbulent diffusion (mainly axial direction diffusion) in the near-axis region (the convection terms are negligibly small because of the proximity of the disc's surface). Lying between these two regions is the developing shear layer which possesses both positive and negative shear stress gradients on either side of the maximum in the axial velocity profile. Further downstream (Fig. 4a,b) the 'three-layer' nature of the flow persists with axial convection becoming more important in the near axis region as the location of the maximum negative velocity is approached. The erosion of the maximum in the inlet velocity profile has led to a simpler shape of the convection term with only two identifiable peaks at $x = 5$ and 10 mm. Downstream of the latter station practically the whole contribution to nett turbulent diffusion is derived from the radial shear-stress gradient and this leads to momentum being transferred from larger to smaller radii. At $x = 10$ mm the favourable pressure gradient near the pipe wall is larger than this diffusive transport, so that the region is still gaining momentum, but by $x = 20$ mm the sign of the pressure gradient has reversed (unfavourable) and nett convection is negative over the whole radius. At the final station shown ($x = 40$ mm, close to the end of the recirculation bubble) turbulent diffusion has become large over the whole flow and, somewhat surprisingly as this is only just at the end of the recirculation bubble, the axial pressure gradient is uniform radially as would occur in a boundary-layer type flow. Even further downstream the flow evolves towards a fully-developed pipe flow with the pressure gradient remaining uniform in the radial direction but becoming progressively smaller and eventually negative. The convection term also becomes smaller while the turbulent diffusion tends to a constant (positive) value in the radial direction, balancing the pressure gradient.

Strictly speaking of course, the truncation error estimation should be performed for all dependent variables; since the major errors with upwind differencing usually occur in regions where the flow is at large angles to the grid-lines, and since in these regions U and V velocities are of the same order of magnitude, then it is probably sufficient just to test the U-momentum equation. If turbulent transport is important in the scalar equations, e.g. for k and ϵ, then the numerical error check should also be employed here in case grid-refinement is required in locations different from those demanded by the U-momentum balance plots. It has sometimes been claimed that in highly turbulent recirculating flows the production and dissipation terms in the k and ϵ equations will dominate and numerical diffusion errors in these equations will be unimportant. To confirm this claim the residual analysis was applied to the turbulence kinetic energy equation and the resulting balances, with the terms normalised using $\rho U_0^3/R$, are presented in Fig. 5 for six axial locations covering a distance of approximately three recirculation lengths. Five terms now appear in each plot corresponding to (1) nett convection, (2) nett diffusion, (3) production, (4) dissipation, and, (5) out-of-balance residual. (Note that the terms have this time been taken over to the right hand side of the equation so that production and dissipation represent positive and negative contributions respectively). The balance at the station 1 mm from the disc has been excluded as all terms are small over most of the pipe radius apart from a narrow region around the disc tip where production is balanced by convection and turbulent diffusion with dissipation and the residual being both small terms. This is thus a region of rapid turbulence production in the developing shear layer and can still be identified in the plots at $x = 5$ and 10 mm although dissipation is now no longer insignificant; the nett source term is still positive, however, so the level of turbulence energy is still increasing at these stations as can be seen from the sign of the nett convection term (the maximum k value occurs at $x \cong 20$ mm). Downstream of $x = 10$ mm up to about $x = 50$ mm it is true that the largest terms in the equation are the source and sink terms for k. It can also be seen, however, that turbulent transport of k is always an appreciable contribution to the balance and thus it is, at least in this flow, certainly important to check the turbulence equations also for numerical diffusion effects.

It is shown below that the turbulence energy solutions are more sensitive to the later stages of grid refinement than the velocity field. It can also be concluded then that high-

Fig. 5 a–f. Radial profiles of terms in the turbulence kinetic energy transport equation for 55 × 50 grid. Abscissa in units of $\rho U_0^3/R$. **(a)** $x = 5$ mm; **(b)** $x = 10$ mm; **(c)** $x = 20$ mm; **(d)** $x = 40$ mm; **(e)** $x = 65$ mm; **(f)** $x = 125$ mm

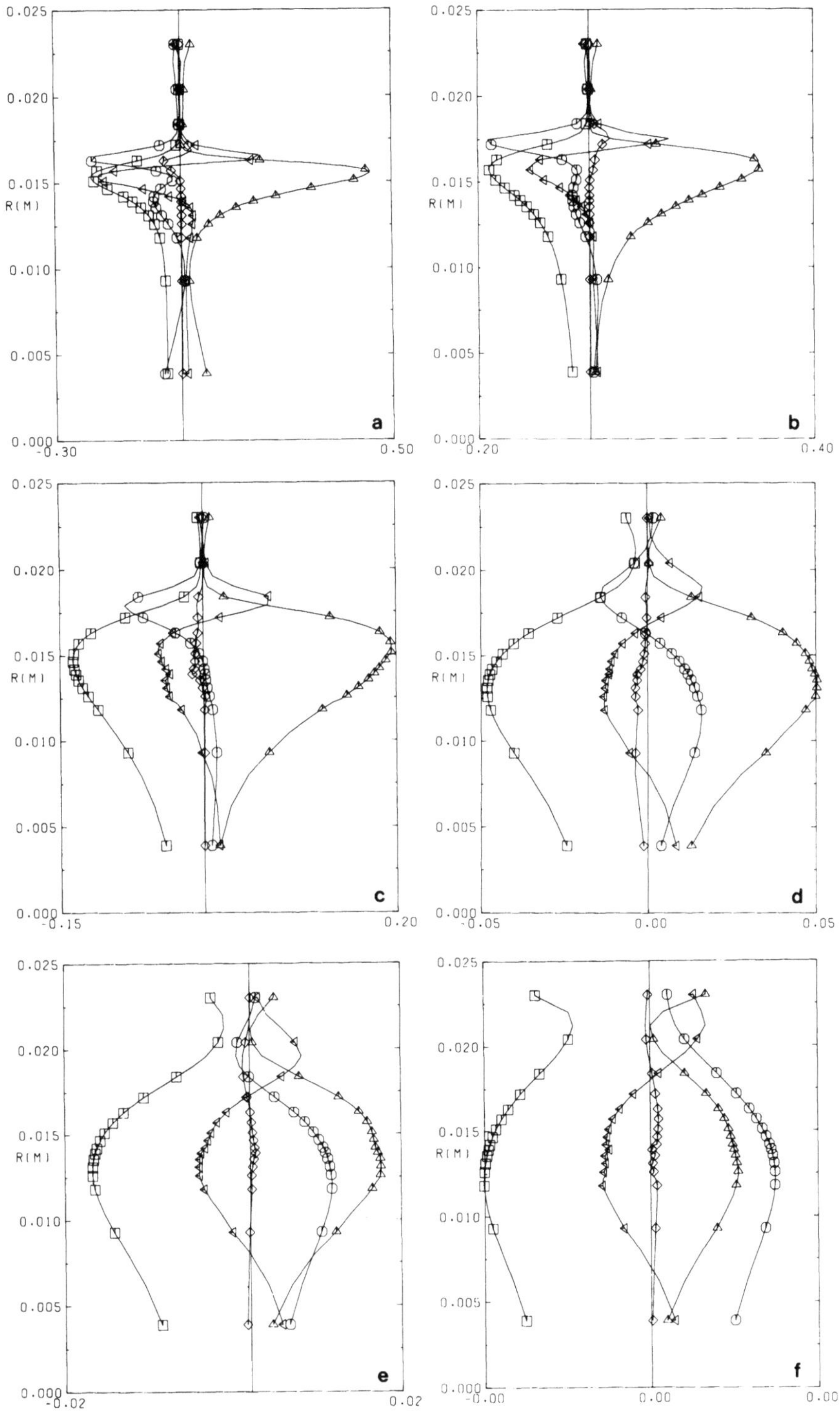

○ Nett convection (by central differencing)

◁ Nett diffusion

△ Turbulent production of turbulence kinetic energy

□ Turbulent dissipation of turbulence kinetic energy

◇ Out-of-balance residual

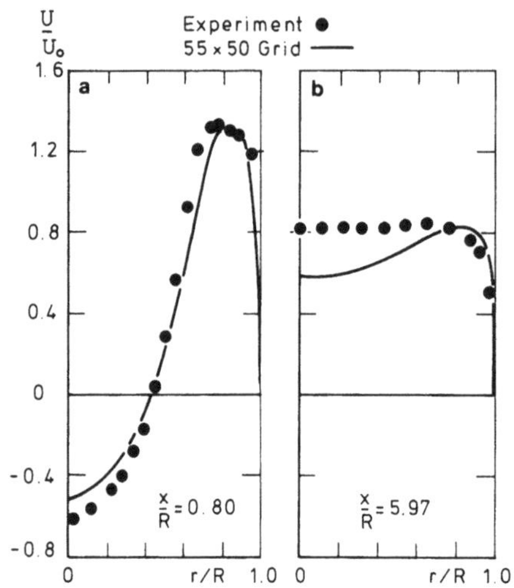

Fig. 6a,b. Radial profile of axial velocity at two axial stations

er-order schemes are in general also required for the turbulence equations, although the practice to date has been to restrict use of the more accurate schemes to the momentum equations and retain upwind differencing in the turbulence equations. Figure 5 indicates that this could prevent grid-independent solutions being obtained unless the selective grid refinement technique is employed. At stations 5 and 10 mm from the disc (Fig. 5a,b) the same relative size of residual can be observed as in the U-momentum plots and at all other stations the residual terms are very small; the solution that has been obtained with the 55 × 50 grid has, therefore, passed the numerical error test also for the turbulence equations (it is unlikely that errors will occur in the ϵ equation that do not show up in the k balance plots). Downstream of $x = 40$ mm (Fig. 5d) the wake-like nature of the flow past the reattachment point can be clearly identified in the importance of the convection term which is even larger than the production term at the furthest downstream station (Fig. 5f) showing the 'weak-shear' characteristic of wake-flows. Only in the near-wall region can the appearance of a local-equilibrium ($P = \epsilon$) region be discerned, and even here at the furthest downstream station, turbulent transport is still important.

The calculated solution is now briefly compared with the experimental results of [13] so as to evaluate the performance of the k-ϵ model. The radial profile of axial velocity within the recirculation zone, Fig. 6a, shows that the calculated value of $\partial U/\partial r$ in the mixing layer is smaller than that measured, suggesting that the calculated diffusion of axial momentum from larger to smaller radii is too great. Far downstream of this region, Fig. 6b, shows that the calculations retain a marked velocity deficit on the centreline which implies that here the radial diffusion of axial momentum is too small in contrast to the result within the recirculation bubble. Figure 7 depicts the centreline development of axial velocity in which the maximum reversed velocity is too small by about one-tenth although the length of the recirculation bubble is too long by about 8%. The difference in the solutions given by the three different grids is also indicated in Fig. 7. The improvement in the recirculation length is particularly obvious between the 20 × 20 and 40 × 40 grid, while the change in the solution on the finest grid is relatively small and apparent only downstream of the recirculation bubble.

The agreement in the radial profiles of turbulent kinetic energy, shown in Fig. 8a, for a location within the recirculation zone is NOT typical. The graph has been included because it emphasizes the difference (about 10% of the profile maximum) in the turbulence kinetic energy solutions obtained with the 40 × 40 and 55 × 50 grids. The disa-

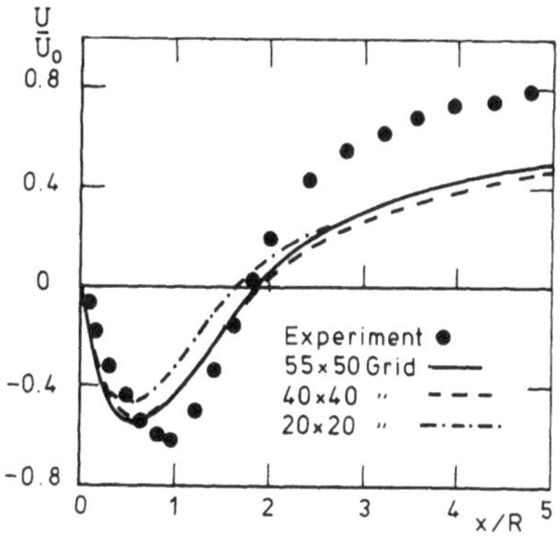

Fig. 7. Centreline development of axial velocity. Comparison between measurement and calculation

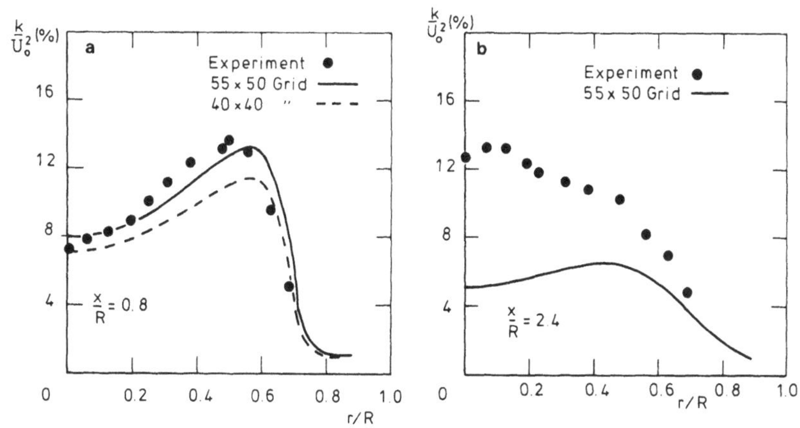

Fig. 8a,b. Radial profiles of turbulence kinetic energy. Comparison between measurement and calculation

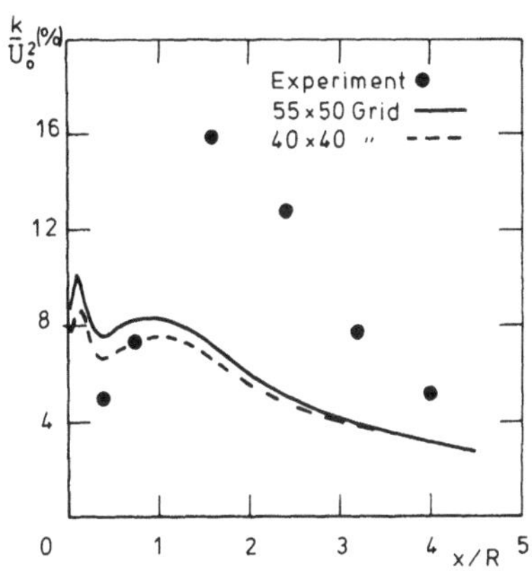

Fig. 9. Centreline development of turbulence kinetic energy. Comparison between measurement and calculation

greement further downstream, Fig. 8b, is much more typical and comparison shows that k is very much smaller than that measured. The centreline profile, Fig. 9, is similarly poor and shows that it is particularly the experimentally found peak in k, at the rear stagnation point, which is poorly represented in the calculations.

The calculations reported here represent an improvement over previously reported calculations of similar near-wake disc flows (e.g. [20, 21]) both because reliable experimental profiles for U and V were available for use as boundary conditions and also because numerical diffusion has in this case been demonstrated to be unimportant. It is noteworthy that the predicted recirculation length is *longer* than that measured, whereas it has generally been observed that the k-ϵ model gives rise to *short* lengths in a wide variety of flows. The principle shortcomings of this model's predictions are that, in the mean flow, the reform of the wake downstream of the recirculation zone is too slow and, in the turbulence field, the levels of turbulence kinetic energy are in error by up to 50%.

Some explanation for the relatively poor performance of the k-ϵ model can be found by examination of the terms in the transport equation for k, shown in Fig. 5 and already presented above. At the first two stations (x = 1.5 mm) the shear flow is far from being in local equilibrium and the change in the shape of the production and dissipation curves implies that it is also not self-preserving. It can, therefore, be expected that the eddy viscosity hypothesis is untenable in the initial region of the shear layer; this is because the success of this hypothesis is usually associated with flows which are at least approximately self-preserving. In addition, the shear layer undergoes substantial streamline curvature over its whole length and this is known (e.g. *Bradshaw*, [22]) to be a further restriction upon the applicability of the effective viscosity hypothesis (see also [21]). A further defect is that the calculated production of k near the centreline up to x = 46 mm is very small whereas the measurements indicate that generation through normal stress/normal strain is important and may be responsible for the large values of k in the vicinity of the free stagnation point, as shown in Fig. 9. The defect arises because the effective viscosity hypothesis is known not to represent normal stresses adequately. Finally, towards the end of the recirculation zone (x = 21 to 46 mm) the major terms in the transport equation are the turbulence production and dissipation but, as the flow passes downstream, these progressively change (x = 72–140 mm) to the characteristic balance of a far-wake, in which dissipation exceeds production and convection is an important term. *Launder* et al. [23] have demonstrated that in such flows an effective viscosity hypothesis can be used but that C_μ can no longer be assigned a constant value, independently of the local ratio of production to dissipation.

The important conclusion to be drawn is that only the use of a Reynolds stress level of closure offers any promise for the removal of the defects which have been discussed above. It is *essential*, however, to ensure that the steps taken in the calculations reported here to minimise numerical error be repeated in any calculations using a higher level of closure to make certain that changes in the turbulence model are not swamped by numerical diffusion.

Summary

The following is a summary of the main results and conclusions:

1. A technique has been described for identifying the location and magnitude of numerical diffusion errors in upwind difference calculations of turbulent recirculating flows.

2. A demonstration was given of the application of this technique to the flow over a confined axisymmetric disc and grid-independent solutions were obtained using a 55×50 grid.

3. The importance of using this technique to examine the occurrence of numerical errors also in the turbulence model equations was emphasised – this may have impli-

cations for the development of higher-order schemes suitable for such positive definite quantities.

4. The performance of the two-equation k-ϵ model in this flow was then critically assessed using experimental data and the model was found to exhibit several notable deficiencies. The recirculation length was predicted to be 8% too long, the maximum reverse velocity was 10% too weak and the recovery in the wake region significantly too slow. More importantly the development of the turbulence energy field was substantially in error, the maximum level of k being too small by about 50%.
5. It is recommended that all future calculations of turbulent recirculating flow which claim to assess turbulence models and which adopt first-order upwind difference approximations for the convective terms *must* demonstrate in some way, perhaps using the method described here, the absence of numerical diffusion.
6. It is also suggested that the technique demonstrated here be used to replace ad hoc grid refinement tests in calculations obtained with *any* higher-order finite-difference scheme.

References

1. Roache, P. J. *Computational Fluid Dynamics* (Hermosa, Albuquerque, NM 1976)
2. Raithby, G. D. A critical evaluation of upstream differencing applied to problems involving fluid flow. Comput. Methods Appl. Mech. Eng. **9**, 75–103 (1976)
3. Castro, I. P. "Numerical Difficulties in the Calculation of Complex Turbulent Flows", in *Turbulent Shear Flows I*, ed. by F. Durst, B. E. Launder, F. W. Schmidt, J. H. Whitelaw (Springer, Berlin, Heidelberg, New York 1979)
4. Leonard, B. P., Leschziner, M. A., McGuirk, J. *Third-order Finite-Difference Method for Steady Two-Dimensional Convection*, Proc. 1st Int. Conf. on Numerical Methods in Laminar and Turbulent Flow, ed. by C. Taylor (Pentech, London 1978)
5. Cheng, S. I., Shubin, G. Computational accuracy and mesh Reynolds number. J. Comput. Phys. **28** 315–326 (1978)
6. Stubley, G. D., Raithby, G. D., Strong, A. B. Proposal for a new discrete method based on an assessment of discretization errors. Numerical Heat Transfer **3**, 411–428 (1980)
7. Gresho, P.M., Lee, R.L. "Don't Suppress the Wiggles – They're Telling you Something", in *Finite Element Methods for Convection Dominated Flows,* ed. by T.J. Hughes (ASME Applied Mechs. Dir. Publication AMD34, 1979)
8. Forester, C. K. Higher order monotonic convective difference schemes. J. Comput. Phys. **23**, 1–22 (1977)
9. Leonard, B. P. A stable and accurate convective modelling procedure based on quadratic upstream interpolation. Comp. Methods Appl. Mech. Eng. **19**, 59–98 (1979)
10. Chan, R. K-C. A balanced expansion technique for constructing accurate finite-difference advection schemes. Int. J. Numer. Methods Eng. **12**, 1131–1150 (1978)
11. McGuirk, J. J., Rodi, W. A depth-averaged mathematical model for the near-field of side-discharges into open-channel flow. J. Fluid Mech. **86**, 761–781 (1978)
12. Krause, E., Kordulla, W. "An Error Analysis of Finite-Difference Approximations for the Boundary-Layer Equations", in *Numerical Methods in Applied Fluid Dynamics,* ed. by B. Hunt (Academic, London 1980)
13. Taylor, A. M. K. P. "Confined, Isothermal and Combusting flows Behind Axisymmetric Baffles"; Ph. D. Thesis, University of London (1981)
14. Jones, W.P., Launder, B.E. The prediction of laminarisation with a two-equation model of turbulence. Int. J. Heat Mass Transfer **15**, 301–314 (1972)
15. Launder, B. E., Spalding, D. B. The numerical computation of turbulent flows. Comput. Methods Appl. Mech. Eng. **3**, 269 (1974)
16. Gosman, A. D., Pun, W. M. "Lecture Notes for Course Entitled – Calculation of Recirculating Flows", Heat Transfer section report HTS/74/2, Dept. of Mech. Eng., Imperial College, London (1973)
17. Patankar, S. V., Spalding, D. B. A calculation procedure for heat, mass and momentum transfer in three-dimensional parabolic flows. Int. J. Heat Mass Transfer **15**, 1787 (1972)

18. Spalding, D. B. A novel finite-difference formulation for differential expressions involving both first and second derivatives. Int. J. Numer. Methods Eng. **4**, 551 (1972)
19. Castro, I. P., Cliffe, K. A., Norgett, M. J. "Numerical Predictions of the Laminar Flow Over a Normal Flat Plate", Theoretical Physics Division Report TP870, AERE, Harwell, England (1981)
20. Pope, S. B., Whitelaw, J. H. The calculation of near-wake flows, J. Fluid Mech., **73**, 1 (1976)
21. Leschziner, M. A., Rodi, W. "Calculation of Annular and Twin Parallel Jets Using Various Descretization Schemes and Turbulence Model Variants", Report SFB80/T/159, Sonderforschungsbereich 80, University of Karlsruhe (1980)
22. Bradshaw, P. "Effects of streamline curvature on turbulent flow", AGARDograph 169 (1973)
23. Launder, B. E., Morse, A. P., Rodi, W., Spalding, D. B.: The prediction of free shear flows – a comparison of six turbulence models. NASA Spec. Pub. **311** (1972)

Turbulent and Mean Flow Measurements in an Incompressible Axisymmetric Boundary Layer with Incipient Separation

P. Dengel, H.H. Fernholz and J.-D. Vagt

Institut für Thermo- und Fluiddynamik, Technische Universität Berlin, D-100 Berlin 10, Fed. Rep. of Germany

Abstract

Measurements have been performed of mean and fluctuating velocities as well as of Reynolds and wall shear stress in an incompressible axisymmetric turbulent boundary layer with a severe adverse pressure gradient. In the flow region where separation was incipient and, further downstream, where reverse flow occurred, only the x-components $\bar{u}$ and u' and the skin friction were measured. The measurements were carried out by using conventional hot wires, pulsed wires and surface fences. A comparison of hot-wire and pulsed-wire measurements showed good agreement also at high local turbulence levels as long as the flow did not reverse; large discrepancies were found, however, close to those positions in the flow where the flow velocity changed sign and in the separated region of the boundary layer.

1. Introduction

Boundary layers with adverse pressure gradients have always been of special interest in fluid flow research since they usually entail flow separation from the wall. *Prandtl* (1935) was apparently the first to conceive the idea that a maximum pressure recovery in a flow is obtained if a sharp increase in pressure is followed by a streamwise pressure distribution allowing the boundary layer to remain on the verge of separation. Such pressure distributions are of great importance in duct flow and aerofoil design.

At least two attempts (*Stratford* 1959, *Fernholz* and *Gibson* 1967) were made in the past to generate a boundary layer with incipient separation and to keep it on the verge of separation over a distance of several boundary layer thicknesses downstream. Mean velocity profiles were measured by means of Pitot tubes and skin friction by means of surface fences. The axisymmetric test section used in the latter investigation was especially well suited for setting up a stable boundary layer on the verge of separation, i.e. one where the separating velocity profiles remained fixed in space and where the mean flow was steady. A similar test section was therefore built in Berlin to perform the experiment described below. It has been the aim of the present investigation to complement the earlier mean flow data by fluctuating flow measurements and to extend measurements into the separation region itself.

So far we have measured mean velocities and the component $\overline{u'^2}$ of the Reynolds normal stress in the separation region. Due to the strong growth of the boundary layer in the vicinity of separation, the boundary layer almost filled the annulus between the inner test wall and the outer perforated cylinder from about 200 mm downstream of separation.

There are few measuring techniques which can be used to perform measurements in boundary layers where the turbulence level is high, say larger than 30%, and where the mean velocity vector changes sign, with the specifically critical region where the mean velocity goes through zero and where the fluctuating velocity is non-zero. In such a case the often used definition of the turbulence level $(\overline{u'^2})^{1/2}/\bar{u}$ becomes meaningless because it tends to infinity.

So far two experiments, both under aerofoil pressure distributions, have been published where the turbulence properties of an adverse pressure gradient boundary layer and its downstream separation region were investigated. *Coles* and *Wadcock* (1979) and *Wadcock* (1980) used a flying hot wire to perform turbulence and mean flow measurements in the boundary layer and in the wake of an aerofoil. No skin-friction measurements were performed. *Simpson* et al. (1980) and *Shiloh* et al. (1980) carried out experiments in an adverse pressure gradient boundary layer with conventional hot-wire anemometry and in the separation region with a directionally-sensitive laser-anemometer system. In the present investigation conventional hot wires were used in the attached flow region and pulsed wires (*Bradbury* and *Castro* 1971) in the reverse flow region. The boundary layer described below, differs from the two cases above in that it had a different upstream history (*Fernholz* and *Vagt* 1975, 1981), a smaller magnitude of the reverse flow velocity and that the separation region was stable. Furthermore the boundary layer was axisymmetric, thus avoiding three-dimensional effects as they become apparent, for example, by the $\overline{u'w'}$ component (*Shiloh* et al. 1980) of the Reynolds stress tensor.

Section 2 of this paper presents an account of the experimental arrangement and the flow regime, while Sect. 3 provides a discussion of the measurements and of some preliminary conclusions. In Sect. 4 some pecularities of the hot-wire and pulsed-wire techniques are discussed as they are relevant to this experiment.

2. Experimental Arrangement and Flow Regime

The wind tunnel used in this investigation (Fig. 1) is a low-speed blower tunnel with a 12 kW motor and centrifugal fan, an airfilter intake and a 2 m long settling chamber with two wire gauzes and a slab of foam rubber, followed by an 11 : 1 axisymmetric contraction. The exit velocity distribution was uniform with local mean velocities $\bar{u}$ in the core of the flow at the nozzle exit varying at most by ± 1.5% and had a turbulence intensity $[(\overline{u'^2})^{1/2}/\bar{u}]$ of 0.10% in a frequency range up to 10^4 Hz.

All measurements were performed at a test section inlet velocity of about 16 m/s with a fixed Reynolds number $(\bar{u}_\infty/\nu)_{ref} = 1.0081 \times 10^6\ m^{-1}$. The flow temperature was held constant at 21 ± 0.3°C. The test section consisted of a sting-mounted horizontal inner circular cylinder (0.25 m diameter, 155 m long and made of Ultramid S) with an elliptical nose cone 0.30 m long and a perforated outer cylinder (38% open area ratio) of 0.60 m diameter.

All probes, mounted on an electrically driven traverse gear, were introduced into the test section through a slot along a generator of the outer cylinder wall. The traverse gear provided precise linear (incremental resolution 0.005 mm) resolution.

Surface fences, protruding approximately 0.10 mm from the surface of the inner cylinder (*Vagt* and *Fernholz* 1973), alternated with static pressure tappings (0.8 mm diameter) along one generator of the inner cylinder. By turning the inner and/or the

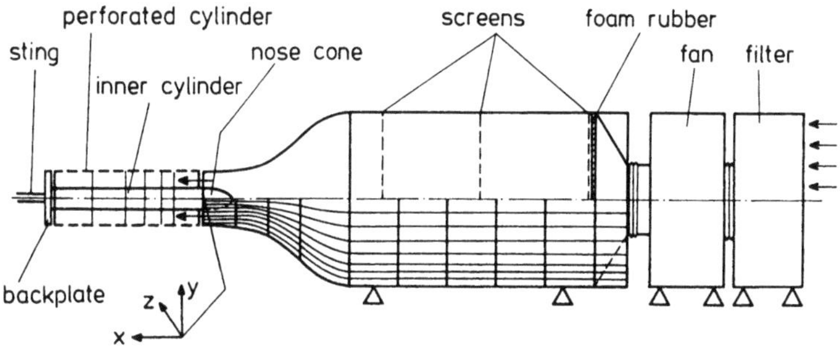

Fig. 1. Wind tunnel and test section

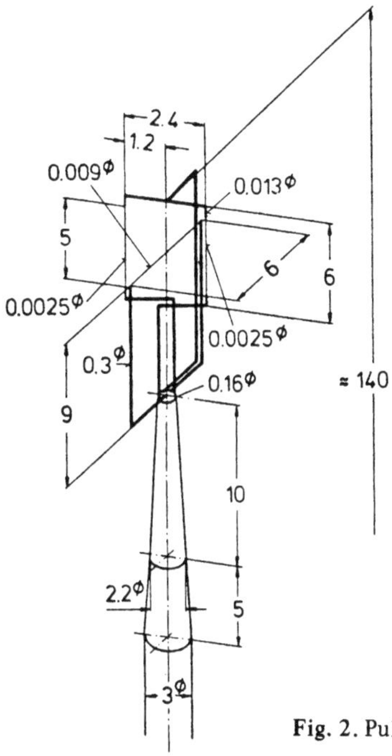

Fig. 2. Pulsed-wire probe

outer cylinder, measurements could be made with the wall probes at fixed positions x along the circumference and with the other probes at any position in the flow field covered by a circumferential angle of about 60°. Mean and fluctuating velocities were measured using probes with single normal and X-wires in the boundary layer upstream of separation (*Vagt* and *Fernholz* 1979) and by means of a pulsed-wire probe in the separation region. The hot-wire probes were operated by DFVLR CTA units and a pulsed-wire anemometer manufactured at the University of Surrey was used for the pulsed-wire measurements.

The pulsed-wire probe was especially developed and tested for this investigation. It was built along the lines of the original Bradbury probe (*Bradbury* and *Castro* 1971) but smaller in size (Fig. 2) in order to get closer to the wall (minimum wall distance is now 3.5 mm). The sensor wires were soldered to the prongs in the same way as the wires of our standard hot-wire probes (*Dahm* and *Vagt* 1977). The distance x between sensor and pulsed wire is 1.2 mm and the length l of the sensor wires 5 mm. With this geom-

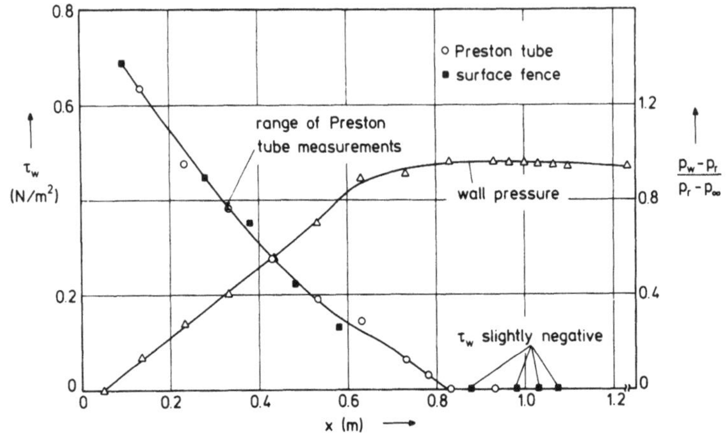

Fig. 3. Skin friction and wall pressure distributions in an adverse pressure gradient turbulent boundary layer

etry the probe has a yaw response of ± 65°. Since the angle between the flow and the probe stem is 90° the stem was conical in the vicinity of the prongs in order to minimize aerodynamic interference effects. For the same reason the wires are 9 mm away from the tip of the probe. The velocity range of this pulsed-wire probe lies between 0.2 m/s and 11 m/s.

Skin friction was measured by means of Preston tubes (0.50 mm and 0.89 mm outer diameter) and by surface fences. The data are shown in Fig. 3.

Measurements from the surface fences agreed well with Preston-tube results over most of the boundary layer, with a repeatability of about 2% for both measuring devices. A check of the skin friction distribution along the circumference for 87.2 mm on both sides of the x-axis showed variations in skin friction of ± 2%. Surface fences have an advantage over Preston tubes in or close to the separation region in that they give the sign and the magnitude of the skin-friction vector. In our case the reverse flow was so weak, however, that the pressure difference across the fence was often smaller than the accuracy of the micromanometer (about 0.005 mm water) and that it was therefore impossible to discern whether the signal was positive or negative. A decision could be made, once the velocity profile measured by means of the pulsed wire was available. So the sign of skin friction was denoted as being negative if the pulsed wire showed reverse flow.

The uncertainty of the skin-friction measurements is ± 5% except close to separation where the absolute values of the pressure become too small.

The measured skin friction distribution was compared with values obtained from a "modified" Ludwig-Tillmann formula (*Fernholz* 1964) which holds also for flows with a form parameter H_{12} larger than 1.80:

$$c_f = 0.058\,\gamma^{1.705} \cdot \mathrm{Re}_{\delta_2}^{-0.268} \tag{1}$$

with

$$\gamma = \log_{10}(8.05/H_{12}^{1.818}), \quad \mathrm{Re}_{\delta_2} = \rho_\delta u_\delta \delta_2/\mu_w \quad \text{and} \quad H_{12} = \delta_1/\delta_2 \; .$$

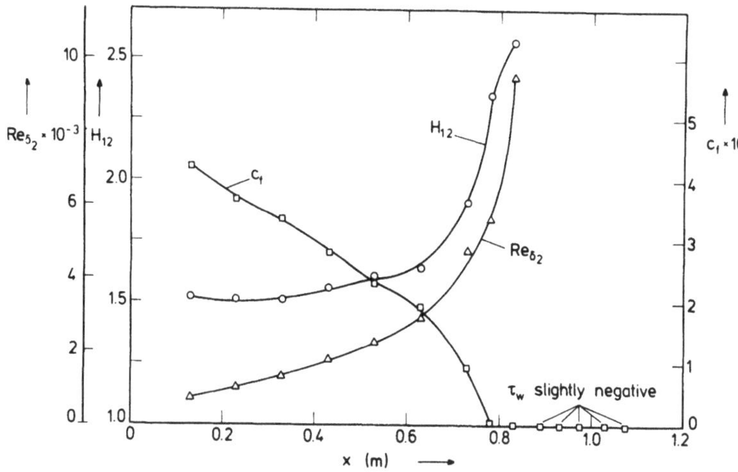

Fig. 4. Streamwise development of skin friction coefficient c_f, shape parameter H_{12} and Reynolds number $Re_{\delta 2}$

The indices δ and w denote the edge of the boundary layer and the wall, respectively. The comparison between c_f from Eq. (1) and from measurements shows that the calculated values agree fairly well with the measured ones.

The pressure distribution which generated the boundary layer on the verge of separation was set up by adjusting the distance between the back plate and the annulus (Fig. 1) and by putting plastic bands of different widths around the circumference of the outer perforated cylinder at different x-positions in order to block the outflow of air until several consecutive surface fences showed nominally zero skin friction.

In Fig. 4 we present the development of the form parameter H_{12}, the Reynolds number $Re_{\delta 2}$ and the skin friction coefficient c_f. Form parameter and Reynolds number show the steep increase towards separation which is characteristic of an adverse pressure gradient boundary layer. Both quantities are, however, not defined in flows where the velocity changes sign which leaves the coefficient of the static pressure (Fig.3) as the only relevant parameter in the separation region. Since the usual scaling quantities u_τ (skin friction velocity) and (ν/u_τ) are not defined either in a boundary layer close to separation, we suggest to replace them by using the maximum shear-stress velocity

$$\overline{u}_s = [(|\overline{u'v'}|)_{max}]^{1/2} \tag{2}$$

and the scaling length $(\nu/\overline{u}_s)$.

3. Discussion of Velocity Measurements and Conclusions

In this section we shall discuss measurements of mean velocity profiles and of Reynolds stresses in an adverse pressure gradient boundary layer and its downstream separation region. These measurements were performed with single-wire probes up to $x = 0.731$ m and with pulsed-wire probes from $x = 0.831$ m downstream. The distributions of mean and fluctuating velocities at $x = 0.781$ m were measured with both probes.

Figure 5 shows the development of the mean velocity across the boundary layer plotted against the wall distance y in a semi-logarithmic manner in order to emphasize

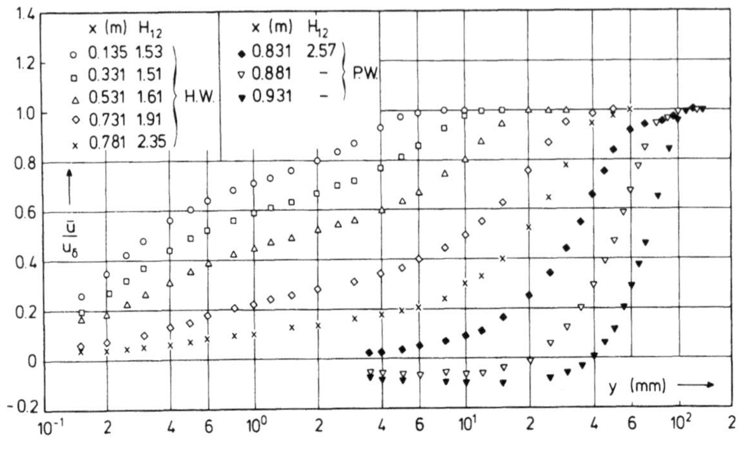

Fig. 5. Development of mean velocity profiles in an adverse pressure gradient turbulent boundary layer with separation

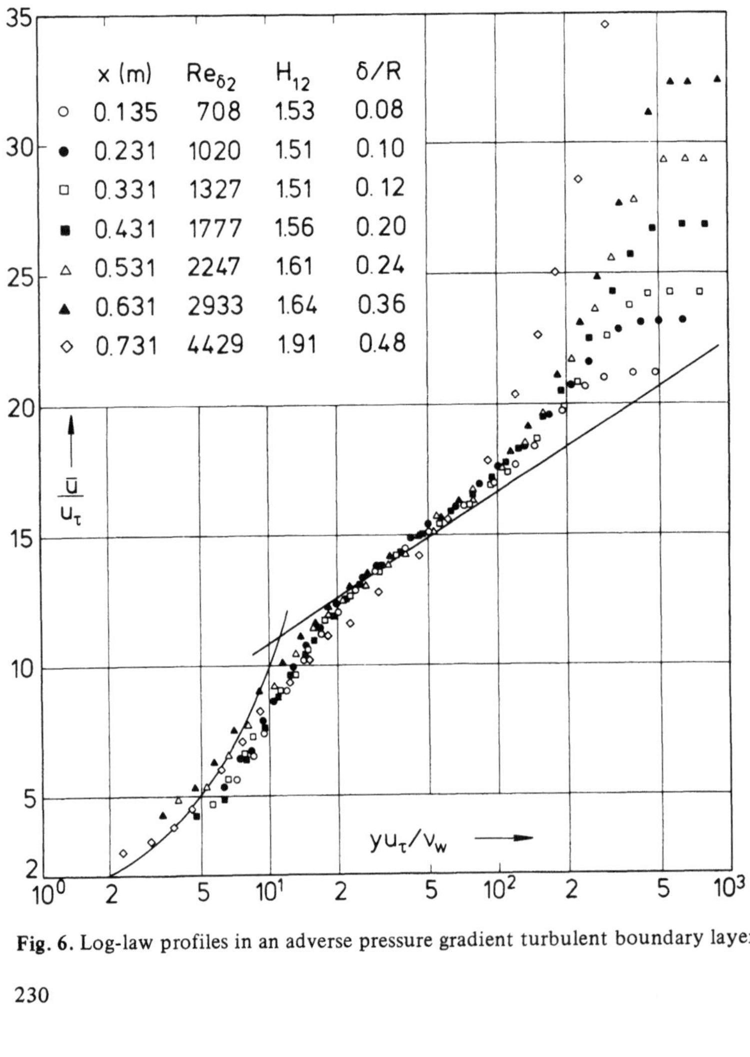

Fig. 6. Log-law profiles in an adverse pressure gradient turbulent boundary layer

the near-wall region. The profiles show the usual adjustment to the adverse pressure with the decreasing slope at the wall. As was mentioned before it has been rather difficult to determine where skin friction became zero. Judging from the mean velocity profiles and the measured skin friction distribution, the separation profile must be close to $x = 0.831$ m. Velocity profiles further downstream show distinct regions of reverse flow over the lower third of the boundary layer thickness although the surface-fence signal was nominally zero.

In Fig. 6 all profiles measured upstream of separation are presented in the usual log-law co-ordinates, showing good agreement with the logarithmic law ($\kappa = 0.40$ and $C = 5.10$) except for the profile at $x = 0.731$ m which has a steeper slope. Agreement between measurements and the linear law of the viscous sublayer is also satisfactory. The slight scatter of the data points is due to small errors (about 3/100 mm) in determining the distance between the hot wire and the wall. In the legend of Fig. 6 we have given the ratio of the boundary layer thickness and the radius of the cylinder in order to show that effects due to transverse curvature are negligibly small. It has been surprising, however, that the pressure gradient hardly affects the length of the range where the measurements agree with the logarithmic law, except for the last profile at $x = 0.731$ m which has a slope different from that of the profiles further upstream.

Figure 7 shows the distribution of the Reynolds normal stress $\overline{u'^2}$ across the boundary layer at different x positions. $\overline{u'^2}$ is non-dimensionalized by the freestream velocity at $x = 0.135$ m in order to show the development in an adverse pressure gradient. The distribution of the Reynolds normal stress has a maximum in the near-wall region ($x = 0.135$ m) which decreases with increasing static pressure, while a second maximum occurs in the outer region of the boundary layer ($x = 0.531$ m). The inner maximum then disappears ($x = 0.731$ m) whereas the level of the outer maximum increases by a factor 4 for the profile at $x = 0.931$ m. So far we cannot explain this shift of $\overline{u'^2}$ and must wait for measurements of $\overline{w'^2}$ and $\overline{v'^2}$ to see whether these components give further insight into the turbulence structure.

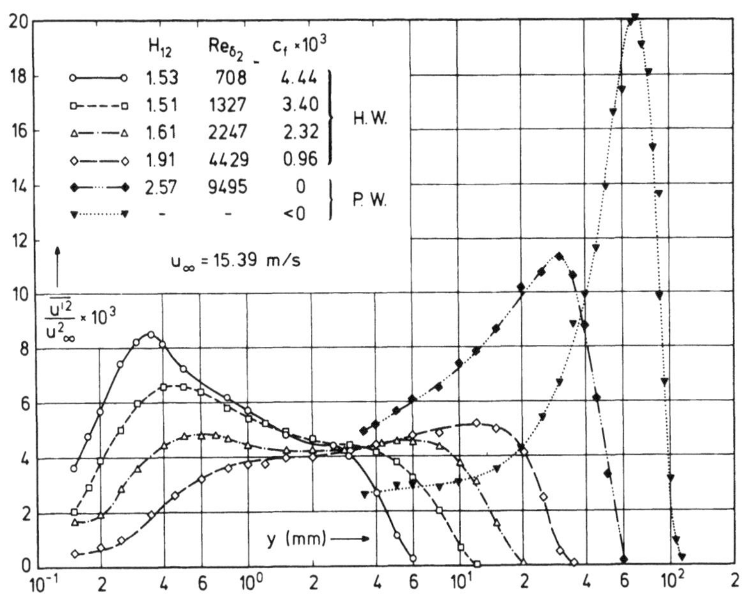

Fig. 7. Development of Reynolds normal stress in an adverse pressure gradient turbulent boundary layer

All measurements are uncorrected although corrections both for higher and lower values could be applied for a number of good reasons. These corrections would affect the velocity profiles in different ways at different locations of x and y. It is conceivable, for example, that the mean velocity is slightly high for y^+ larger than about 40 (Fig. 6) and that the measurements would then follow the log-law along a longer range the length of which would perhaps be determined by the increase in static pressure. At the present stage of our knowledge, both of the hot-wire and the pulsed-wire technique, we estimate the following uncertainty:

Quantity	Pulsed wire	Hot-wire
$\bar{u}$	about ± 10%;	± 2% (without reverse flow)
$\overline{u'^2}$	about ± 10%;	± 10% (without reverse flow)

Here we mean the value between the "true" value and the measured deviation. For all pulsed-wire measurements a sample rate of 4000 was used.

The distributions of Reynolds shear stress at four positions of x upstream of separation are presented in Fig. 8. Unfortunately, the boundary layer is only twice as thick as the x-wire probe at position $x = 0.135$ m where the first profile was taken so that the maximum of $|\overline{u'v'}|$ could not be determined. With increasing pressure the maximum of the Reynolds shear stress moves towards the outer region of the boundary layer. Reynolds shear stress profiles could not yet be measured in the reverse-flow region since a pulsed-wire probe for this purpose is still being developed.

At $x = 0.731$ m the measured shear stress data had to be corrected for misalignment of the probe after an upwash angle of about 4° had been measured by means of a slanted-wire probe. The corrected data turned out to be 35% higher near the maximum and are given as the broken line in Fig. 8. Taking into account a slight misalignment, say ± 1°, of the probe and a high level of turbulence the uncertainty of the Reynolds shear stress measurements is about ± 15% in the boundary layer upstream of separation with a repeatability of ± 3%.

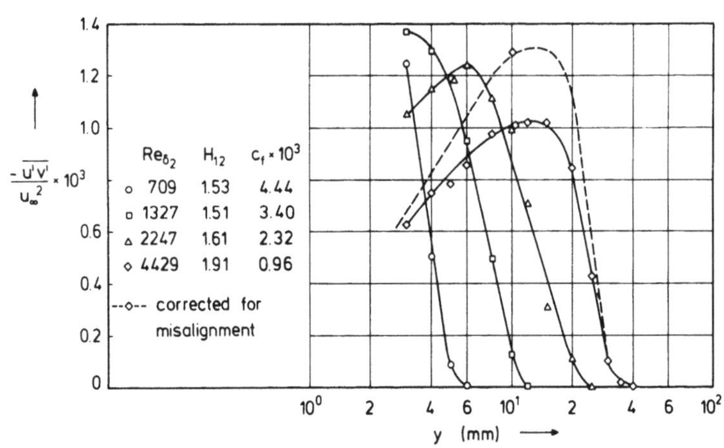

Fig. 8. Development of Reynolds shear stress in an adverse pressure gradient turbulent boundary layer

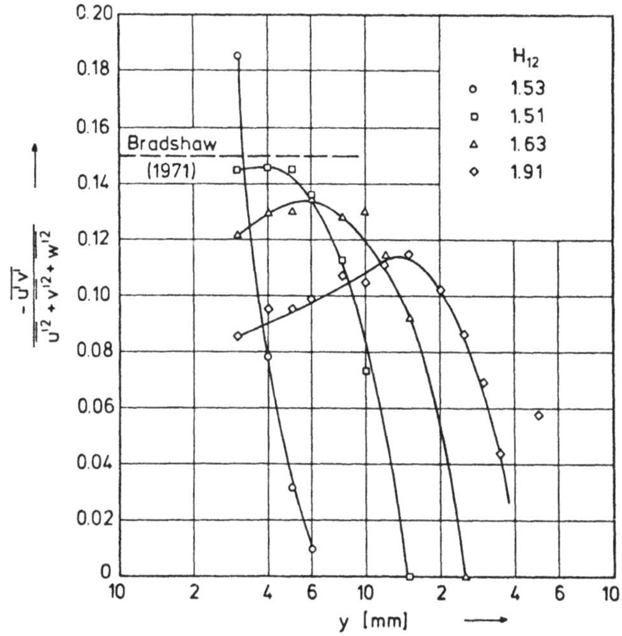

Fig. 9. Distribution of the ratio of Reynolds shear stress and kinetic energy of the fluctuating motion in an adverse pressure gradient boundary layer

In addition to $\overline{u'^2}$ we have also measured the Reynolds normal stress components $\overline{v'^2}$ and $\overline{w'^2}$ in the boundary layer upstream of separation in order to show the ratio (Fig. 9) of the Reynolds shear stress $\bar{\rho}\,\overline{u'v'}$ and the kinetic energy of the fluctuating motion $(\overline{u'^2}+\overline{v'^2}+\overline{w'^2})$. Although these measurements could only be performed in the outer region of the boundary layer, the distribution of this ratio shows quite distinctly that it is not constant in an adverse pressure gradient boundary layer as assumed for example by *Bradshaw* (1971).

The pulsed-wire technique offers the possibility of determining which percentage of the velocity samples has a positive and a negative sign. We have therefore defined a „back flow" parameter $\gamma = (S_t - S_n)/S_t$ where S is the number of samples and where the indices t and n denote total and negative, respectively. This parameter is one if there

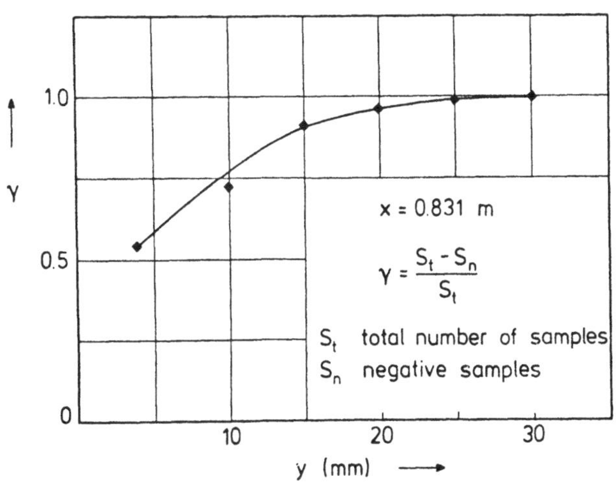

Fig. 10. Back-flow parameter γ

is no reverse flow and 0.5 if half the number of samples is negative. As an example, we show the distribution of γ for the velocity profile at $x = 0.831$ m where the skin friction is assumed to be zero (Fig. 10) (see also *Simpson* et al. 1980).

4. Some Pecularities of the Measurements

Vagt and *Fernholz* (1979) have discussed extensively aerodynamic interference effects which may occur if conventional boundary layer probes are used in the vicinity of a wall. During the present investigation which was performed in an annulus where most of the surface of the outer cylinder was impermeable, we found that the stem (3 mm diameter) of a hot-wire probe changed the pressure distribution so that a mean velocity was measured which was too high by as much as 20% in the viscous sublayer and by about 8% in the region of the logarithmic law. In Fig. 11 we have plotted measurements taken by means of a flattened Pitot probe and by a hot-wire probe the stem of which was normal and inclined by 45° towards the wall, respectively. Agreement between the inclined hot-wire probe and the Pitot tube was good in the log-law and in the outer region whereas the hot-wire probe with the stem normal to the wall read consistently high except towards the edge of the boundary layer. Such differences were not observed in the same test section if the pressure could adjust itself more easily due to the perforated outer cylinder (*Vagt* and *Fernholz* 1979, Fig. 6). Pitot tube and inclined hot-wire probe agreed well with each other, except in the viscous sublayer, because no negative influence of the stem occurred.

Near the separation region the velocity vector in the boundary layer is directed relatively strongly away from the wall. "Upwash angles" of up to 16° were found at position $x = 0.831$ m which means that the boundary layer assumption $\overline{v}/\overline{u} \ll 1$ may be violated.

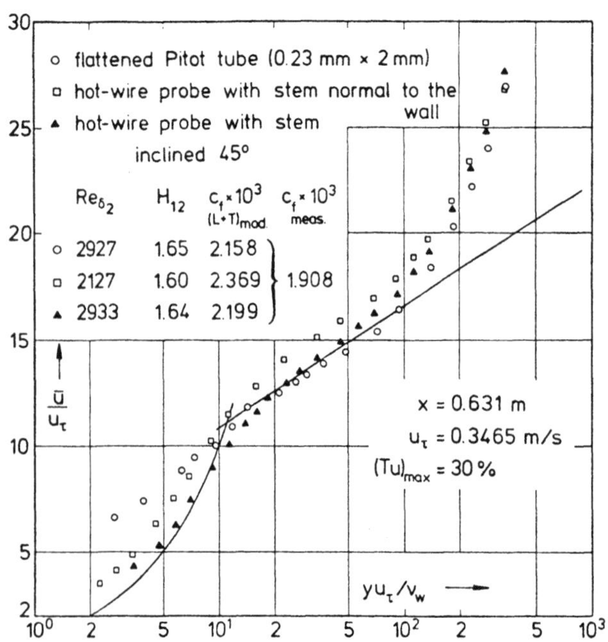

Fig. 11. Comparison of hot-wire and flattened Pitot probe measurements in an adverse pressure gradient boundary layer

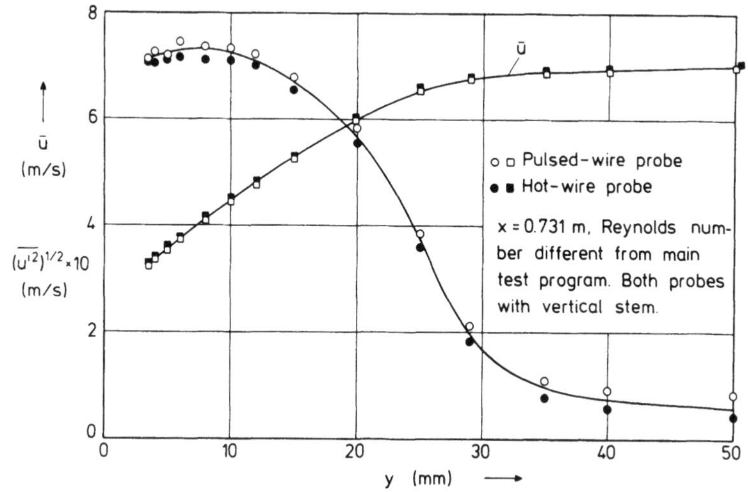

Fig. 12. Comparison between pulsed-wire and hot-wire measurements in an adverse pressure gradient turbulent boundary layer with no reverse flow

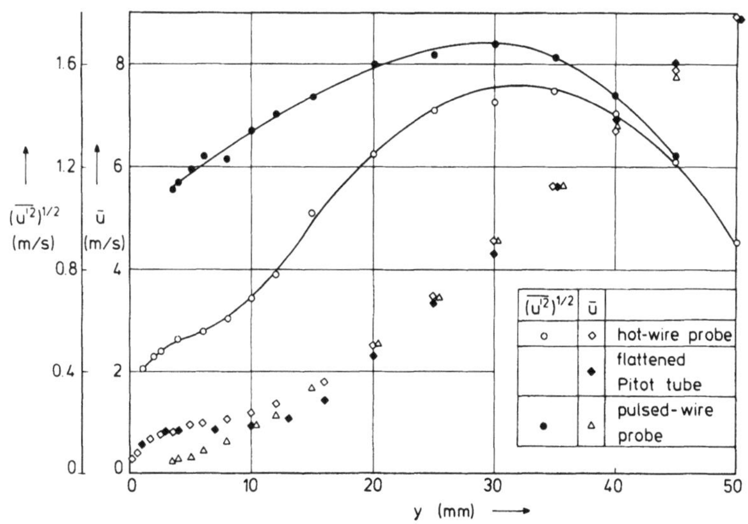

Fig. 13. Comparison between pulsed-wire and hot-wire measurements in an adverse pressure gradient turbulent boundary layer

In Fig. 12 and 13 pulsed wire and hot-wire measurements are compared, first at an x-station without reverse flow and then at a position where instantaneous reverse flow occured although the mean velocity was either still positive or zero. At $x = 0.731$ m (no reverse flow) agreement between the pulsed-wire and hot-wire data is good, both for the mean and the fluctuating velocities which are shown here in dimensionalized form plotted against the distance y normal to the wall (Fig. 12). Differences do occur, however, when the turbulence level is small (about 8%). This is due to pecularities of the pulsed-wire technique.

At $x = 0.831$ m the back flow parameter has values smaller than one (Fig. 10), i.e. there is instantaneous reverse flow in the wall region. Since a hot wire cannot distinguish between positive and negative velocities, its integrated signal must be too high compared with the pulsed-wire signal. The agreement between the flattened Pitot tube data and those of the hot wire is purely coincidental (Fig. 13) in this case. There is, however, a large discrepancy between the turbulence signals obtained from the hot wire and the pulsed wire and this discrepancy increases as the wall is approached. This is due to the probability distribution seen by the respective system. The hot wire sees the negative events which increase in number closer to the wall as positive and adds them up. The hot-wire signal is then too high.

A detailed description of the changes in the probability distributions between free jets and boundary layers and the ensuing differences in signal interpretation wil be given in a paper by *Vagt* and *Dengel.*

References

1 Bradbury, L. J. S., Castro, I. P.: A pulsed-wire technique for velocity measurements in highly turbulent flows. J. Fluid Mech. **49**, 657–691 (1971)

2 Bradshaw, P.: Calculation of three-dimensional turbulent boundary layers. J. Fluid Mech. **46**, 417–445 (1971)

3 Coles, D., Wadcock, A. J.: Flying hot-wire study of two-dimensional turbulent separation on a NACA 4412 airfoil at maximum lift. AIAA J. **17**, 321–329 (1979)

4 Dahm, A., Vagt, J.-D.: Entwicklung und Herstellung interferenzarmer Hitzdrahtsonden. HFI Inst. Ber. 01/77 (1977)

5 Fernholz, H. H.: Halbempirische Gesetze zur Berechnung turbulenter Grenzschichten nach der Methode der Integralbedingungen. Ing. Arch. **33**, 384–395

6 Fernholz, H. H., Gibson, P.: Experiments on Turbulent Boundary Layers Along a Circular Cylinder With and Without Separation. Gas Turbine Lab., M.I.T. Rep. No. 91 (1967)

7 Fernholz, H. H., Vagt, J.-D.: Measurements in an Axisymmetric Turbulent Boundary Layer With Weak and Strong Three-Dimensional Disturbances, in: *Structure and Mechanisms of Turbulence I,* ed. by H. Fiedler, Lecture Notes in Physics, Vol. 75 (Springer, Berlin, Heidelberg, New York, 1975) pp. 222–233

8 Fernholz, H. H., Vagt, J.-D.: Turbulence measurements in an adverse-pressure gradient three-dimensional turbulent boundary layer along a circular cylinder. J. Fluid Mech. **111**, 233–269 (1981)

9 Prandtl, L.: In: *Aerodynamic Theory*, ed. by Durand, Vol. 3 (Springer, Berlin 1935)

10 Shiloh, K., Shivaprasad, B. G., Simpson, R. L.: Measurements of the Transverse Velocity of a Separating Turbulent Boundary Layer, Project SQUID. Tech. Rep. SMU-5-PU (1980)

11 Simpson, R. L., Chew Y.-T., Shivaprasad, B. G.: Measurements of a Separating Turbulent Boundary Layer, Project SQUID. Tech. Rep. SMU-4-PU (1980)

12 Stratford, B. S.: An experimental flow with zero skin friction throughout its region of pressure rise. J. Fluid Mech. **5**, 17–35 (1959)

13 Vagt, J-D., Fernholz H. H.: The surface fence as a means of measuring the magnitude and the direction of wall shear stress in boundary layers. Aeronaut. Q. **XXIV**, 87–91 (1973)

14 Vagt, J.-D., Fernholz, H. H.: A Discussion of Probe Effects and Improved Measuring Techniques in the near-wall Region of an Incompressible Three-Dimensional Turbulent Boundary Layer. AGARD Conf. Proc. 271 (1979) pp. 10.1–10.17

15 Wadcock, A. J.: Simple Turbulence Models and Their Application to Boundary Layer Separation. NASA CR **3283** (1980)

Part IV

Fundamentals

Introductory Remarks

J. L. Lumley
Cornell University, Ithaca, NY, 14853 USA

The first paper in this section, *Pressure Effects on Triple Correlations in Turbulent Convective Flows*, by André, Lacarrère, and Traoré, considers the rate equations for velocity/scalar triple correlations in a purely convective turbulent flow. In particular, following a general discussion of the effect of pressure on the triple correlations, the triple correlations involving the pressure gradient are broken into a combination of rapid and relaxation terms (expressed as suitable groups of triple correlations) similar to the combinations that are customary in the second order equations, with unknown constants. Only situations in which transport is negligible are considered. The equations are solved to compare with laboratory data on unsteady thermal turbulent convection, the unknown constants being evaluated by optimization. Finally, sensitivity studies are carried out on the constants.

This is an interesting and useful study, although the authors give the misleading impression that there is a fundamental difference between the ideas proposed here, and those developed in *Zeman* and *Lumley* (1976). In fact, the modeling of the various terms in these two papers is the same in principle (and very similar in detail), the only substantive difference in the two approaches being that in the latter case, the time derivatives are neglected, and the equations solved algebraically, while here the equations are solved as rate equations. In the model of *Zeman* and *Lumley* (1975), the generalization of the Boussinesq eddy diffusivity assumption is an interesting result of these more sophisticated modeling assumptions, and not an independent assumption. Finally, the authors suggest that the model proposed here in some sense includes non-linear effects that are excluded in the model of *Zeman* and *Lumley*; this is not the case – the models are virtually identical, and both are linear.

It seems to be a moot point whether it is easier to use the rate equations for the triple correlations, or to neglect the time derivatives (Markovianize) and solve for the triple correlations algebraically. In flow situations such as the one modeled here, the time derivatives are in any event small; the principal advantage is the avoiding of the necessity to invert a complicated matrix. Earlier workers (*Deardorff*, 1978) have reported instability of the rate equation for triple moments in the absence of diffusive terms, inclusion of which which would considerably complicate matters.

Although the authors refer to *Zeman* and *Lumley* (1976), they do not note that an approximate inclusion of the rapid terms in the expressions for the third moments was carried out there (a more accurate inclusion of the rapid terms in the equations for the third moments was presented in *Lumley*, 1975). The alternative derivation of *Lumley* (1978), which the authors cite, produces terms similar to the authors' and has the advantage that no additional constants are introduced. It would be interesting to compare the values of the constants obtained here with those derived in these papers.

The transport forms derived by *Zeman* and *Lumley* (1976) have been very successful in modeling the buoyancy-driven mixed layer of the atmosphere. The transport forms derived by *Lumley* (1978) have given very satisfactory results in isothermal flows (*Janicka* and *Lumley*, 1981 a,b; *Taulbee* and *Lumley*, 1981 a,b; *Sief*, 1981) and have the property of resolving the plane-round jet spreading-rate anomaly (*Pope*, 1977).

The second paper, *A Model of Three-dimensional Transfer in Non-isotropic Homogeneous Turbulence,* by Bertoglio, closes the equation for the spectral tensor by developing a model for the spectral transfer reminiscent of the models that are used in real space, but of course more complex. The model is calibrated by comparison with calculations of the transfer essentially using the *E*ddy-*D*amped *Q*uasi-*N*ormal *M*arkovian hypothesis. The only constant appearing in the equation is the usual E.D.Q.N.M. constant, determined in isotropic turbulence. The resulting model is then applied to the calculation of uniform shear and pure plane strain, giving excellent results. There is a difficulty, of course: the method requires the initial value of the entire spectral density tensor, which is never measured. The author has been able here to assume that the flows initially were isotropic, and has either had available measurements of the spectrum, or has used values obtained in other experiments. In more complex situations, however, this would present a real difficulty; in work reported elsewhere (*Cambon* et al. 1981) this group has devised clever ways to overcome this problem in part.

We anticipate that methods of this kind will be useful in situations in which the turbulence is very far from equilibrium, in which it cannot be assumed to have a spectrum which can be parameterized by the scales of the energy-containing range, an assumption that is central to ordinary second-order modeling. A notable exception is the work of *Hanjalic* et al. (1979), in which the spectrum is broken into two parts, with transfer between them proceeding at a different rate from that into the low-wavenumber end of the spectrum. The present method may be considered as the logical extension of that, when trransfer at all wavenumbers proceeds at its own rate. An example of a situation in which a complex method of this type might be useful, is the prediction of the behavior of the wing boundary layer under a leading-edge slat, in which the boundary layer is affected by the pressure field of the periodic wake shed by the slat.

The third paper, *A Theoretical Study of Radiative Cooling in Homogeneous and Isotropic Turbulence*, by Schertzer and Simonin, addresses a problem of considerable importance in the marine atmospheric surface layer, and in the nocturnal planetary boundary layer, where the radiative destruction rate of temperature variance can be a substantial fraction of the molecular one. Despite the general recognition of its importance, the influence of radiation is not routinely taken into account in modeling the planetary boundary layer (although a few efforts have been made), essentially because no simple parameterizations exist. In this paper, the authors define a radiative Prandtl number; at a critical value of this, the advective and radiative time scales are equal. Above this critical value, radiative cooling has only a slight effect on the spectrum. Below this critical value, an inertial-radiative subrange appears, and effects are strong, the slope of the spectrum changing markedy. The authors work out the form of the temperature spectrum and the spectral transfer in these cases. This paper represents a useful further step toward the ultimate inclusion of radiation effects in modeling the planetary boundary layer, although more work will be required to produce simple parameterizations that can be used in practice.

The next two papers, *Second-Order Closure for Variable Density Free Shear Layer*, by Vandromme and Kollman, and *A New Approach to the Analysis of Turbulent Mixing in Variable-Density Flows,* by Ha Minh, Launder, and MacInnes, both address the same problem in different ways. The modeling of mixing of non-reactive gases of different densities is important in its own right, of course, and has many applications (e.g.-cryogenic spills); it is generally regarded as a preliminary to the modeling of combustion, however, which is characterized by large density fluctuations. These two papers represent quite different approaches to the problem. The first is a relatively straightforward extension of customary second-order modeling techniques to Favre (or mass-weighted) averaging. Of course, despite the simplification (in some respects) introduced by the Favre averaging, there are a number of new problems to be addressed in extending modeling to this situation. The second paper uses a mix of Favre-averaged and Reynolds-averaged variables, and again extends more-or-less customary second-order modeling ideas, giving particular emphasis to correlations involving the pressure. While the former paper is essentially complete, giving predictions for inhomogeneous flows and drawing direct comparison with data, the second restricts its attention to a local-equilibrium treatment and a necessarily less detailed comparison with experiment. Both these papers should be compared with *Janicka* and *Lumley* (1981a), in which a model is developed entirely using Reynolds averaging, by a quite different technique. *Janicka* and *Lumley* (1981b) successfully extend this technique to diffusion flames with fast chemistry. It is not yet clear whether it will be useful, or necessary, to use Favre averaging, or a combination of Favre and Reynolds averaging, in modeling these flows. It has been generally assumed that Favre averaging will be necessary to describe combustion; however, the success of the technique of *Janicka* and *Lumley* (1981b) raises doubts about this. All that can be said at present is that both Favre and Reynolds averaging appear to give equally good results, while the combination of the two suggests interesting possibilities.

The final paper in this section is *Direct Simulation of Homogeneous Turbulent Shear Flows on the Illiac IV Computer: Applications to Compressible and Incompressible Modeling*, by Feiereisen, Shirani, Ferziger, and Reynolds. These exact simulations have now reached a state such that they can be considered on an equal footing with experiment. Unfortunately, with the destruction of the Illiac IV, time will probabily not be available on comparable machines, and we may have seen the last of these simulations. The value of these particular computations lies in the comparison with the various models. With regard to the return-to-isotropy and the anisotropic dissipation, the authors seem to be unaware that *Lumley* and *Newman* (1977) showed that the customary modelling did *not* assume that the dissipation was isotropic, but rather combined the anisotropy of the dissipation with the return-to-isotropy term (even though the authors did not realize that this is what they were doing). In addition to this reference, *Lumley* made this same point with emphasis (but evidently not clearly) at the Round Table Discussion on modeling at the London Shear Flow Conference, where he shared the platform with one of the co-authors of this paper. Of course, the results of the simulation support the necessity of combining the terms in this way. The *Lumley* and *Newman* (1977) model for the return-to-isotropy term, which the authors found produced large scatter in this flow, was optimized for return-to-isotropy of axisymmetric turbulence with negative values of the third invariant. It was later found that the form suggested gave absurd values for flows with zero or positive values of the third invariant (such as the flows considered here), and in *Lumley* (1978) an alternate form was presented which appeared to be better. However, there is relatively little direct experimental information available on return to isotropy (as opposed to shear) with zero or positive values of the third invariant, and proper calibration must await new experiments (underway at Cornell University and the Ecole Centrale de Lyon).

Acknowledgements. Supported in part by the U.S. National Science Foundation under Grants No. ATM 79–22006 and CME 79–19817, in Part by the U.S. Office of Naval Research under the following programs: Fluid Dynamics (Code 438), Power (Code 473) and Physical Oceanography (Code 481), and in part by the U.S. NASA-Ames Research Center under Grant No. MSG-2382.

References

Cambon, C., Bertoglio, J.P., Jeandel, D. "Spectral Closure of Homogeneous Turbulence", in Proceedings 1980–81 AFOSR-HTTM-Stanford Conference on complex turbulent flows: Comparison of computation and experiment, ed. by S.J. Kline, Palo Alto, 1981

Deardorff, J.W. Closure of second- and third-moment rate equations for diffusion in homogeneous turbulence. Phys. Fluids **21**, 525–530 (1978)

Hanjalic, K., Launder, B.E., Schiestel, R. "Multiple Time Scale Concepts in Turbulent Transport Modeling", in *Turbulent Shear Flows II*, ed. by L.J.S. Bradbury, F. Durst, B.E. Launder, F.W. Smidt, J.H. Whitelaw (Springer, Berlin, Heidelberg, New York 1980) pp. 36–49

Janicka, J., Lumley, J.L. "Second Order Modeling in Non-Constant Density Flows", Sibley School of Mechanical and Aerospace Engineering Rpt. No. FDA 81–01, Cornell University (1981a) J. Fluid Mech. (to be published)

Janicka, J., Lumley, J.L. "Computation of Density Fluctuations in Diffusion Flames (with Implications for Combustion Noise)", in *Euromech 142, Acoustics of Turbulent Flows*, ed. by G. Comte-Bellot, M. Sunyach (Ecole Centrale, Lyon 1981b) pp. 4.3–4.3C

Lumley, J.L. "Prediction Methods for Turbulent Flows, Introduction", in *Prediction Methods for Turbulent Flow* (Von Karman Institute, Rhode-St-Genese, Belgium 1975)

Lumley, J.L. "Computational Modeling of Turbulent Flows", in *Advances in Applied Mechanics*, Vol. 18, ed. by C.-S. Yih, (Academic, New York 1978) pp. 123–176

Lumley, J.L., Newman, G.R. The return to isotropy of homogeneous turbulence. J. Fluid Mech. **84**, 581–597 (1977)

Pope, S.B. "An Explanation of the Turbulent Round Jet/Plane jet Anomaly", Rpt. No. FS/77/12, Imperial College, London (1977)

Seif, Ali A. "Higher Order Closure Model for Turbulent Jets", Tech. Rpt. No. TRL-110, Turbulence Research Laboratory, Suny/Buffalo (1981)

Taulbee, D.B., Lumley, J.L. "Prediction of the Turbulent Wake with a Second Order Closure Model", Sibley School of Mechanical and Aerospace Engineering Rpt. No. FDA 81–04 Cornell University (1981)

Taulbee, D.B., Lumley, J.L. "Implications Regarding Large Eddy Structures from Attempts to Model Turbulent Transport", in *Applications of Fluid Mechanics and Heat Transfer to Energy and Environmental Problems*, ed. by D.D. Papilou, D.K. Papailou (Springer, Berlin, Heidelberg, New York 1981)

Zeman, O., Lumley, J.L. Modeling buoyancy-driven mixed layers. J. Atmos. Sci. **33**, 1974–1988 (1976)

Pressure Effects on Triple Correlations in Turbulent Convective Flows

J.-C. André[1,2], P. Lacarrère[1], K. Traoré[1,3]

1 Direction de la Météorologie, EERM/GMD, Boulogne, F–92100, France
2 Department of Atmospheric Sciences, Oregon State University, Corvallis, U.S.A.

Abstract

The parameterization of pressure effects on triple correlations is shown to be an important part of the closure problem.

Based on arguments relevant to homogeneous and isotropic turbulence, it is argued that pressure effects are not only of a relaxative nature. In the case of turbulent thermal convection, a parameterization scheme including non-linear and rapid effects is proposed and used within the framework of a third-order model. Comparison with experimental results by *Ferreira* [10] indicates that the rapid part of pressure effects represents an important contribution to the budget of triple correlations and that it prevents the development of spurious stable stratification in the interior of the convective layer.

Nomenclature

C	Dimensionless constants
$\overline{e}$	Eddy kinetic energy (per unit mass)
h	Depth of the turbulent layer
i	Square root of (-1)
$\boldsymbol{k}, \boldsymbol{p}, \boldsymbol{q}, \boldsymbol{r}, \boldsymbol{s}$	Wave-vectors in Fourier space
k, p, q, r, s	Wave-numbers in Fourier space
t	Time
T	Temperature
T_*	Convective temperature scale
$\boldsymbol{u}' =$	$(u'_1, u'_2, u'_3) = (u', v', w') =$ Fluctuating velocity
$\hat{\boldsymbol{u}} =$	$(\hat{u}_\alpha, \alpha = 1$ to $3)$ Fourier transform of the fluctuating velocity field
w_*	Convective velocity scale
$\boldsymbol{x} =$	$(x_1, x_2, x_3) = (x, y, z)$ Position
z	Height
α	Coefficient of thermal expansion
β	Buoyancy parameter
δ_{ij}	Kronecker delta
ϵ	Eddy kinetic energy dissipation rate
ν	Kinematic viscosity
ρ_0	Constant density

1. Introduction and Motivations

It is now widely known that the turbulent transfers which take place in convectively driven flows do not follow the simple Boussinesq eddy viscosity assumption, i.e. in many cases the turbulent flux of a property 'a' is oriented from regions with low 'a' toward re-

3 *Present address:* Météorologie Nationale, Division Agrométéorologique, Bamako, Mali

gions with high 'a'. This "counter-gradient" turbulent transport has been documented experimentally for the case of a mean property like the mean temperature $\overline{T}$ [1] and also for the case of a second-order quantity like the eddy kinetic energy $\overline{e} = \frac{1}{2} \rho_0 \overline{u'_k u'_k}$ [2]. Parameterization methods and numerical simulation techniques for convective flows have consequently undergone a significantly different development as compared to the one which occurred in the case of pure shear flows. It is indeed known that even simplified second-order models, like the so-called "k-ϵ" model, can be used quite successfully for the simulation of a variety of shear flows in different geometrical configurations (e.g. [3]), while it has been shown that the numerical simulation of the most simple convective flows requires the use of sophisticated turbulence models (e.g. [4, 5]).

These models are presently based:

either: (I) on Lumley's idea that the Boussinesq eddy diffusivity assumption can be properly generalized in order to account for buoyancy effects and can then be used to relate, in matrix form, third-order correlations to the gradients of second-order ones, in the framework of a more or less complex model [6, 7]

or: (II) on the explicit use of rate equations for the triple correlations [8, 9].

Both these methods require knowledge and parameterization of quadruple correlations, molecular and pressure effects on third-order terms. The two former problems are usually solved by retaining respectively the quasi-normal approximation and the assumption of isotropic dissipative scales, while the solution of the latter is based on the idea that pressure effects are indeed relaxative processes. This assumption of pressure relaxation has been originally proposed by *Rotta* [25] for the return-to-isotropy of double correlations like Reynolds stresses. It has been further extended by *Launder* [18] to other double correlations like heat fluxes and generalized to account for shear- and buoyancy-induced effects (see also *Reynolds* [26] for a review). When turning back to triple correlations, one is then led to the introduction of a relaxation time which directly determines the value of triple terms in Lumley's method, while the use of rate equations involves not only the above linear relaxation but also allows for more sophisticated, possibly non-linear, pressure-relaxative effects. The purpose of this paper is to study these pressure effects on third-order correlations.

Up to now similar studies have been based either on theoretical considerations or on indirect experimental evidence since the only available laboratory measurements were quite scarce and restricted to the so-called case of penetrative convection, where another different physical phenomenon (i.e. the entrainment of an overlying stably stratified fluid into an underlying turbulent convective layer) leads to additional complexity and does not allow for unambiguous determination of pressure effects. In this respect, the recent experiment by *Ferreira* [10], which we shall use in the present study, does not suffer from the same difficulties since it deals with non-penetrative convection between rigid boundaries. This experimental study has the further advantage of including more precise thermometric and velocimetric measurements of numerous triple correlations. Unfortunately, the price to pay for the reduction of complexity due to penetrative convection is the introduction of another viscous layer at the upper boundary. Our study will consequently be restricted to the interior of the convective layer, where it can be safely assumed that turbulence is fully developed, and will not deal with the adjacent viscous sublayers.

2. Pressure Relaxation in Homogeneous Turbulence

Before turning to the subject of convective flows, it is of interest to recall and summarize some general properties of the effects of pressure fluctuations on triple correlations.

The closure problem associated with pressure effects is mainly due to the fact that these effects are non-local in $\boldsymbol{x}$-space and involve convolution integrals over the whole domain of the flow. It is consequently of much help to introduce two-point correlations in order to solve, at least formally, the above closure problem. Due to its inherent complexity, it is unfortunately almost impossible to use this formalism in any situation other than the simplest, albeit theoretically interesting, case of homogeneous turbulence. In this case, and using Fourier transform $\hat{\boldsymbol{u}}(\boldsymbol{k})$ of the velocity field $\boldsymbol{u}(\boldsymbol{x})$, the exact rate equations for double and triple correlations read (e.g. [11, 12])

$$\left(\frac{\partial}{\partial t}+2\nu k^2\right)<\hat{u}_\alpha(\boldsymbol{k})\hat{u}_\beta(-\boldsymbol{k})> = -\frac{i}{2}D_{\alpha\rho\sigma}(\boldsymbol{k})\int_{\boldsymbol{p}+\boldsymbol{q}=\boldsymbol{k}}<\hat{u}_\rho(\boldsymbol{p})\,\hat{u}_\sigma(\boldsymbol{q})\,\hat{u}_\beta(\boldsymbol{k})>d\boldsymbol{p}+\begin{bmatrix}\alpha\rightleftharpoons\beta\\ \boldsymbol{k}\rightleftharpoons-\boldsymbol{k}\end{bmatrix} \tag{1}$$

$$\begin{aligned}\left(\frac{\partial}{\partial t}+\nu(k^2+p^2+q^2)\right)&\underset{\boldsymbol{k}+\boldsymbol{p}+\boldsymbol{q}=0}{<\hat{u}_\alpha(\boldsymbol{k})\hat{u}_\beta(\boldsymbol{p})\hat{u}_\gamma(\boldsymbol{q})>}\\ =-\frac{i}{2}D_{\alpha\rho\sigma}(\boldsymbol{k})&\int_{\boldsymbol{r}+\boldsymbol{s}=\boldsymbol{k}}<\hat{u}_\rho(\boldsymbol{r})\hat{u}_\sigma(\boldsymbol{s})\hat{u}_\beta(\boldsymbol{p})\hat{u}_\gamma(\boldsymbol{q})>_c d\boldsymbol{r}\\ &-iD_{\alpha\rho\sigma}(\boldsymbol{k})<\hat{u}_\beta(\boldsymbol{p})\hat{u}_\rho(-\boldsymbol{p})><\hat{u}_\gamma(\boldsymbol{q})\hat{u}_\sigma(-\boldsymbol{q})>+\begin{bmatrix}\alpha\rightleftharpoons\beta\\ \boldsymbol{k}\rightleftharpoons\boldsymbol{p}\end{bmatrix}+\begin{bmatrix}\alpha\rightleftharpoons\gamma\\ \boldsymbol{k}\rightleftharpoons\boldsymbol{q}\end{bmatrix}\end{aligned} \tag{2}$$

where $<uuuu>_c = <uuuu> - \Sigma<uu><uu>$ is fourth-order cumulant,

$$\begin{bmatrix}\alpha\rightleftharpoons\beta\\ \boldsymbol{k}\rightleftharpoons\boldsymbol{p}\end{bmatrix}$$

a symbol representing additional terms obtained by the indicated permutations of indices and wave-vectors, and

$$D_{\alpha\beta\gamma}(\boldsymbol{k}) = k_\beta\delta_{\alpha\gamma}+k_\gamma\delta_{\alpha\beta}-2k_\alpha k_\beta k_\gamma/k^2 \tag{3}$$

a geometrical tensor taking into account advective and pressure effects, the former being traced to the Kronecker symbol δ and the latter described by the term involving k^{-2} (inverse Laplacian operator in $\boldsymbol{k}$-space). The fact that pressure effects are always associated with k^{-2} makes it easy to trace their effects. By considering Eq. (2) one is then led to the following conclusions concerning the effect of pressure fluctuations:

(I) the term involving fourth-order cumulants has been shown [11] to be of a relaxative nature and approximated [13] by

$$\begin{aligned}-\frac{i}{2}D_{\alpha\rho\sigma}(\boldsymbol{k})\int_{\boldsymbol{r}+\boldsymbol{s}=\boldsymbol{k}}<\hat{u}_\rho(\boldsymbol{r})\hat{u}_\sigma(\boldsymbol{s})\hat{u}_\beta(\boldsymbol{p})\hat{u}_\gamma(\boldsymbol{q})>_c d\boldsymbol{r}+\begin{bmatrix}\alpha\rightleftharpoons\beta\\ \boldsymbol{k}\rightleftharpoons\boldsymbol{p}\end{bmatrix}+\begin{bmatrix}\alpha\rightleftharpoons\gamma\\ \boldsymbol{k}\rightleftharpoons\boldsymbol{q}\end{bmatrix}\\ = -\{\mu(k)+\mu(p)+\mu(q)\}<\hat{u}_\alpha(\boldsymbol{k})\hat{u}_\beta(\boldsymbol{p})\hat{u}_\gamma(\boldsymbol{q})>\quad,\end{aligned} \tag{4}$$

where the eddy damping rate $\mu(k)$ is proportional to a typically positive geometrical coefficient $B(k, r, s)$ constructed from the geometrical tensor D of Eq. (3)

$$B(k,r,s) = -D_{\alpha\beta\gamma}(\boldsymbol{k})\, D_{\beta\epsilon\alpha}(\boldsymbol{r})\, (\delta_{\gamma\epsilon} - s_\gamma s_\epsilon / s^2) \tag{5}$$

This coefficient is typically positive since

$$B(k,r,s) + B(k,s,r) = \Delta\,[k^2(r^2 + s^2) + (r^2 - s^2)^2]/4\,k^2 r^2 s^2 \geqslant 0$$

where k, r and s are the lengths of the three interacting wave vectors $\boldsymbol{k} = \boldsymbol{r} + \boldsymbol{s}$ and where Δ is a fourth-order polynomial totally symmetric with respect to k, r and s which takes positive values only when k, r and s are the sides of a triangle

$$\Delta = -k^4 - r^4 - s^4 + 2\,k^2 r^2 + 2\,k^2 s^2 + 2\,r^2 s^2 .$$

Distinction between advective (A) and pressure (P) effects in Eqs. (4) and (5) leads to the partition of the geometrical coefficient B into

$$B(k,r,s) = B^{(A)}(k,r,s) + B^{(P)}(k,r,s) \quad , \tag{6}$$

where the advective part is given by

$$\begin{aligned} B^{(A)}(k,r,s) &= -(\delta_{\alpha\beta} k_\gamma + \delta_{\alpha\gamma} k_\beta)\, D_{\beta\epsilon\alpha}(\boldsymbol{r})\, (\delta_{\gamma\epsilon} - s_\gamma s_\epsilon / s^2) \\ &= \Delta[k^4 + k^2(r^2 - 3s^2) - 2s^2(r^2 - s^2)]/4k^2 r^2 s^2 \end{aligned} \tag{7}$$

and the pressure part by

$$\begin{aligned} B^{(P)}(k,r,s) &= 2\,\frac{k_\alpha k_\beta k_\gamma}{k^2}\, D_{\beta\epsilon\alpha}(\boldsymbol{r})\, (\delta_{\gamma\epsilon} - s_\gamma s_\epsilon / s^2) \\ &= \Delta[-k^4 + k^2(r^2 + 2s^2) + s^2(r^2 - s^2)]/4k^2 r^2 s^2 \end{aligned} \tag{8}$$

Careful examination of Eq. (7) shows that $B^{(A)}$ is negative in the largest part of the three-dimensional domain Γ within which Δ is positive, that is within which k, r and s can effectively be taken as the three legs of a triangle. This indicates that advective effects are most likely to give rise to a growth of triple correlations [see Eq. (4)], as could be expected. On the other hand, the same kind of analysis applied to Eq. (8) shows that $B^{(P)}$ is typically positive within Γ; that is, that this pressure effect leads very probably to a linear relaxation of the triple correlations.

(II) the term involving products of second-order cumulants can also be split into advective and pressure parts, the latter being typically negative, i.e. leading to a decrease of triple correlations.

In other words, the above arguments confirm that pressure fluctuations mainly destroy triple correlations. They also show that (unfortunately!) pressure effects are more complicated than the usually assumed linear relaxation to isotropy.

In the more general case of convective flows, the need for simplicity will nevertheless lead us to adopt a parameterization of pressure fluctuations based on a relaxation assumption, but we shall keep in mind that it would not be reasonable to ascribe every possible unphysical growth of triple correlations to an underestimated pressure-induced linear relaxation to isotropy.

3. Pressure Parameterization for Single-Point High-Order Modeling

The simulation of turbulent convective flows which are inhomogeneous along a vertical cannot, at least at the present time, be undertaken in the above presented two-point closure framework. One has then to turn to the much more manageable single-point closures.

Unfortunately, these are less satisfactory from the theoretical point of view. A possible alternative would be to rely on the so-called "modal method" (e.g. [14]), in which the simplification due to the use of a relatively crude and prescribed horizontal structure allows for high resolution along the vertical. There is then no need for sophisticated turbulence closure, turbulent transfer being then explicitly described.

The model we shall be dealing with is a third-order model already used for the simulation of various turbulent flows [15, 16]. It should be noted that the realizability conditions [8] which were used in the earlier simulations are not necessary in the present study. Indeed, the refinements we shall introduce below for the closure assumptions prevent triple correlations from becoming too large and overshooting their maximum permissible values derived from generalized Schwarz's inequalities.

We shall not go into all details of the model here, its description for the simulation of convective flows being given in *André* et al. [16], to which the interested reader is referred. It is simply recalled that the rate equations for triple correlation with the usual assumptions and standard notation read

$$\frac{\partial \overline{u_i'u_j'w'}}{\partial t} = -\overline{w'^2}\frac{\partial \overline{u_i'u_j'}}{\partial z} - \overline{u_j'w'}\frac{\partial \overline{u_i'w'}}{\partial z} - \overline{u_i'w'}\frac{\partial \overline{u_j'w'}}{\partial z} - \epsilon_{uuu}$$
$$+ \beta(\overline{u_i'u_j'T'} + \overline{u_i'w'T'}\delta_{3j} + \overline{u_j'w'T'}\delta_{3i})$$
$$- \frac{1}{\rho_0}\left(\overline{u_i'u_j'\frac{\partial p'}{\partial z}} + \overline{u_i'w'\frac{\partial p'}{\partial x_j}} + \overline{u_j'w'\frac{\partial p'}{\partial x_i}}\right) \tag{9}$$

$$\frac{\partial \overline{u_i'u_j'T'}}{\partial t} = -\overline{u_i'u_j'w'}\frac{\partial \overline{T}}{\partial z} - \overline{u_i'w'}\frac{\partial \overline{u_j'T'}}{\partial z} - \overline{u_j'w'}\frac{\partial \overline{u_i'T'}}{\partial z} - \overline{w'T'}\frac{\partial \overline{u_i'u_j'}}{\partial z}$$
$$+ \beta(\overline{u_i'T'^2}\delta_{3j} + \overline{u_j'T'^2}\delta_{3i}) - \epsilon_{uuT} - \frac{1}{\rho_0}\overline{T'\left(u_i'\frac{\partial p'}{\partial x_j} + u_j'\frac{\partial p'}{\partial x_i}\right)} \tag{10}$$

$$\frac{\partial \overline{w'T'^2}}{\partial t} = -2\overline{w'^2T'}\frac{\partial \overline{T}}{\partial z} - \overline{w'^2}\frac{\partial \overline{T'^2}}{\partial z} - 2\overline{w'T'}\frac{\partial \overline{w'T'}}{\partial z} - \epsilon_{uTT} + \beta\overline{T'^3} - \frac{1}{\rho_0}\overline{T'^2\frac{\partial p'}{\partial z}} \tag{11}$$

$$\frac{\partial \overline{T'^3}}{\partial t} = -3\,\overline{w'T'^2}\frac{\partial \overline{T}}{\partial z} - 3\,\overline{w'T'}\frac{\partial \overline{T'^2}}{\partial z} - \epsilon_{TTT}\;, \tag{12}$$

in which the parameterization of molecular terms ϵ... is based on the assumption of isotropic dissipative scales

$$\epsilon_{uuu} = \epsilon_{uTT} = 0$$

$$\epsilon_{uuT} = c_{10} \frac{\epsilon}{\bar{e}} \frac{\overline{u'_k u'_k T'}}{3} \delta_{ij}$$

$$\epsilon_{TTT} = c_{10} \frac{\epsilon}{\bar{e}} \overline{T'^3} \tag{13}$$

Turning now to the pressure terms, these are split as in [15], into three parts: diffusive, diagonal, and trace-free parts. We shall neglect the diffusive part since it is very likely to be of small importance in the interior of convective layers, pressure transport being significant only close to walls or in stably stratified layers. The diagonal part will be modeled as a source term following *Launder*'s [17] recommendation. Finally the trace-free part will be modeled as a redistribution term including a 'slow' return-to-isotropy contribution and a 'rapid' contribution, proportional to the anisotropic part of the production tensor. These two contributions come respectively from the non-linear turbulent and buoyant parts of the Poisson's equation for fluctuating pressure (see [18]).

With all these assumptions, and in the particular case of convection in the absence of mean shear, the pressure terms will be given by, in general agreement with the developments of proceeding section:

$$-\frac{1}{\rho_0} \overline{\left(u'_i u'_j \frac{\partial p'}{\partial z} + u'_i w' \frac{\partial p'}{\partial x_j} + u'_j w' \frac{\partial p'}{\partial x_i}\right)} = -c_8 \frac{\epsilon}{\bar{e}} \overline{u'_i u'_j w'}$$

$$-c_{11}\, \beta (\overline{u'_i u'_j T'} + \overline{u'_i w' T'}\, \delta_{3j} + \overline{u'_j w' T'}\, \delta_{3i}) \tag{14}$$

$$-\frac{1}{\rho_0} \overline{T'\left(u'_i \frac{\partial p'}{\partial x_j} + u'_j \frac{\partial p'}{\partial x_i}\right)} = c_9 \frac{\epsilon}{\bar{e}} \overline{u'_k u'_k T'}\, \delta_{ij} - c_8 \frac{\epsilon}{\bar{e}} \left(\overline{u'_i u'_j T'} - \frac{\overline{u'_k u'_k T'}}{3} \delta_{ij}\right)$$

$$-c_{11}\, \beta (\overline{u'_i T'^2}\, \delta_{3j} + \overline{u'_j T'^2}\, \delta_{3i} - \tfrac{2}{3} \overline{w' T'^2}\, \delta_{ij}) \tag{15}$$

$$-\frac{1}{\rho_0} \overline{T'^2 \frac{\partial p'}{\partial x_i}} = -c_8 \frac{\epsilon}{\bar{e}} \overline{u'_i T'^2} - c_{11}\, \beta\, \overline{T'^3}\, \delta_{3i} \tag{16}$$

where the 'c_9-term' corresponds to diagonal part and the 'c_8-' and 'c_{11}-terms' correspond respectively to the relaxative and rapid parts.

Deardorff [27] has pointed out that the present parameterization of the rapid part bears some resemblance to the one obtained by assuming that the fluctuating pressure in a Boussinesq fluid satisfies the hydrostatic relation $\partial p'/\rho_0 \partial x_i = -(\rho'/\rho_0)\, g\, \delta_{3i} = \beta T' \delta_{3i}$, in which case one would have, for example, $c_8 = 0$ and $c_{11} = 1$ in Eqs. (14) and (16).

4. Numerical Results and Sensivity Studies

Some "computer optimization" and sensivity studies, which will be reported later, have been used to determine the 'best' values of the constants c_8, c_9, c_{10}, and c_{11}:

$$c_8 = 8.0; \quad c_9 = -0.67; \quad c_{10} = 6.0; \quad c_{11} = 0.2, \tag{17}$$

where the value of c_9 is determined from the realizability constraint [15],

$$3\,c_9 = c_{10} - c_8 . \tag{18}$$

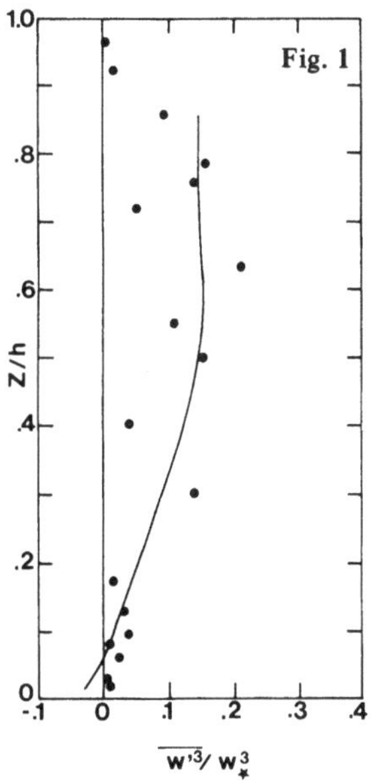

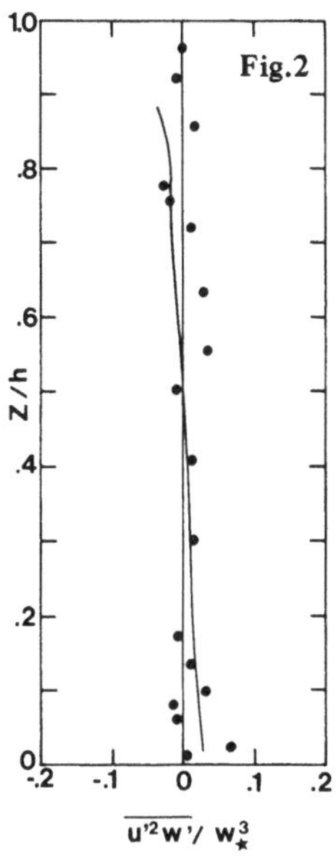

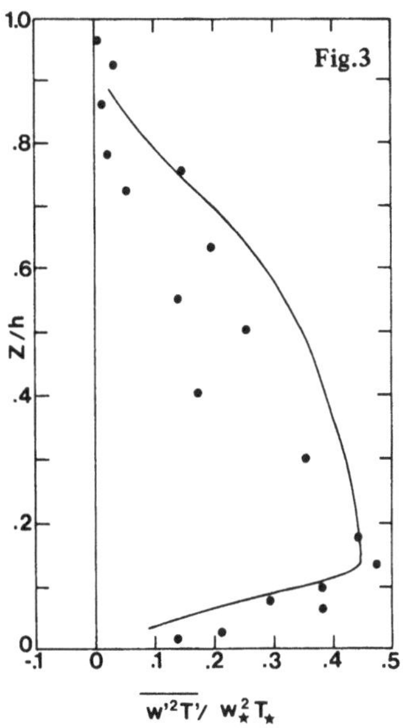

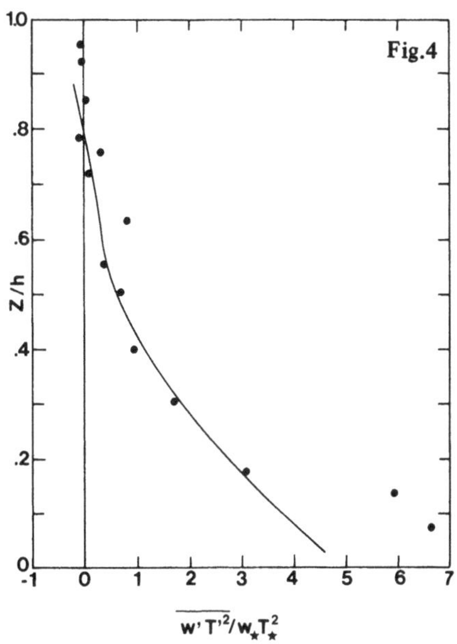

Fig. 1. Dimensionless profile of the eddy flux of vertical component of kinetic energy. (*Points*) laboratory measurements; (*curve*) numerical results

Fig. 2. Dimensionless profile of the eddy flux of horizontal component of kinetic energy. Symbols as in Fig. 1

Fig. 3. Dimensionless profile of the eddy flux of kinematic turbulent heat flux. Symbols as in Fig. 1

Fig. 4. Dimensionless profile of the eddy flux of temperature variance. Symbols as in Fig. 1

The corresponding results, normalized with the aid of *Deardorff's* [19] convective scaling, are shown in Figs. 1–4 for $\overline{w'^3}$, $\overline{u'^2 w'}$, $\overline{w'^2 T'}$ and $\overline{w'T'^2}$ respectively. It must be noted here that the experimental results by *Ferreira* [10] which appear on these figures have been rescaled. The velocity scale w_* used in convective scaling depends in fact on the coefficient of thermal expansion α. Since this coefficient varies significantly (almost linearly) with temperature in the range we are concerned with , we used the exact value of α corresponding to each particular experiment instead of a mean value as in the results shown by Ferreira. This led to a change in the dimensionless results by as much as 20% in the cases of $\overline{w'^3}$ and $\overline{u'^2 w'}$.

It can be seen that the agreement between our numerical results and the "corrected" measurements by *Ferreira* [10] is quite satisfactory in the convective layer interior, with the only possible exception that $\overline{w'^2 T'}$ is slightly overestimated. This effect could possibly be ascribed to the fact that parameterizations (13) and (15) of molecular and pressure effects in Eq. (10) are not completely consistent with realizability conservation. Indeed, the realizability constraint (18) is sufficient to ensure that $\partial \overline{w'^2 T'} / \partial t = 0$ whenever $w' = 0$ but it does not ensure that $\partial \overline{u'^2 T'} / \partial t = 0$ when $u' = 0$.

These parameterizations consequently induce an overestimation of $\overline{u'^2 T'}$ and $\overline{v'^2 T'}$ which in turn, through the c_8-term describing return-to-isotropy, [see Eq. (15)], leads to an overestimation of $\overline{w'^2 T'}$. Although a plausible explanation has been found, it has not been possible to derive, at least up to now, a completely satisfactory modification of parameterization (15).

It should also be noticed from Figs. 1–4 that, despite the fact that viscous wall layers are not taken into account in the simulation, the numerical results are still in agreement with experimental data close to the lower boundary.

The sensitivity of model results to a +20% modification of constants c_8, c_{10} and c_{11} is shown in Table 1 [c_9 being always determined from (18)].

Table 1. Sensitivity of model results to changes in the constants

+ 20%	c_8	c_{11}	c_{10}
$\overline{w'^3}$ $(z/h = 0.5)$	- 28%	- 5%	- 2%
$\overline{w'^2 T'}$ $(z/h = 0.5)$	- 3%	- 2%	- 1%
$\overline{w'T'^2}$ $(z/h = 0.5)$	- 20%	- 1%	- 5%
$\overline{T'^3}$ $(z/h = 0.5)$	Less than 1%		- 7%

It can first be checked that an increase in the constants leads to a decrease of triple correlations (and vice-versa), in agreement with the general idea of damping by pressure effects. It should further be noticed that c_8 is the easiest constant to determine since it strongly influences the results, with the only exception that any significant increase of c_8 would lead to an important deterioration of $\overline{w'^3}$ and $\overline{w'T'^2}$ profiles without improving significantly the $\overline{w'^2 T'}$ profile. Such an increase of c_8, i.e. of the linear relaxative effect, would furthermore be in contradiction to the developments of Sect. 2 and would also lead to much underestimated calculated triple correlations in the case of other turbulent flows like the asymetric channel flow of *Hanjalic* and *Launder* [20], see [16]. Any significant increase of c_8 being ruled out, one has then to look for another pressure mechanism to efficiently destroy triple correlations. This is the main reason for the introduction of the rapid linear part described by the 'c_{11}-term'.

Another justification for the introduction of rapid linear pressure effects is as follows. For stationary conditions, and near the middle of the convective layer where $\overline{w'^2}$ reaches a maximum, the budget for $\overline{w'^2T'}$ reduces to [see Eqs. (10) with $i=j=3$ and (13, 15) and (18)]:

$$c_8 \frac{\epsilon}{\bar{e}} \overline{w'^2T'} = -\overline{w'^3} \frac{\partial \bar{T}}{\partial z} - 2\overline{w'^2} \frac{\partial \overline{w'T'}}{\partial z} + 2\beta \left(1 - \frac{2}{3} c_{11}\right) \overline{w'T'^2} \tag{19}$$

Adapting to the triple correlation $\overline{w'^2T'}$ an argument developed by *Wyngaard* and *Coté* [21] for the double correlation, or kinematic heat flux, $\overline{w'T'}$, it can be said that the introduction of rapid effects (the c_{11}-term) decreases the magnitude of the buoyant contribution in Eq. (19). This allows in turn the budget to balance with larger values of $-\overline{w'^3}$ $\partial\bar{T}/\partial z$, i.e. with smaller values of $\partial\bar{T}/\partial z$ or, in other words, with enhanced unstable stratification. This argument has been verified by running numerical experiments with $c_{11} = 0$, and by noting that, in such cases, spurious stable stratification (positive $\partial\bar{T}/\partial z$) did indeed appear in the lower half of the convective layer.

5. Concluding Remarks

It has been shown in the present study that an accurate simulation of turbulent convective flows, necessarily using higher-order-closure assumptions, requires a careful parameterization of pressure effects on triple correlations, including linear relaxation-to-isotropy by turbulent fluctuations and 'rapid' destruction by gravitationally-induced effects. This last mechanism could possibly be further documented by employing the linear technique proposed by *Cambon* and *Jeandel* [28] or an alternative decomposition of the Poisson equation for pressure fluctuations due to *Lumley* [29] or *Domingos* [30].

More generally, convective turbulent flows exhibit many peculiar features: the existence of countergradient transfer, the presence of organized large-scale structures, the importance of pressure effects. It is believed that these features are closely related to each other, the key mechanism being probably the appearance of quite regular mean motion at larger scales, even at very high Rayleigh number. These large structures develop in already turbulent flows and are probably the result of a secondary instability mechanism (see e.g. [22] for a review). They bear some resemblances, from the morphological and possibly also the mathematical point of view, to the more familiar convective patterns observed for much smaller Rayleigh number. Their physics and dynamics are however basically different since in this case the dissipative phenomena are due to the 'background' smaller scale turbulent motions, instead of being due directly to viscosity. One can then question the relevance of 'direct' numerical simulations for the study of these convective turbulent flows. Can they shed some light on flow dynamics? Can they be of interest for the development of closure schemes and parameterization techniques? Whatever their precise type may be, whether "modal" [14] or "three-dimensional" with subgrid-scale parameterization (e.g. [23]) or large eddy simulation techniques [24], they are restricted to relatively small equivalent Rayleigh numbers. They can indeed only deal with narrow turbulence spectra due to the limited memory size of presently available computers. Despite this limitation concerning their true physical significance, it is believed (or at least hoped!) that they will promote an understanding of many of the effects associated with large-scale convective structures.

References

1. Deardorff, J.W.: The counter-gradient heat flux in the lower atmosphere and in the laboratory. J. Atmos. Sci. **23**, 503–506 (1966)
2. Willis, G.E., Deardorff, J.W.: A laboratory model of the unstable planetary boundary layer. J. Atmos. Sci. **31**, 1297–1307 (1974)
3. Launder, B.E., Spalding, D.B.: The numerical computation of turbulent flows. Comput. Methods Appl. Mech. Eng. **3**, 269–289 (1974)
4. André, J.C., De Moor, G., Lacarrère, P., du Vachat, R.: Turbulence approximation for inhomogeneous flows – Part II. The numerical simulation of a penetrative convection experiment. J. Atmos. Sci. **33**, 482–491 (1976)
5. Lumley, J.L., Zeman, O., Siess, J.: The influence of buoyancy on turbulent transport. J. Fluid Mech. **84**, 581–597 (1978)
6. Zeman, O., Lumley, J.L.: Modeling buoyancy driven mixed layers. J. Atmos. Sci. **33**, 1974–1988 (1976)
7. Sun, W.Y., Ogura, Y.: Modelling the evolution of the convective planetary boundary layer. J. Atmos. Sci. **37**, 1558–1572 (1980)
8. André, J.C., De Moor, G., Lacarrère, P., du Vachat, R.: Turbulence approximation for inhomogeneous flows – Part I. The clipping approximation. J. Atmos. Sci. **33**, 476–481 (1976)
9. Warn-Varnas, A.C., Piacsek, S.A.: An investigation of the importance of third-order correlations and choice of length scale in mixed layer modelling. Geophysical and Astrophysical Fluid Dynamics, **13**, 225–243 (1979)
10. Ferreira, R.T.S.: "Unsteady turbulent thermal convection"; Ph. D. Thesis, University of Illinois, Urbana Champaign (1978)
11. André, J.C.: Irreversible interaction between cumulants in homogeneous, isotropic, two-dimensional turbulence theory. Phys. Fluids **17**, 15–21 (1974)
12. Orszag, S.A., Kruskal, M.D.: Formulation of the theory of turbulence. Phys. Fluids **11**, 43–60 (1968)
13. Orszag, S.A.: Analytical theories of turbulence. J. Fluid Mech. **45**, 363–386 (1970)
14. Latour, J., Spiegel, E.A., Toomre, J., Zahn, J.P.: Stellar convection theory I. The anelastic modal equations. Astrophys. J. **207**, 233–243 (1976)
15. André, J.C., De Moor, G., Lacarrère, P., Therry, G., du Vachat, R.: Modeling the 24-hour evolution of the mean and turbulent structures of the planetary boundary layer. J. Atmos. Sci. **35**, 1861–1883 (1978)
16. André, J.C., De Moor, G., Lacarrère, P., Therry, G., du Vachat, R.: "The Clipping Approximation and Inhomogeneous Turbulence Simulations", in: *Turbulent Shear Flows I*, ed. by F. Durst, B.E. Launder, F.W. Schmidt, J.H. Whitelaw (Springer, Berlin, Heidelberg, New York 1979) pp. 307–318
17. Launder, B.E.: Private communication (1976)
18. Launder, B.E.: On the effects of a gravitational field on the turbulent transport of heat and momentum. J. Fluid Mech. **67**, 569–581 (1975)
19. Deardorff, J.W.: Convective velocity and temperature scales for the unstable planetary boundary layer and for Rayleigh convection. J. Atmos. Sci. **27**, 1211–1213 (1970)
20. Hanjalic, K., Launder, B.E.: Fully developed asymmetric flow in a plane channel. J. Fluid Mech. **51**, 301–335 (1972)
21. Wyngaard, J.C., Coté, O.R.: The evolution of a convective planetary boundary layer – A higher-order-closure model study. Boundary-Layer Meteorol. **7**, 289–304 (1974)
22. Brown, R.A.: Longitudinal instabilities and secondary flows in the planetary boundary layer: a review. Rev. Geophys. Space Phys. **18**, 683–697 (1980)
23. Deardorff, J.W.: A numerical study of three-dimensional turbulent channel flow at large Reynolds numbers. J. Fluid Mech. **41**, 453–480 (1970)
24. Kwak, D., Reynolds, W.C., Ferziger, J.H.: "Three-Dimensional Time-Dependent Computation of Turbulent Flow", TF-5, Department of Mechanical Engineering, Stanford University CA (1975)
25. Rotta, J.C.: Statistische Theorie nichthomogener Turbulenz. Z. Phys. **129**, 547–572 (1951)
26. Reynolds, W.C.: Computation of turbulent flows. Annu. Rev. Fluid Mech. **8**, 183–208 (1976)
27. Deardorff, J.W.: Private communication (1980)
28. Cambon, C., Jeandel, D.: "Approach of Non-isotropic Homogeneous Turbulence 'Submitted' to Mean Velocity Gradients", Proc. 3rd Symp. on Turbulent Shear Flows, University of California, Davis, Sept. 1981, pp. 17.7–17.11
29. Lumley, J.L.: Computational modeling of turbulent flows. Adv. Appl. Mech. **18**, 123–176 (1978)
30. Domingos, J.J.D.: "Pressure Strain: Exact Results and Models", Proc 3rd Symp. on Turbulent Shear Flows, University of California, Davis, Sept. 1981, pp. 19.26–19.30

A Model of Three-Dimensional Transfer in Non-Isotropic Homogeneous Turbulence

Jean-Pierre Bertoglio
Ecole Centrale de Lyon, 36 Avenue Guy de Collongue, F–69130 Ecully, France

Abstract

A model for predicting homogeneous anisotropic turbulence submitted to uniform mean velocity gradients is presented. Equations of second order correlations are used. They are closed in the frame of the E.D.Q.N.M. theory. Since the use of the fully three-dimensional expression for the tranfer results in cumbersome computations, a modeled form is proposed to interpolate this term. Then only a few complete calculations are required and computations can be carried out. Our results agree fairly well with the trends predicted by the experiments.

Nomenclature

D'_{ijklmn}	function of $\boldsymbol{K}, \boldsymbol{P}$, and $\boldsymbol{Q}$, depending on the geometry of the triad
D'_{ijml}, D''_{ijml}	idem
$E(K,t)$	kinetic energy spectrum
κ	structural parameter $\equiv \dfrac{\overline{u_2^2} - \overline{u_3^2}}{\overline{u_2^2} + \overline{u_3^2}}$
$\boldsymbol{K}$	wave vector
K	wave number, $= \lvert\boldsymbol{K}\rvert$
$\boldsymbol{P}$	wave vector
P_{ij}^{L}	linear part of the velocity pressure corelation: $\frac{1}{\rho}\overline{\left(u_i \frac{\partial p}{\partial X_j} + u_j \frac{\partial p}{\partial X_i}\right)}$
$\boldsymbol{Q}$	wave vector
$\overline{q^2}$	kinetic turbulent energy, $= 1/2\,(\overline{u_1^2} + \overline{u_2^2} + \overline{u_3^2})$
S	shear, $= \partial \overline{U}_2 / \partial X_3$
t	time
$T_{ij}(\boldsymbol{K},t)$ or $\underline{\underline{T}}$	three-dimensional transfer, $= T_{ij}^{+} + T_{ij}^{-}$
$T_{ij}^{+}(\boldsymbol{K},t)$	part of T_{ij} corresponding to the backscatter
$T_{ij}^{\mathrm{M}+}(\boldsymbol{K},t)$	modeled form for T_{ij}^{+}
$T_{ij}^{-}(\boldsymbol{K},t)$	part of T_{ij} corresponding to the drain
$\mathrm{tr}(\lvert\boldsymbol{K}\rvert,t)$	isotropic transfer associated with an isotropic turbulence, $= \mathrm{tr}^{+} + \mathrm{tr}^{-}$
$\mathrm{tr}^{+}(\lvert\boldsymbol{K}\rvert,t)$	isotropic backscatter
$\mathrm{tr}^{-}(\lvert\boldsymbol{K}\rvert,t)$	isotropic drain
$u_i(\boldsymbol{X})$	velocity fluctuation
u'_i	$\sqrt{\overline{u_i^2}}$
$\overline{u_i u_j}$	Reynolds stress tensor
$\overline{U}_i(\boldsymbol{X})$	mean velocity
ϵ	dissipation
$\eta_{ij}(\boldsymbol{K},\boldsymbol{P},\boldsymbol{Q},t)$	$= -\Theta_{KPQ}\,[D'_{ijml}\Phi_{ml}(\boldsymbol{Q}) + D''_{ijml}\Phi_{ml}(\boldsymbol{K})]$
η_{ij}^{M}	modeled form for η_{ij}
$\Phi_{ij}(\boldsymbol{K},t)$, or $\underline{\underline{\Phi}}$	spectral tensor, Fourier transform of the second order velocity correlation at two points

μ	damping factor	Θ_{KPQ}	E.D.Q.N.M. characteristic time
ν	kinematic viscosity		

1. Introduction

The use of two-point correlations has often been very helpful in predicting turbulent flows. In particular, concerning homogeneous anisotropic turbulent fields, the Rapid-Distortion Theory has provided satisfactory information about the interaction between turbulence and mean velocity gradients, which is a linear process. Unfortunately, at the present time, the use of analytical theories, such as D.I.A., T.F.M. and E.D.Q.N.M., to take into account the nonlinear action of turbulence on itself, has essentially been restricted to the area of isotropic turbulence. The main reason for this restriction is that using these theories without simplification to close the rate equation for the three-dimensional second order spectral tensor $\Phi_{ij}(\boldsymbol{K},t)$ would result in too cumbersome computations.

One is therefore led to introduce hypotheses to obtain a reduction in computational labor. Two interesting attempts should be mentioned at this point. The first is due to *Herring* [1, 2] who developed the $\Phi_{ij}(\boldsymbol{K},t)$ tensor into a spherical harmonic expansion and retained only low order terms. Computations of the return to isotropy of turbulence then become tractable and have been carried out by using the equations of the D.I.A.. Agreement with direct simulation is fairly good. This study is nevertheless limited to the particular case of an axisymmetrical homogeneous turbulence, and a much larger number of terms would probably be needed if turbulent fields generated by mean velocity gradients were to be studied in that way.

Another interesting approach has recently been developed by *Cambon* [3]. The idea is to use the equation governing the average of $\Phi_{ij}(\boldsymbol{K},t)$ over a sphere of radius K and to close this equation with the E.D.Q.N.M. assumptions. When performing the integration over the sphere, information concerning the "directivity" of $\Phi_{ij}(\boldsymbol{K},t)$ is lost, and a modeled form of Φ_{ij} is needed. $\Phi_{ij}(\boldsymbol{K})$ is then expressed as an isotropic function of its average over the sphere and of $\boldsymbol{K}$. Good agreement with experiment has been obtained when predicting the classical homogeneous flows with this model, in particular the case of plane strain. However the model does not seem to fit every kind of anisotropy, in particular we have pointed out that the presence of rotation affects its validity [4].

In the present paper, we develop a different kind of approach. We propose a model in which hypotheses are not made on the $\Phi_{ij}(\boldsymbol{K},t)$ tensor itself, but on the three-dimensional transfer $T_{ij}(\boldsymbol{K}, t)$. The advantage is that the linear terms are then exactly taken into account. Values of $T_{ij}(\boldsymbol{K}, t)$ computed by using the complete three-dimensional E.D.Q.N.M. closure are needed only for a restricted number of wave vectors and for a few values of time. In between, a modeled form of T_{ij} is adopted, based on an interpolation between the computed values. The interpolated form is chosen to depend on Φ_{ij} and on the "isotropic" transfer, which is easier to compute.

In order to find a suitable form for the interpolation, which preserves the most important properties of the "true" three-dimensional E.D.Q.N.M. transfer, a preliminary study is made: T_{ij} is computed in particular cases of anisotropy resulting from the action of rapid-distortion on an initially isotropic turbulence. The main characteristics of transfer are then examined. A modeled form is thereafter proposed. Agreement between the model and the computed values of $T_{ij}(\boldsymbol{K})$ is found to be satisfactory.

Computations using the modeled $T_{ij}(\boldsymbol{K})$ are then carried out. The first results and comparisons with experiment are presented here.

2. The Closure of the Equation Governing $\Phi_{ij}(K,t)$ in the Frame of the E.D.Q.N.M.

We chose the E.D.Q.N.M. theory (*Orszag* [5]) as it is the simplest analytical closure and because it has been successfully employed in isotropic turbulence, in particular by *Lesieur* [6].

Putting an isotropic Θ_{KPQ} in the equation of the Test Field Model given by *Kraichnan* [7], assuming furthermore that Θ_{KPQ} can be expressed by using the same function of $|K|$ and $E(|K|)$ as in isotropic E.D.Q.N.M., give:

$$\frac{\partial}{\partial t}\Phi_{ij}(K,t) + 2\nu K^2 \Phi_{ij}(K,t) = \underbrace{\int_\Delta \Theta_{KPQ} D_{ijklmn}\,\Phi_{kl}(P,t)\,\Phi_{mn}(Q,t)\,dP\,dQ}_{=T^+_{ij}(K,t)}$$

$$\underbrace{-\int_\Delta [\eta_{ik}(K,P,Q)\,\Phi_{kj}(K,t) + \eta^*_{jk}(K,P,Q)\,\Phi_{ik}(K,t)]\,dP\,dQ}_{=T^-_{ij}(K,t)} \qquad (1)$$

where the integrals extend over the triads $K + P + Q = 0$, and where

$$\eta_{ij}(K,P,Q) = -\Theta_{KPQ}[D'_{ijml}\Phi_{ml}(Q) + D''_{ijml}\Phi_{ml}(P)] \quad ,$$

D, D' and D'' being coefficients depending on the geometry of the triad.

Equation (1), in which the terms containing mean velocity gradients have been omitted for the sake of brevity, can be considered as the equation of E.D.Q.N.M. for a non-isotropic turbulence.

The assumptions made on Θ_{KPQ} imply that the damping factor μ of the E.D.Q.N.M. is not affected by anisotropy, and furthermore that the mean gradients are neglected in the equation of third-order correlations. From the point of view of the D.I.A., an equivalent of those assumptions is to assume, as suggested by *Leslie* [8], that Green's function is affected neither by the anisotropy of turbulence, nor by the presence of mean velocity gradients. If we have an idea of the validity of the first part of the proposition, which is probably acceptable (see *Schumann* and *Herring* [2]), the influence of the mean gradients is not clear (see *Kraichnan* [7] and *Cambon* [16]). In the present paper we make the simplest hypothesis on Θ_{KPQ} from practical considerations.

The right hand side of Eq. (1) is the 3D transfer, and we shall call it $T_{ij}(K, t)$.

3. Study of the Three-Dimensional Transfer and Phenomenological Approach of Its Modeled Form

In order to study the main properties of $T_{ij}(K, t)$ we have computed the values of this tensor in particular cases of anisotropic turbulence in which $\Phi_{ij}(K,t)$ can easily be obtained. For practical considerations we have chosen turbulent fields resulting from the action of rapid-distortion on the initially isotropic turbulence measured by *Comte-Bellot* [9]. Obviously the non linear processes are not taken into account when "building" the Φ_{ij} tensor in that way, but it is hoped that the observed properties of the computed form of the 3D transfer will still be valuable for real homogeneous flows, i.e. for weakly anisotropic turbulence.

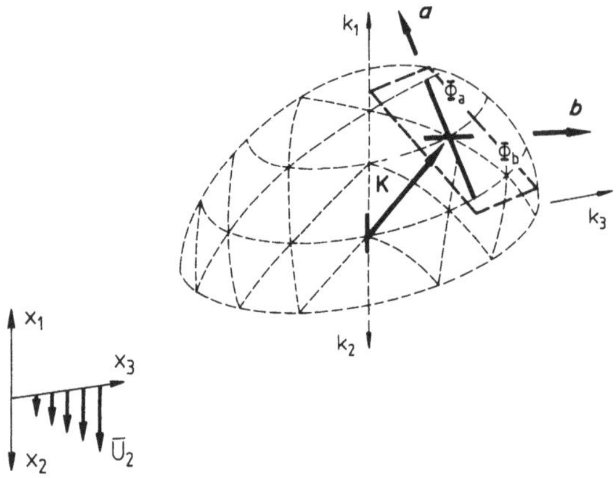

Fig. 1. (*a, b*) Eigen vectors of Φ_{ij}, (Φ_a, Φ_b) eigen values of Φ_{ij}

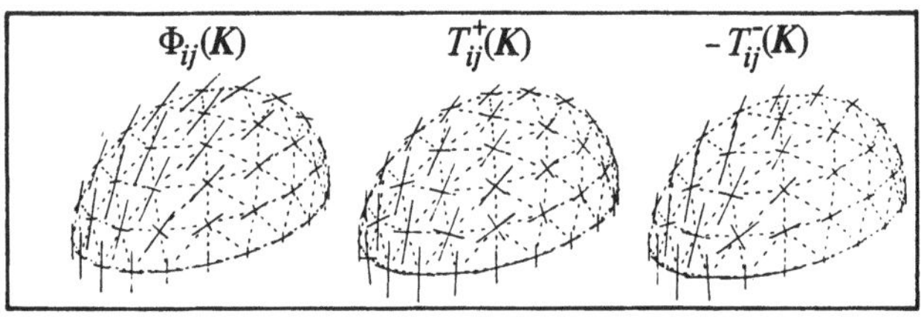

Fig. 2. Visualization of Φ_{ij}, T^+_{ij} and T^-_{ij} on a sphere of radius $K = 0.45\ \mathrm{cm}^{-1}$, in the case of a mean shear $S = 13.6\ \mathrm{s}^{-1}$ and for a time $t = 0.07$ s (same conventions as in Fig. 1)

The results are presented by using a visualization over spheres of radius $|\boldsymbol{K}|$. The principal axes of the tensors are visualized and segments proportional to the eigen values are plotted, as shown on Fig. 1. T_{ij} has been split into two parts: $T_{ij}^{+}(\boldsymbol{K}, t)$ usually called backscatter, and $T_{ij}^{-}(\boldsymbol{K}, t)$ called drain. Each part is visualized, as well as the $\Phi_{ij}(\boldsymbol{K},t)$ tensor itself.

The most important results are that, on the one hand, the distributions on the sphere of both T_{ij}^{+} and T_{ij}^{-} are strongly influenced by the distribution of $\Phi_{ij}(\boldsymbol{K},t)$, and, on the other hand, that the "local" anisotropy of T_{ij}^{-}, at a given point of the sphere, is strongly influenced by the local anisotropy of Φ_{ij}. Moreover, it appears that T_{ij}^{+} is less connected to $\Phi_{ij}(\boldsymbol{K})$ than T_{ij}^{-} is, which is suggested by Eq. (1). These considerations can be deduced from Fig. 2 in the case of a uniform shear flow; they remain true for a pure plane strain.

We now try to find a modeled form for T_{ij} which respects those main features. We try to express $T_{ij}(\boldsymbol{K}, t)$, for any values of $\boldsymbol{K}$ and t, as a functional of $\Phi_{ij}(\boldsymbol{K},t)$, of the isotropic transfer tr $(|\boldsymbol{K}|, t)$ and of $T_{ij}(\boldsymbol{K}', t')$ computed for $\boldsymbol{K}'$ describing a small set of spheres and for a few values of time. We still make the split $T_{ij} = T_{ij}^{+} + T_{ij}^{-}$ and then model each part separately.

Calling $(K_1, \ldots |K_\alpha| \ldots, K_\gamma)$ the set of wave numbers corresponding to the spheres on which the 3D transfer is computed by E.D.Q.N.M. for the set of times $(t_1, \ldots |t_\beta| \ldots, t)$ we write the functional F^{+} such that:

$$T_{ij}^{\mathrm{M}+}(\boldsymbol{K},t) = F^{+}\left[\left(\underline{\underline{T}}^{+}\left(K\frac{K_\alpha}{|\boldsymbol{K}|}, t_\beta\right), \quad \alpha = 1 \ldots \gamma\,;\ \beta = 1 \ldots \delta\right)\right., \\ \left. \boldsymbol{K}, \underline{\underline{\Phi}}(\boldsymbol{K},t),\ \mathrm{tr}^{+}(|\boldsymbol{K}|,t), E(|\boldsymbol{K}|,t)\right] \tag{2}$$

For the drain, it appears that it is better to model η_{ij}^{-} than to directly express T_{ij}^{-} by using a functional like Eq. (2).

We finally adopt, for $\eta_{ij}(\boldsymbol{K}, t)$:

$$\eta_{ij}{}^{\mathrm{M}}(\boldsymbol{K},t) = \frac{\mathrm{tr}^-(K,t)}{E(K,t)} \left[\eta_{ij}\left(\boldsymbol{K}\frac{K_\alpha}{K}, t_\beta\right) \frac{E(K_\alpha, t_\beta)}{\mathrm{tr}^-(K_\alpha, t_\beta)} \frac{K_{\alpha+1} - K}{K_{\alpha+1} - K_\alpha} + \eta_{ij}\left(\boldsymbol{K}\frac{K_{\alpha+1}}{K}, t_\beta\right) \frac{E(K_{\alpha+1}, t_\beta)}{\mathrm{tr}^-(K_{\alpha+1}, t_\beta)} \frac{K - K_\alpha}{K_{\alpha+1} - K_\alpha} \right] \quad (3)$$

t_β being the value of the set such that $t_\beta < t < t_{\beta+1}$, and K_α such that $K_\alpha < K < K_{\alpha+1}$. $E(K, t)$, $\mathrm{tr}^+(K, t)$ and $\mathrm{tr}^-(K, t)$ are respectively the integral of Φ_{ij} over the sphere of radius K, and the two parts of the isotropic E.D.Q.N.M. transfer associated with $E(K, t)$.

A similar form of $T_{ij}^{\mathrm{M}+}$ is adopted, which is based on the principal directions of Φ_{ij}.

For different values of K, which do not belong to the set of K_α, the modeled forms $T_{ij}^{\mathrm{M}+}(\boldsymbol{K}, t)$ and $T_{ij}^{\mathrm{M}-}(\boldsymbol{K}, t)$ are tested by comparison with the true three-dimensional results of the E.D.Q.N.M. computation. An example of such a comparison is given on Fig. 3. The model can be considered as satisfactory, and no optimization of constants is found to be necessary.

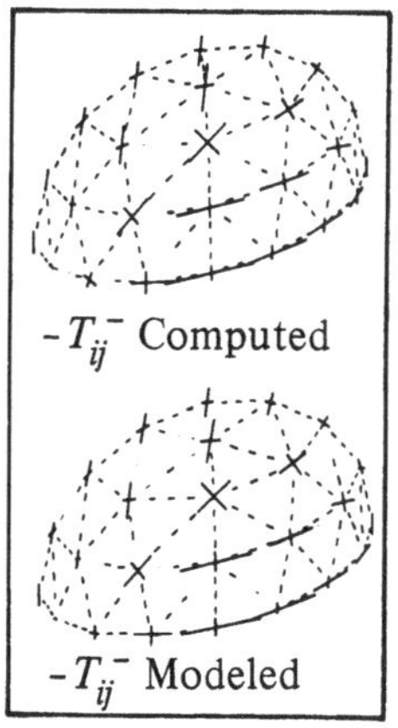

Fig. 3. Comparison between the computed form of T_{ij}^-, at $K = 0.2\ \mathrm{cm}^{-1}$, and its modeled form, interpolated from the results of computations at $K = 0.1\ \mathrm{cm}^{-1}$ and $K = 0.5\ \mathrm{cm}^{-1}$ ($S = 13.9\ \mathrm{s}^{-1}$, $t = 0.1$ s); (same conventions as in Fig. 1)

4. Results

We now use the modeled form of the transfer in the equation governing Φ_{ij}. The only constant appearing in the closure is then the usual E.D.Q.N.M. constant determined in isotropic turbulence. The computation time remains tractable since only a few complete 3D values of T_{ij} have to be computed. As typical values corresponding to the results presented here, we can indicate that T_{ij} is computed over 4 spheres of radius K, and at intervals of 0.02 s. Since the modeled form of the transfer is not exactly conservative, the backscatter is adjusted in the code to ensure energy conservation.

The computing method has been tested in the case of isotropic turbulence. It gives good results when compared to the experiment of *Comte-Bellot* and *Corrsin* [9].

Two non-isotropic cases are presented. They concern the uniform shear flow and the pure plane strain. Figures 4 and 5 show a comparison between our results and those of Rapid-Distortion Theory in the case of a shear flow acting on an initially isotropic turbulence supposed to be identical to the one measured by Comte-Bellot. The evolution

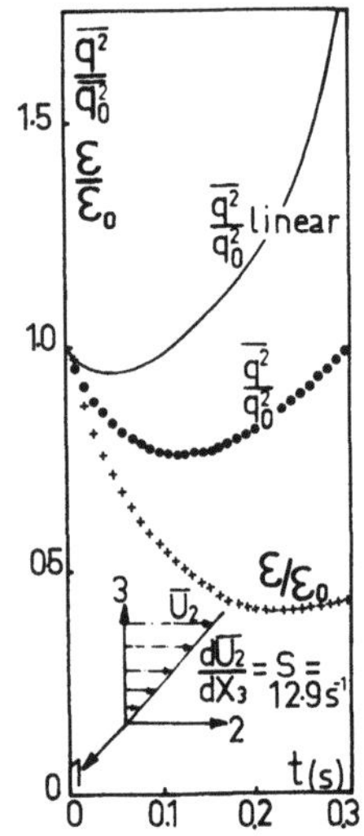

Fig. 4. Evolution with time of the turbulent kinetic energy $\overline{q^2}$, (···) present model, (—) rapid distortion; and evolution with time of the dissipation ϵ, (+) present model

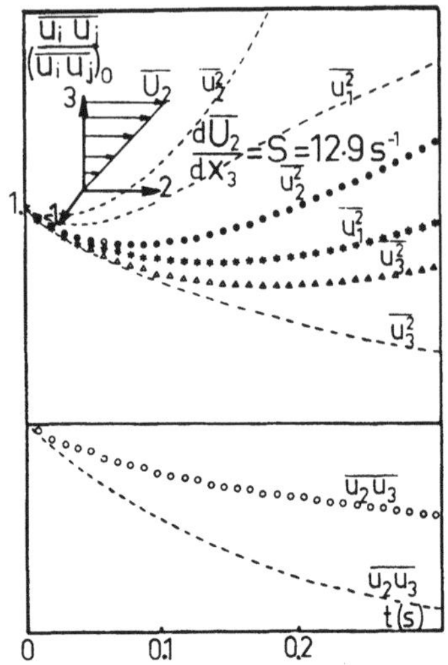

Fig. 5. Evolution with time of the components of the Reynolds stress tensor. (●★△○) Present model, (---) rapid-distortion theory

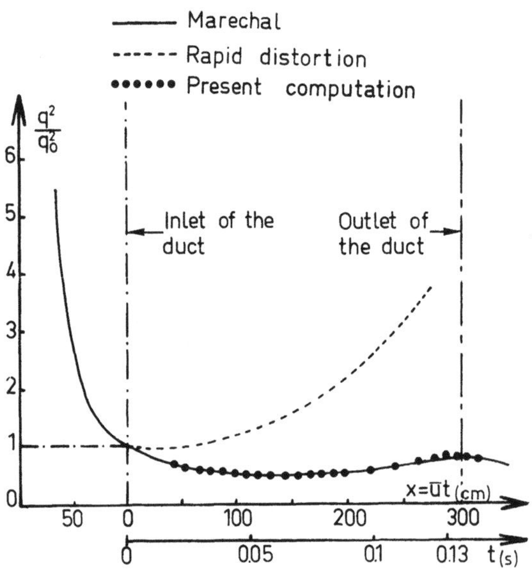

Fig. 6. Evolution with time of the turbulent kinetic energy the case of a plane strain distortion. ($a = \partial\bar{U}_3/\partial X_3 = -\partial\bar{U}_2/\partial X_2 = 19\ \mathrm{s}^{-1}$). Comparison with the experimental values of Maréchal

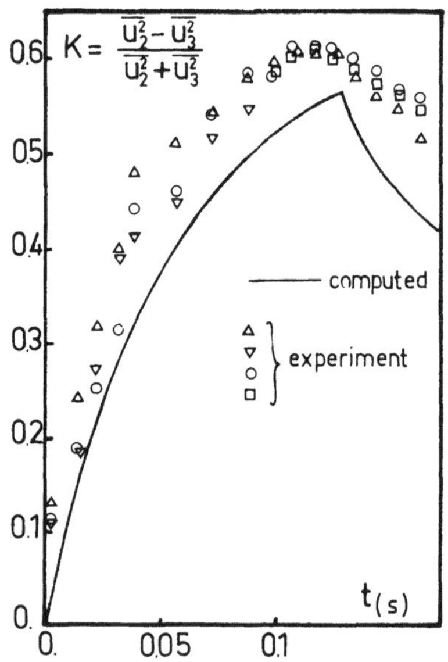

Fig. 7. Evolution with time of the structural parameter in the case of a plane strain distortion ($a = \partial\bar{U}_3/\partial X_3 = -\partial\bar{U}_2/\partial X_2 = 19\ \mathrm{s}^{-1}$)

with time of the turbulent kinetic energy (Fig. 4) indicates that, as expected, the growth of turbulence is reduced by the presence of non-linear terms. The evolution of the components of the Reynolds stress tensor, plotted in Fig. 5, also confirms a common idea: anisotropy is smaller when the action of the turbulence motion on itself is taken into account.

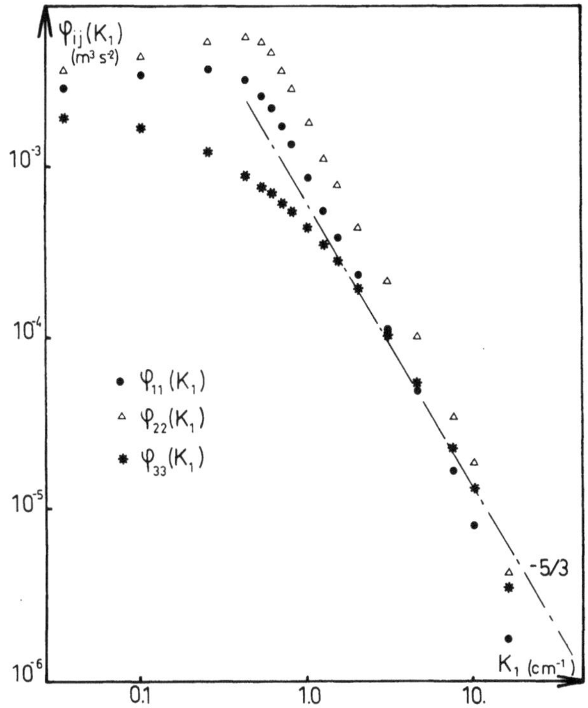

Fig. 8. One-dimensional spectra of the components of the Reynolds stress tensor (t = 0.13 s, a = 19 s^{-1})

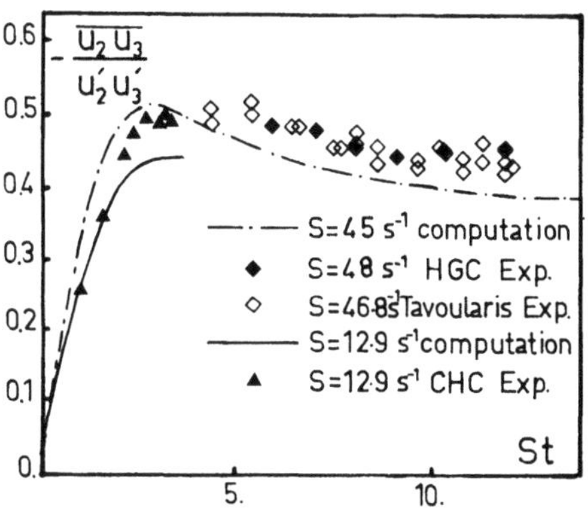

Fig. 9. Evolution with St of the correlation coefficient. Comparison with the experiments of C.H.C., H.G.C. and *Tavoularis* and *Corrsin* (St has been evaluated with the mean value S = 45 s^{-1} for both H.G.C. and Tavoularis experiments)

Comparison with experiment has also been carried out for one of the few cases in which an initial spectrum is known: the pure strain distortion of *Maréchal* [10]. Figure 6 shows that agreement on the evolution of the turbulent kinetic energy is good. As for the structureal parameter κ, it reaches 0.56 (Fig. 7) whereas values of about 0.6 were obtained by Maréchal. The corresponding one-dimensional spectra are plotted in Fig. 8.

The lack of initial experimental data concerning homogeneous shear flow badly restrains the validity of comparison with experiments in the shear case. A comparison is nevertheless attempted on the correlation coefficient $\overline{u_2 u_3}/\sqrt{\overline{u_2^2}}\sqrt{\overline{u_3^2}}$ which seems to be less sensitive to initial conditions than the other quantities. Two values of shear are examined, S = 12.9 s^{-1} and S = 45 s^{-1}, respectively corresponding to the experiments of *Champagne, Harris* and *Corrsin* [11] and *Harris, Graham* and *Corrsin* [12] (or more recently *Tavoularis* and *Corrsin* [13]). In both cases we assume that turbulence is isotropic

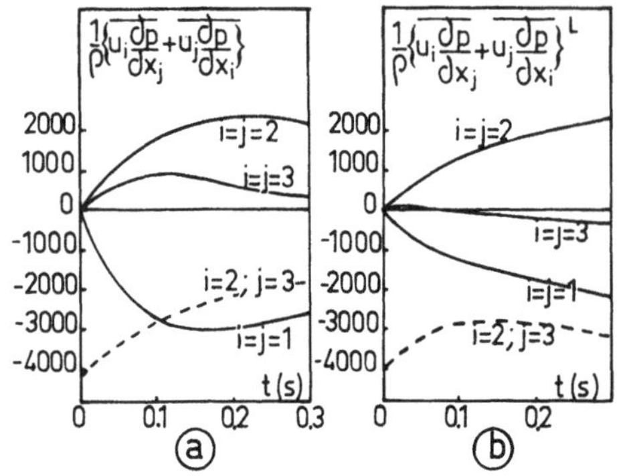

Fig. 10 a, b. Evolution of the linear part of the pressure velocity correlation, for $S = 13.6\ s^{-1}$. **(a)** Rapid-Distortion; **(b)** Present model

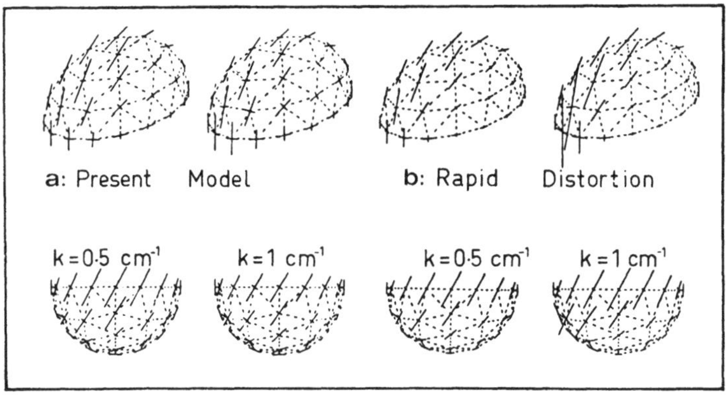

Fig. 11. Visualization of $\Phi_{ij}(\bar{K})$. Comparison between the present model **(a)**, and Rapid distortion **(b)**, in the case of shear $S = 13.6\ s^{-1}$, $t = 0.1$ s, for two values of $K = 0.5\ cm^{-1}$ and $1\ cm^{-1}$

at the entrance of the wind tunnel and we use the initial spectrum deduced from Comte-Bellot's experiment. In the C.H.C. case, our results tend to a nearly asymptotic value of about 0.44 (Fig. 9). The experimental results reach a slightly higher value, but the evolution with time can be considered to be fairly well predicted. In the high shear case, after reaching a maximum, our values are decreasing with time, which seems to agree with the experiments, especially with the measurements of Tavoularis.

Several other features, concerning the behaviour of different turbulent quantities, can be deduced from our computation. As an example, we can mention the results concerning the pressure velocity correlations in the presence of shear, because of the importance of carefully modeling these correlations when doing one point closure (*Launder* [14], *Lumley* [15]). Figure 10 shows that the behaviour of the linear part of the pressure velocity tensor P_{ij}^L is found to be very different from what it is in Rapid-Distortion Theory. All the components are found to be increasing with time, for large values of time. Moreover P_{33}^L soon becomes negative, which corresponds to a supply of energy to $\overline{u_3^2}$, while it remains positive in Rapid-Distortion Theory. The discrepancy between the behaviour of P_{33}^L in our computation and in linear studies is directly connected to the difference of the distributions of Φ_{ij} on the spheres of radius $|K|$ (Fig. 11).

These remarks concerning P_{ij}^L are important because they point out the danger of modeling the linear part of pressure velocity correlations only with reference to Rapid-Distortion Theory when effecting one-point closure.

5. Conclusions

The present paper has shown that a good qualitative and even quantitative prediction of homogeneous turbulent flows can be derived from the analytical theories of turbulence by modeling the non-linear transfer term. This conclusion results from experimental comparisons whose number is restricted by the fact that only isotropic initial spectra were used in our computations. We must mention that a more complete comparison with experiments has been made, by using non-isotropic initial data (*Cambon* [17]). To a certain extent, the results confirm the validity of our model, but the technique for "building" initial anisotropic values for $\Phi_{ij}(\boldsymbol{K})$ can be very questionable.

As no modeled form of the $\Phi_{ij}(\boldsymbol{K},t)$ tensor itself is employed, our method is thought to be particularly appropriate for predicting flows in which the use of isotropic functions is inadequate: rotating flows, for example. The results can be directly used to predict flows, or can provide information to improve one-point closures or other type of model.

References

1. Herring, J.R.: Approach of axisymmetric turbulence to isotropy. Phys. Fluids **17** (1974)
2. Schumann, H., Herring, J.R.: Axisymmetric homogenous turbulence: a comparison of direct spectral simulations with the direct-interaction approximation. J. Fluid Mech. **76**, 755–782 (1976)
3. Cambon, C., Jeandel, D., Mathieu, J.: Spectral modeling of homogenous non-isotropic turbulence. J. Fluid Mech. **104**, 247–262 (1981)
4. Bertoglio, J.P.: Homogenous Turbulent Field Within a Rotating Frame, AIAA 13th Fluid and Plasma Conference, paper 80–1353 (1980)
5. Orszag, S.A.: Analytical theories of turbulence. J. Fluid Mech. **41**, 363–386 (1970)
6. Lesieur, M., Schertzer, D.: Dynamique des gros tourbillons et décroissance de l'énergie cinétique en turbulence tridimensionelle isotrope à grand nombre de Reynolds. J. Méc. **225**, 464 (1978)
7. Kraichnan, R.H.: Test-field model for inhomogenous turbulence. J. Fluid Mech. **56**, 287–304 (1972)
8. Leslie, D.C.: *Developments in the Theory of Turbulence* (Clarendon Press, Oxford 1973)
9. Comte-Bellot, G., Corrsin, S.: Simple Eulerian time correlation of full and narrow band velocity signals in grid-generated isotropic turbulence. J. Fluid Mech. **48**, 273–337 (1971)
10. Marechal, J.: Etude expérimentale de la déformation plane d'une turbulence homogène. J. Méc. **11**, 263–294 (1972)
11. Champagne, F.H., Harris, V.G., Corrsin, S.: Experiments on nearly homogenous turbulent shear flow. J. Fluid Mech. **41**, 81–139 (1970)
12. Harris, V.G., Graham, J.A.H., Corrsin S.: Further experiments in nearly homogenous turbulent shear flow. J. Fluid Mech. **81**, 657–687 (1977)
13. Tavoularis, S., Corrsin, S.: Experiments in nearly homogenous turbulent shear flow with a uniform mean temperature gradient – Part I. J. Fluid Mech. **104**, 311–347 (1981)
14. Launder, B.E., Reece, G.J., Rodi, W.: Progress in the development of a Reynolds stress turbulence closure. J. Fluid Mech. **68**, 537–566 (1975)
15. Lumley, J.L.: *Prediction Methods for Turbulent Flows*, Lecture Series 76 (Karman Institute 1975)
16. Cambon, C., Jeandel, D.: "Approach of Non-Isotropic Homogenous Turbulence Submitted to Mean Velocity Gradients", 3rd Symp. on Turbulent Shear Flows, Davis, Sept. 1981
17. Cambon, C., Bertoglio, J.P., Jeandel, D.: "Spectral Closure of Homogenous Turbulence", The 1980 –81 Afosr Httm Stanford Conference on Complex Turbulent Flows, Stanford, Sept. 1981

A Theoretical Study of Radiative Cooling in Homogeneous and Isotropic Turbulence

D. Schertzer[1] and O. Simonin[2,3]

1 Direction de la Météorologie, EERM/CRMD, Rue de Sevres, F–92 100 Boulogne, France
2 Institut de Mécanique Statistique de la Turbulence, F–13 200 Marseille, France
3 Ecole Nationale Supérieure de Techniques Avancées, F–75 015 Paris, France

Abstract

We apply two-point closure to study radiative cooling in homogenous and isotropic turbulence which should provide a systematic tool for examining this problem. The radiative dissipation of temperature inhomogeneities intervenes as a linear dissipative term in the equation of advection of temperature fluctuations.

A critical regime is studied which could not be analyzed by phenomenological models. In that regime, a balance occurs between the radiative dissipative term and a non-local part of the transfer term. It creates an "intertial-radiative" subrange, where the slope of the temperature spectrum is steeper than -3.

Nomenclature

$B_\nu(T)$	Planck function at frequency ν
$C_{c.o}$	Corrsin-Obukhov constant
C_K	Kolmogorov constant
C_P	Specific heat at constant pressure
D	Kinematic viscosity
$D_R(k) = N(k)/k^2$	Radiative diffusivity
D_θ	Molecular thermal diffusivity
D_R^r	Renormalized radiative diffusivity
D_θ^e	Eddy-diffusivity
E	Turbulent energy spectrum
E_θ	Turbulent temperature variance spectrum
$\boldsymbol{F}$	Net flux vector of radiative energy
G_ν	Green's function of the radiative energy transfer equation
i	Square-root of (-1)
$I_\nu(\boldsymbol{x},\boldsymbol{n},t)$	Specific intensity of radiation
$\boldsymbol{k},\boldsymbol{p},\boldsymbol{q}$	Wave-vectors in Fourier space
$k_{c.o}$	Corrsin-Obukhov wave-number
k_I	Wave-number of injection
k_K	Kolmogorov wave-number
$\boldsymbol{n}$	Unit vector specifying the travelling direction of a ray
$N(k)$	Radiative dissipative term
$P_R = D/D_R(0)$	Radiative Prandtl number
$P_T = D/D_\theta$	Thermal Prandtl number
$R = \chi v^2/\epsilon\langle\theta^2\rangle$	Ratio of dissipation time scales
$R_e = v/(k_I D)$	Reynolds number
t	Time
$t_\nu(r) = \exp(-\beta_\nu r)$	Transmission for radiation of frequency ν
$t_\Delta(r) = \frac{1}{\Delta\nu}\int_{\Delta\nu} t_\nu(r)d\nu$	Mean transmission on the the range of frequency $\Delta\nu$
$T(\boldsymbol{x},t)$	Temperature
$T_\theta(k,t)$	Temperature-variance transfer term
$T_\theta^{NL}(k,t)$	Non local temperature-variance transfer term
$\boldsymbol{u}(\boldsymbol{x},t)$	Fluctuating velocity vector
$v^2 = \langle\boldsymbol{u}\cdot\boldsymbol{u}\rangle$	Eddy kinetic energy
β_P	Planck mean absorption coefficient
β_R	Rosseland mean absorption coefficient
β_ν	Absorption coefficient for radiation of frequency ν

ϵ	Eddy kinetic energy dissipation rate	$\chi(k,t)$	Non-linear flux of temperature variance
ζ_R	Characteristic number of radiative influence	χ_D	Molecular diffusion rate of temperature variance
$\theta(\boldsymbol{x},t)$	Temperature fluctuation	χ_I	Injected flux of temperature variance
ν	Frequency of radiation		
$\pi(k,t)$	Non linear flux of kinetic energy	χ_R	Radiative destruction rate of temperature variance
ρ	Density	$\Phi_\theta(k,t) = k^2 E_\theta(k,t)$	Spectrum of the temperature gradient variance
ρ_a	Density of the absorber		
$\tau_e(k), \tau_D(k), \tau_R(k)$	Characteristic times of advection, molecular diffusion, radiative destruction	$\Omega_\theta(k,t) = \int_0^k dp\, p^2 E_\theta(p,t)$	Flux of temperature gradient variance
$\tau_{k,p,q}$	Characteristic time of triple correlation relaxation	$<\cdot>$	Average value of $\cdot$
		$\hat{\cdot}$	Fourier transform of $\cdot$

1. Introduction

Two-point analytical closures have been successfully developed for the study of a passively convected scalar as the temperature field, see *Leslie* [1] for a review, and [2–5] for new developments. These closures should provide a systematic tool for the study of interaction between turbulence and long-wave radiation. *Towsend* [6] and later *Coantic* [7] have analyzed extensively the importance of this interaction in the context of geophysical flows. This interaction seems especially important in nocturnal planetary boundary layers (e.g., *André* et al. [8]), since the radiative destruction rate of temperature variance χ_R is then of the same order as the molecular one χ_D.

We first discuss a direct derivation in Fourier space for the divergence of the radiative flux. Due to the fact that the radiative transfer equation is linear, Fourier space is rather convenient for obtaining such a result, a feature which has been first exploited by *Spiegel* [9].

A phenomenological approach, slightly different from the one used by *Schved* [10], points out the existence of a critical regime. We characterize it with the help of a "radiative number" ζ_R, which is expressed in terms of the global radiative and kinematic characteristics of the flow and is equal to 1 in the case of a critical regime. For a radiative number ζ_R smaller than this critical value, radiative dissipation becomes of prime importance.

Local and non-local interactions, as analyzed by *Lesieur* and *Schertzer* [11], are discussed relative to their influence. For ζ_R larger than 1, local interactions are dominant as in the usual "inertial-convective" range (i.e., inertial for the temperature and velocity fields). Simple phenomenological models of non-linear transfer of temperature variance may be used, as in *Simonin* et al. [12], to model the pertubed temperature spectrum.

In contrast, for lower ζ_R, non-local interactions become predominant and lead to the existence of an "inertial-radiative" subrange. In this subrange, which is included within the velocity inertial range, a balance occurs between radiative destruction (which is no longer of a perturbative nature) and a non-local part of the non-linear transfer of temperature variance.

The temperature variance spectrum then has a slope notably steeper (< -3) than in the usual "inertial-convective" range ($-5/3$). Some consequences of the appearance of such a subrange are discussed.

2. Radiative Dissipative Term

Long-wave radiative colling is described (e.g. [7]) by the divergence of the radiative flux $\boldsymbol{F}$:

$$\left[\frac{\partial}{\partial t} T(\boldsymbol{x},t)\right]_{\mathrm{R}} = -\nabla \boldsymbol{F}/\rho C_P \tag{1}$$

and introduces a linear dissipative term in the equations for the rate of change of temperature fluctuations $\theta(\boldsymbol{x},t)$.

In Fourier space, this equation and that for the spectrum of temperature variance $E_\theta(k,t)$ in homogeneous and isotropic turbulence read:

$$(\partial/\partial t + D_\theta k^2)\,\hat{\theta}(\boldsymbol{k},t) = -N(k)\,\hat{\theta}(\boldsymbol{k},t) - \mathrm{i}\boldsymbol{k} \int_{\boldsymbol{k}=\boldsymbol{p}+\boldsymbol{q}} \hat{u}(\boldsymbol{p},t)\,\hat{\theta}(\boldsymbol{q},t)\,d\boldsymbol{p} \tag{2}$$

$$(\partial/\partial t + 2D_\theta k^2)\,E_\theta(k,t) = -2N(k)\,E_\theta(k,t) + T_\theta(k,t) \tag{3}$$

where $T_\theta(k,t)$ is the non-linear transfer term of scalar variance, D_θ the molecular diffusivity, and $N(k)$ is the radiative dissipative term. It is worth-while noticing that $N(k)$ modifies the Green's function of Eq. (2), so that it intervenes implicitly in the expression of $T_\theta(k,t)$. We will discuss this point later.

The reason for this relatively simple radiative term is that the radiative transfer equation (e.g. [13]), considering only absorption, is linear with respect to $I_\nu(\boldsymbol{x},\boldsymbol{n},t)$,† the specific intensity at point $\boldsymbol{x}$ of rays travelling in the direction $\boldsymbol{n}(|\boldsymbol{n}| = 1)$:

$$\boldsymbol{n} \cdot \nabla I_\nu = -\beta_\nu I_\nu(\boldsymbol{x},\boldsymbol{n},t) - B_\nu[T(\boldsymbol{x},t)] \tag{4}$$

where β_ν, the coefficient of absorption at frequency ν, is proportional to the concentration ρ_{a} of the absorber, and has the dimension of a wave-number. B_ν is the Planck function which is the appropriate source term in the case of local thermodynamical equilibrium (L.T.E.).

Spiegel [9] has shown that, to first order, the effect of variation over the fluid of absorption coefficient with temperature can be neglected. For similar reasons, we will neglect the fluctuations of the concentration of the absorber. We will therefore confine our attention to a medium with constant β_ν.

The Fourier transform of Eq. (4) with respect to $\boldsymbol{x}$, leads to:

$$\hat{I}_\nu(\boldsymbol{k},\boldsymbol{n},t) = G_\nu(\boldsymbol{k},\boldsymbol{n})\,\hat{B}_\nu(\boldsymbol{k},t) \tag{5}$$

G_ν being the Green's function:

$$G_\nu(\boldsymbol{k},\boldsymbol{n}) = \beta_\nu/(\mathrm{i}\,\boldsymbol{n}\cdot\boldsymbol{k} + \beta_\nu) \tag{6}$$

---

†*Note added in proof:* nevertheless, it must be stressed that the temperature field intervens in a highly non-linear way, due to non-linear coupling between the velocity field and the temperature.

From the spectral intensity I_ν, it is then possible to compute the radiative flux and its divergence according to:

$$\mathrm{i}\,\boldsymbol{k}\cdot\hat{\boldsymbol{F}}(\boldsymbol{k},t) = \mathrm{i}\int_0^\infty d\nu \int_{|\boldsymbol{n}|=1} \boldsymbol{k}\,\boldsymbol{n}\,\hat{I}_\nu(\boldsymbol{k},\boldsymbol{n},t)\,d\boldsymbol{n} \quad . \tag{7}$$

Equation (2) is then obtained by using a linear approximation for the Planck function B_ν:

$$T(\boldsymbol{x},t) = \langle T\rangle + \theta(\boldsymbol{x},t)$$

$$B_\nu[T(\boldsymbol{x},t)] = B_\nu(\langle T\rangle) + \frac{dB}{dT}(\langle T\rangle)\,\theta(\boldsymbol{x},t) \quad , \tag{8}$$

where the temperature $T(\boldsymbol{x},t)$ is the sum of a mean part $\langle T\rangle$ and a fluctuating part $\theta(\boldsymbol{x},t)$. Since we are concerned with homogeneous turbulence, $\langle T\rangle$ is independent of the coordinates of the point $\boldsymbol{x}$ considered and makes no contribution to the evolution of fluctuations. From Eqs. (5–8) it follows that:

$$N(k) = \int_0^\infty d_\nu A_\nu \beta_\nu \left[1 - \frac{\beta_\nu}{k}\arctan\left(\frac{k}{\beta_\nu}\right)\right] \quad A_\nu = \frac{4\pi}{\rho C_P}\,\frac{dB_\nu}{dT}(\langle T\rangle) \quad . \tag{9}$$

In the "grey atmosphere" approximation ($\beta_\nu = \beta$), this expression reduces to [9]:

$$N(k) = A\beta\left[1 - \frac{\beta}{k}\arctan\left(\frac{k}{\beta}\right)\right] \quad A = \frac{4\pi}{\rho C_P}\,\frac{4\sigma\langle T\rangle^3}{\pi} \quad . \tag{10}$$

Though the approximation of a grey atmosphere is satisfactory in the case of stellar atmospheres [13], it is not accurate enough for the earth's atmosphere. In the latter case, the dependence on frequency ν of the absorption coefficient β_ν may classically be taken into account by statistical models of absorption bands [14]. The frequency spectrum is then divided into independent bands of frequencies $\Delta\nu$, narrow compared with the frequency-dependence of the Planck function but broad compared with the spacing between absorption lines. In each band, β_ν is considered as the cumulative absorption due to an infinite array of randomly spaced absorption lines with an arbitrary statistical distribution of line intensities. Accurate two-parameter models of the mean transmission t_Δ, in each band of frequencies $\Delta\nu$, can thus be performed [14] for atmospheric gaseous absorbers. In order to use these band models, following *Simonin* [15], we can express $N(k)$ in terms of the monochromatic transmission, $t_\nu(r) = \exp(-\beta_\nu r)$, to obtain finally:

$$N(k) = -\sum_j A_j\left[\frac{dt_{\Delta j}}{dr}(0) + \int_0^\infty \frac{d^2 t_{\Delta j}}{dr^2}(r)\,\frac{\sin kr}{dr}\,dr\right] \tag{11}$$

$$A_j = \frac{4\pi}{\rho C_P}\int_{\Delta\nu_j}\frac{dB_\nu}{dT}\,d\nu \quad .$$

In the lower atmosphere, the major contribution to the radiative dissipative term $N(k)$, arises from the presence of water vapour [14–15]. In the rest of the text, we will refer, for the expression of a non-grey atmosphere, to a computation of $N(k)$ from the above approximation [11] for a dilute mixture water vapour in air [12].

In every case, it is possible to define the two classical limits:

(I) the opaque limit: $k \to 0 \ N(k) \to D_R(0) k^2$

$$D_R(0) = \frac{1}{3} \int_0^\infty d\nu \frac{A_\nu}{\beta_\nu} = \frac{1}{3} \frac{A}{\beta_R} \tag{12}$$

(II) the transparent limit: $k \to \infty \ N(k) \to N(\infty)$

$$N(\infty) = \int_0^\infty d\nu \, A_\nu \beta_\nu = A\beta_P \tag{13}$$

In the case of a grey atmosphere, the Planck mean absorption coefficient β_P and the Rosseland one, β_R, are both equal to β. In contrast, *Goody* [14] indicates that for the case of gaseous absorption in the lower atmosphere, one has: $\beta_P/\beta_R \sim 10^4$.

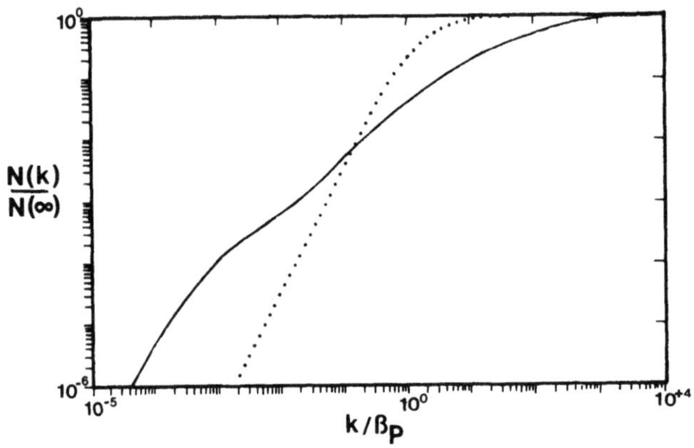

Fig. 1. Radiative dissipative term: (...) grey atmosphere; (—) non grey atmosphere (water vapour dilute with air)

One of the consequences of the large difference between β_P and β_R is that there exists a large subrange of wave-numbers such that:

$$N(k) \sim \alpha k^c \qquad c \geqslant 2/3 \quad . \tag{14}$$

As we will see later, the indicated value (2/3) will be of prime importance.
We can put the radiative dissipative term $N(k)$ in dimensionless form:

$$\psi_P(k/\beta_P) = N(k)/N(\infty) \quad . \tag{15}$$

If the medium contains only one kind of absorber, β_P is proportional to the mean concentration ρ_a of the absorber and the form of the function ψ_P depends only on the mean temperature and pressure [15].

The last expression for $N(k)$ is particulary well adapted to the study of the radiative dissipation in the range of wave-numbers near the transparent limit. For the more global studies, we can use a similar form which depends on the same mean quantities:

$$\psi_M(k/\beta_M) = N(k) \Big/ \left[D_R(0)\,\beta_M^2\right] \qquad \beta_M^2 = \beta_P \beta_R \quad . \tag{16}$$

3. Phenomenology

Classically (e.g. [3]), the local characteristic time of turbulent advection is:

$$\tau_e(k) \sim \left(\epsilon^{1/3} k^{1/3}\right)^{-1} \tag{17}$$

while the molecular-conductive and radiative times are respectively:

$$\tau_D(k) \sim (D_\theta k^2)^{-1} \quad ; \quad \tau_R(k) \sim [N(k)]^{-1} \tag{18}$$

where D_θ is the molecular diffusivity. Recalling that the thermal Prandtl number is defined as:

$$P_T = D/D_\theta \tag{19}$$

(D being the kinematic viscosity) we may define a "radiative Prandtl number" P_R by [see Eq. (12)]:

$$P_R = D/D_R(0) \quad . \tag{20}$$

The inverse of these times is plotted in Fig. 2. A critical regime is obtained when the two curves $\tau_e(k)$ are approximately tangent or intersect. When an intersection occurs, we obtain a subrange $\{k_o, k_t\}$ inside which radiative effects are presumably of prime importance, because $\tau_R(k)$ is the smaller characteristic time. The first point of intersection k_o is on the side of the opaque limit (small k), the second one, k_t, on the side of the transparent limit (large k). Of course, if the intersection of $\tau_D(k)$ and $\tau_R(k)$ occurs at a wave-number k'_t, k'_t characterizes the end of the aforesaid subrange.

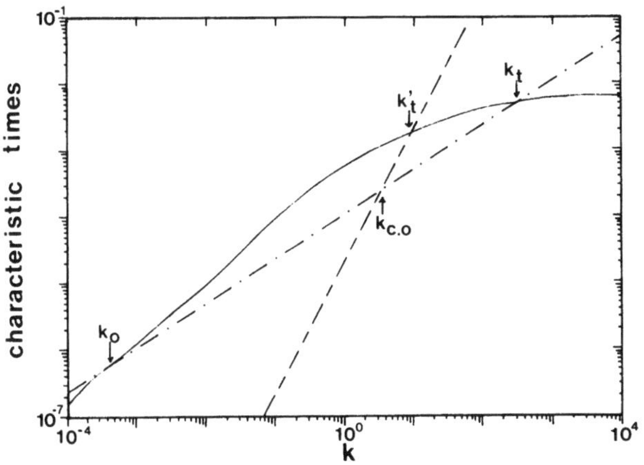

Fig. 2. Characteristic times in s^{-1}, $\zeta_R \cong 0.2$: (- - -) τ_D^{-1}; (—·—·—) τ_e^{-1}; (—) τ_R^{-1}. (non-grey atmosphere

In the case of a grey atmosphere, using crude (but fairly accurate) approximations of $N(k)$ corresponding to its two limits (Eq. 12 and 13), we may approximate k_o and k_t by:

$$k_o \cong \epsilon^{1/4}/D_R(0)^{3/4} \quad ; \quad k_t \cong [3D_R(0)\beta^2]^{3/2}/\epsilon^{1/2} \quad . \tag{21}$$

In any case (whether for grey or non-grey atmospheres), the possible point of tangency gives a certain critical wave-number k_c:

$$k_c d \ln [N(k_c)]/dk = 2/3 \tag{22}$$

while the radiative dissipative term can be written in the form [Eq. (15)]:

$$N(k_c) = \alpha_R N(\infty) \left[\frac{k_c}{\beta_P} \right]^{2/3} \tag{23}$$

in the case of a grey atmosphere: $\alpha_R = 0.284$, for the water vapour dilute mixture with air (see Fig. 1): $\alpha_R \cong 4 \cdot 10^{-2}$.

The case of tangency between τ_e and τ_R is obtained for the critical value ϵ_c of the rate of destruction of turbulent energy satisfying:

$$\epsilon_c^{1/3} \, k_c^{2/3} = N(k_c) \tag{24}$$

and for ϵ lower than ϵ_c, we have a critical regime.

These last Eq. (23, 24) lead us to construct a "radiative number" ζ_R characteristic of the radiative influence on the turbulent temperature variance spectrum:

$$\zeta_R = \frac{\epsilon^{1/3} \beta_P{}^{2/3}}{\alpha_R N(\infty)} \tag{25}$$

ζ_R is equal to 1 when the two curves $\tau_e(k)$ and $\tau_R(k)$ are tangent at a point. And the subrange $\{k_o, k_t\}$ exists when ζ_R is lower than this critical value.

Using Eq. (16) and the classical relations (e.g. [1]):

$$\mathrm{Re} \propto (k_K/k_I)^{4/3} \quad ; \quad k_K = (\epsilon/D^3)^{1/4}$$

(where Re is the Reynolds number, k_K is the Kolmogorov wave-number, k_I is the wave-number of injection of turbulent kinetic energy), one obtains for ζ_R:

$$\zeta_R = \frac{P_R}{\alpha_M} \left(\frac{k_K}{\beta_M} \right)^{4/3} \propto \frac{P_R \mathrm{Re}}{\alpha_M} \left(\frac{k_I}{\beta_M} \right)^{4/3} \tag{26}$$

$$\alpha_M = 3\alpha_R \left(\frac{\beta_M}{\beta_P} \right)^{2/3} .$$

4. Local Interactions

For wave-numbers smaller than k_o or for a characteristic number of radiative influence ζ_R larger than its critical value of unity (i.e. for "small radiative effects") it can be expected that the transfer of temperature variance in Fourier space is dominated by local interactions, as in the case of a simply advected scalar (e.g. [3] or [5]). That is to say, triad interactions such as $k \sim p \sim q$ give the predominant contribution for the non-linear transfer term $T_\theta(k,t)$ and the non-linear flux of temperature variance in scales smaller than k^{-1}:

$$\chi(k,t) = \int_k^\infty dk' \, T_\theta(k',t) \tag{27}$$

Dimensional analysis, or examination of the local divergence of this flux in the framework of two-point closures [5], leads to:

$$\chi(k,t) = C_{c.o}^{-1} k \quad E_\theta(k,t)\, \tau_e^{-1}(k) \tag{28}$$

or:

$$E_\theta(k,t) = C_{c.o}\, k^{-5/3}\, \chi(k,t)\epsilon^{-1/3} \tag{29}$$

$C_{c.o}$ being the Corrsin-Obukhov constant. The variation of the flux $\chi(k)$, with respect to k, is due to the radiative dissipation as a quasi-equilibrium is obtained by an injection of a flux χ_I of temperature variance at a wave-number k_I:

$$d\chi(k) = -\, 2N(k)\, E_\theta(k) dk \quad . \tag{30}$$

It follows that:

$$\chi(k) = \chi_I \exp\left[-\, 2C_{c.o}\, \epsilon^{-1/3} \int_{k_I}^{k} dk' \;\, N(k') k'^{-5/3}\right] \tag{31}$$

i.e., a slight modification, with respect to the usual inertial-convective range.

A further approximation may be made, for the non-critical regime, on noting that in a certain subrange $\{k_1, k_2\}$ around k_c we have [due to the definition of this wave-number, see Eq. (22)]:

$$N(k) \cong \epsilon_c^{1/3}\, k^{2/3} \quad . \tag{32}$$

Thus

$$\chi(k) \cong \chi(k_1)(k_1/k)^{2C_{c.o}(\epsilon_c/\epsilon)^{1/3}} \quad . \tag{33}$$

As noted earlier (14), in the case of a non-grey atmosphere, the approximation (32) can be satisfactory over a large subrange of wave-number, so k_1 may be approximately equal to k_I. This leads a quasi-linear dependence of $\chi(k)$ on $\log(k/k_I)$ as observed on Fig. 3.

In the case of a non-grey atmosphere, the particular form (32) of the radiative dissipative term $N(k)$ leads to a rather flat radiative dissipation spectrum (30) in an "area-preserving" plot (Fig. 4). Thus, for rather large ζ_R the radiative spectral dissipation is

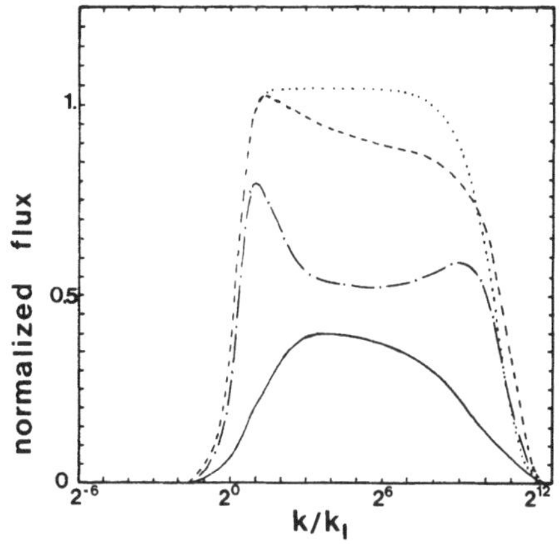

Fig. 3. Weak radiative effect (grey atmosphere) $k_K/k_I \cong 2.5\ 10^3$, $\beta = 2.k_I$, $P_R = 10^{-3} \Rightarrow \zeta_R \cong 16.$: (....) $\pi(k)$, global non-linear flux of kinetic energy; (– – –) $\chi(k)$, global non-linear flux of temperature variance; (——) non-local flux

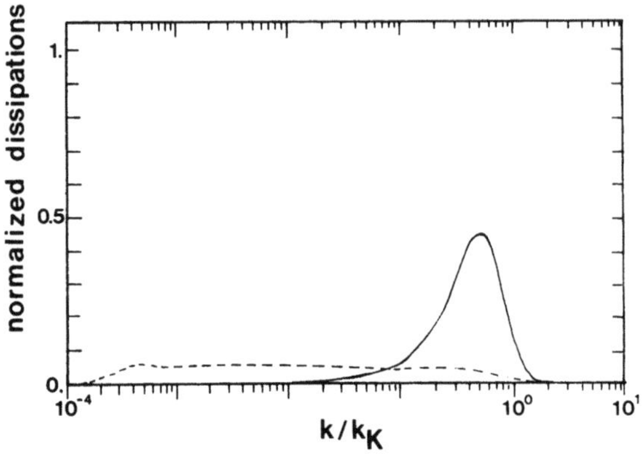

Fig. 4. Weak radiative effect (non-grey atmosphere) $k_K/k_I \cong 2.5 \times 10^3$, $\beta_M = 0.11\ k_I$, $P_R = 4.6 \times 10^{-8} \Rightarrow \zeta_R \cong 20$: (—) $2D_\theta k^2 E_\theta(k,t)$, (---) $2N(k)E_\theta(k,t)$ [area preserving plot $(k/k_K)g(k)$ vs $\log(k/k_K)$]

weak. However, it stretches over a large subrange of wave number and its global contribution χ_R to the global destruction rate of temperature variance χ_I can be important (e.g. [12]: $\zeta_R \cong 20$ and $\chi_R/\chi_I \cong 0.353$).

As soon as ζ_R is less than 1 [or ϵ less than ϵ_c, $\zeta_R = (\epsilon/\epsilon_c)^{1/3}$] Eq. (31) or (33) shows that the exponent of the spectrum may largely change. There is consequently no more guarantee that the local interactions are predominant. For instance, if the slope of the spectrum becomes steeper than −3, the spectrum of the temperature gradient $\Phi_\theta(k)$ is then concentrated in scales larger than k_o^{-1}. Thus, some integrals intervening in $\chi(k)$ are no longer locally divergent. Loosely speaking, what is on the left of the spectrum may weight more than the remainder on the right.

This point is totally missed by phenomenological models leading to (28), even if τ_e is replaced by a time scale taking into account the dissipative one [16].

5. Non-Local Interactions: An "Inertial-Radiative" Subrange

In physical space, non-local interactions mean interactions between very different scales. In Fourier space, a precise definition can be given (e.g. [11], i.e. a triad $(\boldsymbol{k}, \boldsymbol{p}, \boldsymbol{q})$ is non local if:

$$[\min\{k,p,q\}/\max\{k,p,q\}] < a << 1 \tag{34}$$

where a is some fixed parameter, defining the amplitude of non-localness.

Suppose now, as the local analysis of the preceding section shows, that around k_o the slope of the temperature spectrum becomes steeper. The main contributions to the non-linear flux $\chi(k, t)$, $k > k_o$, are given by non-local triads of interactions $(\boldsymbol{k}, \boldsymbol{p}, \boldsymbol{q})$ corresponding to triple correlations $<\hat{\theta}(\boldsymbol{k})\hat{\theta}(\boldsymbol{p})\hat{\boldsymbol{u}}(\boldsymbol{q})>$ with:

$$p << k_o << q \sim k \tag{35}$$

Using the technique developed in *Lesieur* and *Schertzer* [11] and applying it to Eq. (3), the most important part χ_D^{NL} of the non-local flux χ^{NL}, created by the non-local interactions, can be put in the form of an "eddy-diffusivity":

$$\chi_D^{NL}(k,t) = (2/3) \int_k^\infty dq\, E(q,t) \int_0^{\inf\{aq,k\}} dp\, \tau_{q,p,q}\, p^2 E_\theta(p,t) \tag{36}$$

where $E(k, t)$ is the energy spectrum and $\tau_{k,p,q}$ is the characteristic time of relaxation of triple correlation $<\hat{\theta}(\boldsymbol{k})\hat{\theta}(\boldsymbol{p})\hat{\boldsymbol{u}}(\boldsymbol{q})>$. In the velocity-inertial range, we have,

$$\tau_{k,p,q} \cong [\tau_e^{-1}(k) + \tau_e^{-1}(p) + \tau_e^{-1}(q) + N(k) + N(q)]^{-1} \tag{37}$$

On taking the derivative with respect to k, (36) leads to the non-local part of the transfer term:

$$T_{\theta D}^{NL} = (2/3)E(k, t) \int_0^{ak} dp\, \tau_{k,p,k}\, p^2 E_\theta(p, t) - 2D_\theta^e(k, t, a)k^2 E_\theta(k, t) \tag{38}$$

$D_\theta^e(k, t; a)$ being the eddy-diffusivity, taking into account the influences of scales o smaller than $(k/a)^{-1}$ on the scale k^{-1}:

$$D_\theta^e(k, t; a) = (1/3) \int_{k/a}^{\infty} dp\, \tau_{p,k,p}\, E(p, t) \tag{39}$$

When k is larger than k_o, for ζ_R larger than 1, i.e. $N(k) > \tau_e^{-1}(k)$, $\tau_{k,p,q}$ may be approximated by:

$$p < k_o < k \qquad\qquad \tau_{k,p,q} \cong N(k)^{-1} \quad . \tag{40}$$

Moreover, we may introduce the "renormalized radiative diffusivity" D_R^r (using renormalization language [18]):

$$D_R^r(k, t; a) = D_R(k) + D_\theta^e(k, t; a) \tag{41}$$

It must be stressed that, Eq. (41) gives a false appearance of a simple additivity between the radiative diffusivity and an eddy-diffusivity, independent of the radiative diffusivity. This is clearly not the case, an increase in D_R implies (37, 40) a decrease in D_θ^e. We will comment later on this important point.

If we suppose that $T_{\theta D}^{NL}$ is the leading part of the transfer term T_θ, its balance with the radiative dissipation can be written, with the help of Eq. (38–40), as:

$$2D_R^r(k, t; a)k^2 E_\theta(k, t) = (2/3)E(k, t)\Omega_\theta(k, t)/N(k) \tag{42}$$

Equation (42) ensures the existence of an "inertial-radiative" subrange between k_o and k_c:

$$E_\theta(k, t) \sim k^{-s} \qquad\qquad s > 3 \tag{43}$$

Because of Eqs. (40–42) may be roughly approximated by:

$$E_\theta(k, t) \sim (1/3)C_K\, \epsilon^{2/3} k^{-5/3-2c}\, \Omega_\theta(k_o) \tag{44}$$

where we use the convergence of Ω_θ, i.e. $\Omega_\theta(k) \sim \Omega_\theta(k_o)$. Beyond k_c, Eq. (44) indicates a return to the $k^{-5/3}$ law which is reached around k_t, if this wave number is far enough into the transparent limit ($c \cong 0$).

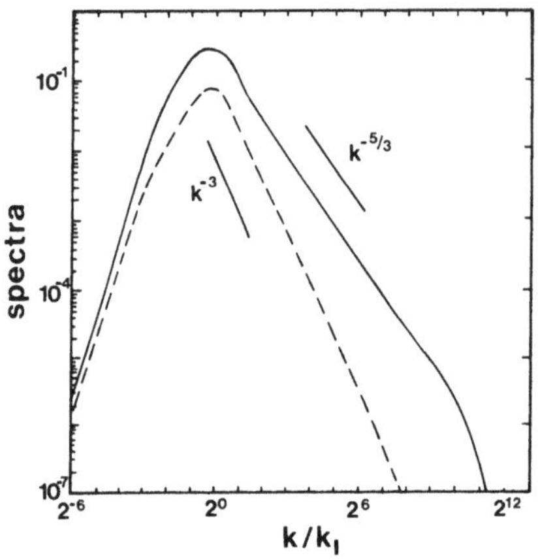

Fig. 5. Strong radiative effect, spectra for non-grey atmosphere: $k_K/k_I \cong 2.5\ 10^3$, $\beta_M = 2k_I$, $P_R = 10^{-7} \Rightarrow \zeta_R \cong 0.2$ (——) $E(k,t)$; (- - - -) $E_\theta(k,t)$

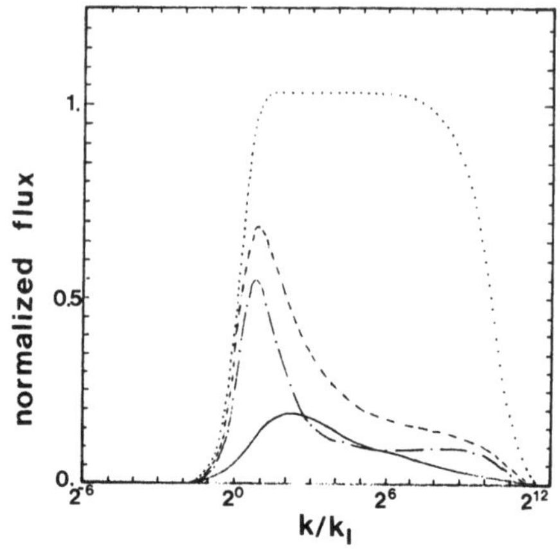

Fig. 6. Strong radiative effect $\zeta_R \cong 0.2$ (non-grey atmosphere), same conditions as Fig. 5 (· · ·) $\pi(k)$, global non-linear flux of kinetic energy; (— — —) $\chi(k)$, global non-linear flux of temperature variance; (–.–.–) local non-linear flux of temperature variance; (——) non-local flux

In the case of a non-grey atmosphere, Fig. 5 shows that in the subrange $\{k_o, k_t\}$ the exponent of the temperature spectrum remains nearly equal to –3, because c itself stays nearly equalt to 2/3 [see (14)].

As noted earlier, Eq. (44), the non-local equilibrium described tends to create an apparent "inertial-convective" subrange near k_t, if the wave number is close enough to the transparent limit, i.e. $N(k_t) \sim N(\infty)$. For $k > k_t$, a real inertial-convective subrange takes place up to $k_{c.o}$, the Corrsin-Obukhov wave-number. In this range the flux χ_D of temperature variance which is cascaking towards $k_{c.o}$ is of course the part of the flux which is left after destruction by radiative effects, and which will be destroyed by molecular diffusivity.

6. Concluding Remarks

Above the critical value of the radiative number ζ_R ($\zeta_R = 1$), for which advective and radiative time scales are tangent, see Eq. (24), radiative cooling has only a slight perturbative effect on the temperature variance spectrum. It introduces a small exponential correction [a small power correction near k_c, $N(k_c) \sim k_c^{2/3}$] to the $k^{-5/3}$ temperature inertial range spectrum. This range survives, though the flux of temperature variance is no longer constant as the radiative destruction occurs through all the range. The analysis made in Sect. 4, see Eq. (31), leads to the following estimation of the molecular dissipation of temperature variance χ_D, compared with its injected flux χ_I:

$$\ln(\chi_D/\chi_I) \cong 2\epsilon^{-1/3} C_{c.o} \int_{k_I}^{k_{c.o}} dk' k'^{-5/3} N(k') \quad . \tag{45}$$

For ζ_R lower than 1, i.e. for larger radiative destruction, the spectrum is largely affected by the appearance of an "inertial-radiative" subrange. Non-local interactions predominate and determine the behaviour of the temperature spectrum, the slope of the spectrum being nearly a constant steeper than -3. In the case of the non-grey atmosphere we have studied (water vapour in air) this exponent remains very close to -3.

For different absorbers, or different conditions, the value of this exponent may be very different because it depends largely, see Eq. (42–44), on the behaviour of $N(k)$, the radiative dissipative term. Moreover, if $N(k)$ is steeper than the one we have used, non-local interactions will be more effective than in our study. In any case, the variance of the temperature gradient is mainly concentrated in scales larger than k_o^{-1} or smaller than k_t^{-1} (provided, of course, an inertial-convective subrange still develops for $k > k_t$, i.e. if $k_t < k_{c.o}$).

On the other hand, the behaviour of the larger scales does not seem greatly affected by radiative transfer in the case of a grey atmosphere. Thus the ratio of dissipation times:

$$R = \chi v^2/\epsilon<\theta^2> \tag{46}$$

remains nearly a constant ($R \cong 2.2$) in the stationary case. This is not the case in a non-grey atmosphere, where R may reach 3.5. This difference seems to rely on the fact that in a non-grey atmosphere the opaque limit (small k) is reached only very slowly. Thus the balance which occurs between the "beating-term" and the eddy diffusivity in the large scales (see [5]) seems strongly affected.

It must be stressed that a cumulative effect of radiative dissipation and molecular dissipation, as modeled by *André* et al. [7], remains doubtful due to our remarks on the way the radiative diffusivity is renormalized (41). That is to say, for large radiative effects, the characteristic times of eddy diffusivity and of radiative diffusivity do not intervene in an additive way. This point of view is supported by the corresponding study made by *Kraichnan* [17] in the case of low thermal Prandtl number.

A first suggestion has been given by *Simonin* in [15] for weak radiative effects ($\zeta_R > 1$) in line with our remarks. Observing that the usual one-point modeling of the destruction rate of temperature variance includes one part of the radiative destruction rate, the other part of the radiative destruction could be taken into account by adding a second term. This term represents only the modification of the large scales due to the radiative destruction of temperature variance. However, this problem needs further work which is in progress.

Also underway is the study of the influence of fluctuations of concentration in the absorber due to the interactions with the fluctuating radiative field generated by the fluctuations of temperature and with a mean anisotropic radiative field.

Finally, we agree with *Schved* [10] that, what we call a critical regime ($\zeta_R < 1$), could probably not be reached in the earth's atmosphere, but rather on planets such as Mars (CO_2 atmosphere and weak pressure). Moreover, it could be worthwhile to study the appearance of this critical regime in reacting flows.

Acknowledgements. The authors would like to thank Prof. M. Coantic for stimulating discussions and Drs J.C. André and U. Frisch for helpful advice.

References

1. Leslie, D.C. : *Developements in the Theory of Turbulence*, 1st ed. (Clarendon Press, Oxford 1973)
2. Newman, G.R., Herring, J.R.: A test field model study of a passive scalar in isotropic turbulence. J. Fluid Mech. **94**, 163–194 (1979)
3. Larchevéque, M., Chollet, J.P., Herring, J.R., Lesieur, M., Newman, G.R., Schertzer, D.: "Two-Point Closure Applied to a Passive Scalar in Decaying Isotropic Turbulence", in *Turbulent Shear Flows II,* ed.by L.J.S. Bradbury, F.Durst, B.E. Launder, F.W. Schmidt, J.H. Whitelaw (Springer Berlin, Heidelberg, New York 1980) pp. 50–66
4. Larchevéque, M., Lesieur, M.: The application of E.D.Q.N.M. to the problem of dispersion of particles in three- and two- dimensional turbulence. J. Méc (1980)
5. Herring, J.R., Schertzer, D., Chollet, J.P., Larchevéque, M., Lesieur, M., Newman, G.R.: A comparative assessment of spectral closures as applied to passive scalar diffusion. J. Fluid Mech. **124**, 411–437 (1982)
6. Towsend, A.A.: The effects of radiative transfer on turbulent flow of a stratified fluid. J. Fluid Mech. **4**, 361–375 (1958)
7. Coantic, M.F.: Interaction Between Turbulence and Radiation. AGARDograph **232**, 175–236 (1978)
8. André, J.C., De Moor, G., Lacarrère, P., Therry, G., Du Vachat, R.: Modeling the 24-hour evolution of the mean and turbulent structures of the planetary boundary layer. J. Atmos. Sci. **35**, 1861–1883 (1978)
9. Spiegel, E.A.: The smoothing of temperature fluctuations by radiative transfer. Astrophys. J. **165**, 202–207 (1957)
10. Schved, G.M.: Influence of radiative transfer on certain types of motion in planetary atmospheres. Adv. Heat Transfer **14**, 249–280 (1978)
11. Lesieur, M., Schertzer, D.: Amortissement autosimilaire d'une turbulence à grand nombre de Reynolds. J. Mec. **17**, 249–280 (1978)
12. Simonin, O., Coantic, M.F., Schertzer, D.: Effet du rayonnement infrarouge sur la turbulence de temperature dans l'atmosphère: structure spectrale et taux de dissipation. C.R. Acad. Sci. Paris **293**, 245–248 (1981)
13. Chandrasekhar, S.: "The equation of Transfer", in: *Radiative Transfer* (Dover, New York 1953) pp. 1–38
14. Goody, R.M.: "Theoretical Basis", in: *Atmospheric Radiation*, Part 1 (Clarendon, Oxford 1964)
15. Simonin, O.: "Influence du transfert radiatif sur la turbulence de temperature: application à la couche limite planétaire": Thèse de Docteur-Ingenieur présenté à l'Ecole Nationale Supérieur de Techniques Avancées, Paris (1981)
16. Hill, R.J.: Models of the scalar spectrum for turbulent advection. J. Fluid Mech **88**, 541–562 (1978)
17. Kraichnan, R.H.: Shear and thermally-driven turbulence. Phys. Fluids **7**, 1048–1062 (1964)
18. Forster, D., Nelson, D.R., Stephen, M.J.: Large-distance and long-time properties of a randomly stirred fluid. Phys. Review A, **16**(2), 732–749 (1977)

Second Order Closure for Variable Density Free Shear Layer

Dany Vandromme[1] and Wolfgang Kollmann[2]
Von Karman Institute, Belgium

Abstract

A full second order closure turbulence model for the prediction of free shear flows with variable density is developed. The importance of density fluctuations due to inhomogeneities in fluid composition is emphasized. Some simplicity is achieved by using density weighted averaging to account implicitly for density fluctuation effects. Transport equations for various second order correlations – such as Reynolds stress, turbulent mass fluxes or density velocity correlations – are derived, modelled and solved, together with the mean equations for momentum and species conservation. Solutions of these equations are compared with the two different sets of experimental data for low speed free mixing layers (helium into nitrogen and freon 12 into air).

Nomenclature

c_I	I-specie mass fraction
D	velocity divergence
k	turbulent kinetic energy
P	pressure
T_c	triple correlation
v_α/v_β	velocity components
α	Falkner Skan parameter
$\delta_{\alpha\beta}$	Kronecker symbol
δ_1, δ_2	characteristic thicknesses
ϵ	kinetic energy dissipation rate
$\epsilon_{\alpha\beta}$	dissipation rate tensor
ϕ_0	velocity ratio
ρ	density
σ	spreading rate parameter
$\tau_{\alpha\beta}$	laminar stress tensor
$(\bar{\ }), (\ )'$	conventional averaging
$(\tilde{\ }), (\ )''$	density weighted averaging
$c_\epsilon, c_{\epsilon 1}, c_{\epsilon 2}, c_{\epsilon 3}, c_{\epsilon 4}, c, c_1, c_2, c_3, c_{\phi 1}, c_{\phi 2}, c_{UDP}, c_{P2}, c_{DS}, c_{c1}, c_{c2}, c_c, c_Q, c_{Q1}, c_{Q2}$	modelling constants

Introduction

The prediction of the mean motion of a chemically homogeneous fluid depends on the evaluation of the Reynolds stress (double velocity fluctuation correlations) that appear in the mean momentum equation for turbulent flow. Similarly, the evaluation of the mean species concentrations or temperatures from their conservation equations requires the definition of the local correlations of fluctuations of velocity and temperature and/or species concentration. Early models of turbulence related these correlations (or fluxes

---

1 *Present address:* NASA-Ames Research Center, Moffett Field, CA 94035, USA
2 *Present address:* Department of Mechanical Engineering, University of California, Davis, CA 95616, USA

of momentum, heat or mass of species) only to the mean values of velocity, temperature or concentration. Many years ago *Boussinesq* [1] related the flux term of the momentum equation to the mean velocity gradient and an eddy viscosity. Such a relationship functionally similar to the Newton's law of viscosity has been used in a large variety of turbulent models. For all of them the main task is to derive a suitable form for the eddy viscosity. Heat and species flux relationship required further assumptions regarding turbulent Prandtl or Schmidt numbers.

The present paper presents a set of turbulence equations closed at a higher level. Transport equations are derived, modelled and solved directly for second order moments (Reynolds stress and the turbulent mass flux) to avoid the questionable concept of eddy viscosity. Several investigations have dealt with this closure level (e.g., [2–8]) but only two [6, 8] have considered variable density flows. This study is restricted to density variations caused by inhomogeneities in fluid composition and does not consider compressibility effects typical of high speed or chemically reacting flows. To achieve checks against experimental data, two different flows are considered:

(I) low speed mixing layer of helium and nitrogen [9]
(II) low speed mixing layer of freon 12 and air [10]

In the next section, equations for the different variables and correlations are commented on. Then the necessary modelling assumptions are presented. The fourth section gives some details about the numerical method used to solve the set of equations, the boundary conditions, and the amount of CPU time for a typical run. Then the results are compared with the experimental data and discussed. The last section summarizes the results obtained so far and suggests future directions for the use of this closure to the solution of more general turbulence problems.

Exact Equations

To account implicitly for variable density effects in all the equations, it is helpful to work with mass weighted variables rather than conventionally (time) averaged quantities. This yields a mean mass conservation equation having a similar form to the instantaneous one, and convection terms of all the transport equations forms similar to instantaneous substantial derivatives. The main disadvantage of that averaging is to produce a more complicated equation for turbulent kinetic energy dissipation rate. Nevertheless, since all the terms of the dissipation equation even when conventionally averaged have to be modelled, the additional complications are merely treated as model modifications. Although the main advantage of the mass weighted averaging is to avoid the use of density velocity correlations, it is necessary to evaluate them to refine the determination of third order correlations and to deal with the mean pressure gradient terms. Thus transport equations for density velocity correlations are included in this model. Nevertheless the convective mean terms keep their exact form and stay density correlation free. The exact equations are not reproduced here because of their length but can be found in reference [11].

Modelling Assumptions

Reynolds Stress

Diffusion Terms

$$-\frac{\partial}{\partial x_\alpha}\left[\bar{\rho}\;\widetilde{v''_\alpha v''_\beta v''_\gamma} + \delta_{\alpha\beta}\;\overline{v''_\gamma P'} + \delta_{\alpha\gamma}\;\overline{v''_\beta P'} - \overline{\tau_{\alpha\beta} v''_\gamma} - \overline{\tau_{\alpha\gamma} v''_\beta}\right] .$$

The diffusive transport exists in three different forms. The first one can be identified as the turbulent diffusion, the second as the pressure diffusion and the last one as the viscous diffusion.

– Turbulent diffusion. To express this third order correlation, we use a procedure similar to *Launder's* [2]. Considering the transport equation for the mass weighted third order velocity correlation, we can simplify it by keeping only the leading terms.

$$\begin{aligned}
&-\frac{\partial}{\partial x_\delta}(\bar{\rho}\;\widetilde{v''_\alpha v''_\beta v''_\gamma v''_\delta}) + \widetilde{v''_\alpha v''_\beta}\;\frac{\partial}{\partial x_\delta}(\bar{\rho}\;\widetilde{v''_\delta v''_\gamma}) \\
&+ \widetilde{v''_\alpha v''_\gamma}\;\frac{\partial}{\partial x_\delta}(\bar{\rho}\;\widetilde{v''_\delta v''_\beta}) + \widetilde{v''_\beta v''_\gamma}\;\frac{\partial}{\partial x_\delta}(\bar{\rho}\;\widetilde{v''_\delta v''_\alpha}) \\
&- \overline{v''_\beta v''_\gamma\;\frac{\partial P'}{\partial x_\alpha}} - \overline{v''_\beta v''_\alpha\;\frac{\partial P'}{\partial x_\gamma}} - \overline{v''_\beta v''_\alpha\;\frac{\partial P'}{\partial x_\beta}} = 0 \;.
\end{aligned} \tag{1}$$

As for constant density flows, use is made of the assumption of quasi-normal velocity distribution, also known as the Millionshtchikov approximation. For velocity distribution nearly Gaussian the triple velocity correlations are small and the fourth order correlation can be expressed in terms of second order correlations only, by the following formulae which is strictly fulfilled for the normal law.

$$\overline{v'_\alpha v'_\beta v'_\gamma v'_\delta} = \overline{v'_\alpha v'_\beta}\cdot\overline{v'_\gamma v'_\delta} + \overline{v'_\alpha v'_\gamma}\cdot\overline{v'_\beta v'_\delta} + \overline{v'_\alpha v'_\delta}\cdot\overline{v'_\beta v'_\gamma} \;. \tag{2}$$

From the definition of the mass weighted averaging the relationship between the fluctuations is:

$$v'_\alpha = v''_\alpha + \frac{\overline{\rho' v'_\alpha}}{\bar{\rho}} \;. \tag{3}$$

This equality allows rewriting the equation (2) in terms of mass weighted correlations. The new expression includes a large collection of density correlations, that are mostly negligible (see [11]). After reduction to the leading terms the fourth order mass weighted correlation can be expressed as

$$\begin{aligned}
\widetilde{v''_\alpha v''_\beta v''_\gamma v''_\delta} = &- k_\alpha\;\widetilde{v''_\beta v''_\gamma v''_\delta} - k_\beta\;\widetilde{v''_\alpha v''_\gamma v''_\delta} - k_\gamma\;\widetilde{v''_\alpha v''_\beta v''_\delta} \\
&- k_\delta\;\widetilde{v''_\alpha v''_\beta v''_\gamma} + \widetilde{v''_\alpha v''_\beta}\cdot\widetilde{v''_\gamma v''_\delta} + \widetilde{v''_\alpha v''_\gamma}\cdot\widetilde{v''_\beta v''_\delta} + \widetilde{v''_\alpha v''_\delta}\cdot\widetilde{v''_\beta v''_\gamma}
\end{aligned} \tag{4}$$

where

$$k_\alpha = \frac{\overline{\rho' v'_\alpha}}{\bar{\rho}} = -\overline{v''_\alpha} \; .$$

In case of constant density flows *Hanjalić* and *Launder* [2] assume that the last three terms of Eq. (1) are "adequately represented" by

$$\overline{v'_\beta v'_\gamma \frac{\partial P'}{\partial x_\alpha}} + \overline{v'_\beta v'_\alpha \frac{\partial P'}{\partial x_\gamma}} + \overline{v'_\gamma v'_\alpha \frac{\partial P'}{\partial x_\beta}} \sim \frac{\epsilon}{k} \overline{v'_\alpha v'_\beta v'_\gamma} \; . \tag{5}$$

Assuming this relationship holds for constant density, we can obtain the corresponding relationship in terms of mass weighted quantities by using Eq. (3). This yields

$$\overline{v''_\beta v''_\gamma \frac{\partial P'}{\partial x_\alpha}} + \overline{v''_\beta v''_\alpha \frac{\partial P'}{\partial x_\gamma}} + \overline{v''_\gamma v''_\alpha \frac{\partial P'}{\partial x_\beta}} \sim \frac{\epsilon}{k} \widetilde{v''_\alpha v''_\beta v''_\gamma} \; . \tag{6}$$

Assembling Eq. (1, 4, 6)

$$\widetilde{v''_\alpha v''_\beta v''_\gamma} = -\bar{\rho} \; c \; \frac{k}{\epsilon} \left[\widetilde{v''_\gamma v''_\delta} \frac{\partial}{\partial x_\delta} (\widetilde{v''_\alpha v''_\beta}) + \widetilde{v''_\beta v''_\delta} \frac{\partial}{\partial x_\delta} \widetilde{v''_\alpha v''_\gamma} + \widetilde{v''_\alpha v''_\delta} \frac{\partial}{\partial x_\delta} \widetilde{v''_\beta v''_\gamma} \right]$$

$$+ \frac{k}{\epsilon} \frac{\partial}{\partial x_\delta} \left[\bar{\rho} \, k_\alpha \widetilde{v''_\beta v''_\gamma v''_\delta} + \bar{\rho} \, k_\beta \widetilde{v''_\alpha v''_\gamma v''_\delta} + \bar{\rho} \, k_\gamma \widetilde{v''_\alpha v''_\beta v''_\delta} \right] \tag{7}$$

c is a constant whose value will be given in the next section. If the density is constant Eq. (7) is strictly identical to the version of [2, 3]. Thus the third order correlations are determined from a set of first order differential equations, which for boundary layer type flow is

$$A_1 \cdot T_c + A_2 \frac{\partial}{\partial y}(T_c) = A_3 \; . \tag{8}$$

A_1, A_2 and A_3 are functions of the turbulence time scale, the second order velocity correlations and the density correlations.

– Diffusive transport by pressure fluctuations. So far no satisfactory model has been used for this term. Most authors consider a combined modelling of the triple correlation with the pressure velocity correlation but none have modelled the pressure correlation separately. *Lumley* [7], *Harlow* and *Nakayama* [16] and *Donaldson* [5] proposed different relationships but none yields improvements of the flow description. Furthermore a number of experiments show that the turbulent energy budget virtually balances without including the unmeasured pressure transport term [17]. Therefore, it is neglected here.

– Diffusive transport by viscous effects. In the assumption of high Reynolds number flow, this contribution is negligible compared to the turbulent diffusive terms. Nevertheless we may ensure a minimal value to the diffusive term by including a constant laminar viscosity in it.

Mean pressure gradient term. As stated previously, the solution of density velocity correlation equations allows dealing with non-zero mean pressure gradient flows. As these terms are exact in the second order moment equations, no modelling assumption is required here.

Pressure-rate of strain correlation. The form to be used here is rewritten from the constant density version of *Launder* [2, 3] in terms of mass weighted averaged quantities. Nevertheless, terms involving the mean dilatation, which do not contribute in constant density flows are retained here. The result is

$$\overline{P'\left(\frac{\partial v''_\beta}{\partial x_\gamma}+\frac{\partial v''_\gamma}{\partial x_\beta}\right)} = - c_1\,\bar\rho\,\epsilon\left[\frac{\widetilde{v''_\beta v''_\gamma}}{k}-\frac{2}{3}\,\delta_{\gamma\beta}\right]+\frac{c_2+8}{11}\,\bar\rho\cdot\left[\widetilde{v''_\beta v''_\delta}\,\frac{\partial\tilde v_\gamma}{\partial x_\delta}+\widetilde{v''_\gamma v''_\delta}\,\frac{\partial\tilde v_\beta}{\partial x_\delta}\right]$$

$$-\frac{6c_2+4}{11}\,\bar\rho\cdot\left[\widetilde{v''_\beta v''_\gamma}\,\tilde D+\delta_{\gamma\beta}\,\widetilde{v''_m v''_l}\,\frac{\partial\tilde v_m}{\partial x_l}\right]$$

$$+\frac{8c_2-2}{11}\,\bar\rho\left[\widetilde{v''_\gamma v''_\delta}\,\frac{\partial\tilde v_\delta}{\partial x_\beta}+\widetilde{v''_\beta v''_\delta}\,\frac{\partial\tilde v_\delta}{\partial x_\gamma}\right]$$

$$-\frac{30c_2-2}{55}\left[\frac{\partial\tilde v_\beta}{\partial x_\gamma}+\frac{\partial\tilde v_\gamma}{\partial x_\beta}\right]\bar\rho\,k-\frac{40c_2+12}{55}\,\delta_{\gamma\beta}\,\bar\rho\,k\,\tilde D\ . \quad (9)$$

Dissipation. The last terms appearing in the Reynolds stress equation have a dissipative character. They correspond to the conversion process of mechanical energy into heat due to viscous effects. Arguing for the locally isotropic nature of dissipative processes at high Reynolds number, these terms can be expressed in terms of the dissipation rate of turbulent kinetic energy for which a transport equation is included in the present set.

$$-\overline{\tau_{\alpha\beta}\,\frac{\partial}{\partial x_\alpha}\,v''_\gamma}-\overline{\tau_{\alpha\gamma}\,\frac{\partial}{\partial x_\alpha}\,v''_\beta}=-\frac{2}{3}\,\bar\rho\,\delta_{\gamma\beta}\,\epsilon\ . \quad (10)$$

Modelled Reynolds stress equation. The previous assumptions lead to the following equation for the Reynolds stress:

$$\frac{\partial}{\partial t}\,(\bar\rho\,\widetilde{v''_\gamma v''_\beta})+\frac{\partial}{\partial x_\alpha}\,(\bar\rho\,\tilde v_\alpha\widetilde{v''_\gamma v''_\beta})=-\bar\rho\,\widetilde{v''_\alpha v''_\beta}\,\frac{\partial\tilde v_\gamma}{\partial x_\alpha}-\bar\rho\,\widetilde{v''_\alpha v''_\gamma}\,\frac{\partial\tilde v_\beta}{\partial x_\alpha}-\frac{2}{3}\,\bar\rho\,\delta_{\gamma\beta}\,\epsilon$$

$$+\frac{\partial}{\partial x_\alpha}\left[\mu_L\,\frac{\partial}{\partial x_\alpha}\,(\widetilde{v''_\beta v''_\gamma})-\bar\rho\,\widetilde{v''_\alpha v''_\beta v''_\gamma}\right]$$

$$-\overline{v''_\beta}\,\frac{\partial\bar P}{\partial x_\gamma}-\overline{v''_\gamma}\,\frac{\partial\bar P}{\partial x_\beta}+\overline{P'\left(\frac{\partial v''_\beta}{\partial x_\gamma}+\frac{\partial v''_\gamma}{\partial x_\beta}\right)} \quad (11)$$

in which the pressure-rate of strain correlation is given by Eq. (9).

Dissipation Rate of Turbulent Kinetic Energy

The form of this equation is somewhat similar to the one used in [12] with the following differences from the standard k-ϵ version.

– The diffusivity coefficients allow non-isotropic processes.

$$-\bar{\rho}\ \overline{v''_\beta \epsilon} = c_\epsilon \cdot \bar{\rho} \cdot \frac{k}{\epsilon} \cdot \widetilde{v''_\alpha v''_\beta} \cdot \frac{\partial \epsilon}{\partial x_\alpha} \quad . \tag{12}$$

– The mean pressure gradients have a form similar to their counterpart in the Reynolds stress equation. S_{p} represents the set of mean pressure gradient terms in the exact equation.

$$S_{\mathrm{p}} = -c_{\epsilon 4} \cdot \frac{\epsilon}{k} \overline{v''_\beta} \frac{\partial \bar{P}}{\partial x_\beta} \quad . \tag{13}$$

– The velocity divergence term, called S_{D} has the form:

$$S_{\mathrm{D}} = -c_{\epsilon 3} \cdot \bar{\rho} \cdot \epsilon \cdot \tilde{D} \quad . \tag{13}$$

In which $\tilde{D}$ is the mass weighted velocity divergence. These assumptions yield the following dissipation equation:

$$\frac{\partial}{\partial t}(\bar{\rho}\epsilon) + \frac{\partial}{\partial x_\alpha}(\bar{\rho}\ \tilde{u}_\alpha \epsilon) = -c_{\epsilon 1} \frac{\epsilon}{k} \bar{\rho}\ \widetilde{v''_\alpha v''_\beta} \frac{\partial \tilde{v}_\alpha}{\partial x_\beta} - c_{\epsilon 2}\ \bar{\rho}\ \frac{\epsilon^2}{k}$$
$$+ \frac{\partial}{\partial x_\alpha}\left(c_\epsilon \cdot \frac{k}{\epsilon} \cdot \widetilde{v''_\alpha v''_\beta} \frac{\partial \epsilon}{\partial x_\beta}\right) - c_{\epsilon 3} \cdot \bar{\rho} \cdot \epsilon \cdot \tilde{D} - c_{\epsilon 4} \frac{\epsilon}{k} \overline{v''_\beta} \frac{\partial \bar{P}}{\partial x_\beta} \quad . \tag{14}$$

Density Velocity Correlation

According to the definition of mass weighted averaging, this correlation corresponds to the mean value of the mass weighted fluctuation.

– The diffusive transport term is modelled according to a constant density model for triple correlation [2]

$$\overline{\rho' v''_\alpha v''_\beta} = -c_{\mathrm{s}} \frac{k}{\epsilon}\left[\widetilde{v''_\alpha v''_\gamma} \frac{\partial}{\partial x_\gamma}(\overline{\rho' v''_\beta}) + \widetilde{v''_\beta v''_\gamma} \frac{\partial}{\partial x_\gamma}(\overline{\rho' v''_\alpha})\right] . \tag{15}$$

This approximation allows permutation of indices of velocity components but no permutation of density and velocity fluctuations.

The next term is approximated by:

$$\overline{v''_\beta \frac{\partial v''_\alpha}{\partial x_\alpha}} = c_{\mathrm{UDP}} \frac{\sqrt{\overline{\rho'^2}}}{\bar{\rho}} \overline{v''_\beta} \frac{\partial}{\partial x_\alpha} \overline{v''_\alpha} \quad . \tag{16}$$

There is no precise way to fix the value of c_{UDP} though it may be presumed of order unity.

The basic idea to approximate the pressure term is to use a procedure similar to the pressure-rate of strain modelling in the Reynolds stress equation. Following *Lumley* [7], the "Rotta" term is approximated by:

$$\phi_{\beta\rho,1} = - c_{\phi 1} \cdot \frac{\epsilon}{k} \cdot \overline{\rho' v''_\beta} - c_{\phi 2} \frac{\epsilon}{k} \left[\frac{\widetilde{v''_\alpha v''_\beta}}{k} - \frac{2}{3} \delta_{\alpha\beta} \right] \overline{\rho' v''_\alpha} \tag{17}$$

allowing for anisotropy effects and the rapid part is

$$\phi_{\beta\rho,2} = c_{\phi 2} \overline{\rho' v''_\beta} \frac{\partial \tilde{v}_\beta}{\partial x_\alpha} \quad . \tag{18}$$

– If the fine scale motion is assumed isotropic the last term

$$\overline{(\rho'/\rho)(\partial \tau_{\alpha\beta}/\partial x_\alpha)}$$

is equal to zero. The modelled transport equation for the density velocity correlation is written as:

$$\frac{\partial}{\partial t} (\bar{\rho}\, \overline{v''_\beta}) + \frac{\partial}{\partial x_\alpha} (\bar{\rho}\, \tilde{v}_\alpha \overline{v''_\beta}) = - \bar{\rho}\, \overline{v''_\alpha} \frac{\partial \tilde{v}_\beta}{\partial x_\alpha} + \widetilde{v''_\alpha v''_\beta} \frac{\partial \bar{\rho}}{\partial x_\alpha} + c_{\mathrm{UDP}} \sqrt{\overline{\rho'^2}}\, \overline{v''_\beta} \frac{\partial}{\partial x_\alpha} (\overline{v''_\alpha})$$

$$- \frac{\partial}{\partial x_\alpha} \left(c_s \frac{k}{\epsilon} \widetilde{v''_\alpha v''_\gamma} \frac{\partial}{\partial x_\gamma} (\overline{\rho' v''_\beta}) + \widetilde{v''_\beta v''_\gamma} \frac{\partial}{\partial x_\gamma} \cdot (\overline{\rho' v''_\alpha}) \right)$$

$$+ c_{\rho 2} \overline{\rho' v''_\alpha} \frac{\partial \tilde{v}_\beta}{\partial x_\alpha} - c_{\phi 1} \frac{\epsilon}{k} \overline{\rho' v''_\beta} - c_{\phi 2} \frac{\epsilon}{k} \left(\frac{\widetilde{v''_\alpha v''_\beta}}{k} - \frac{2}{3} \delta_{\alpha\beta} \right) \overline{\rho' v''_\alpha} \tag{19}$$

Scalar Flux

The modelling assumptions used to close this equation are similar to those for the Reynolds stress equation. Again the production terms are exact.

– The whole diffusive transport process is represented by the following group of terms:

$$\frac{\partial}{\partial x_\gamma} (\bar{\rho}\, \widetilde{v''_\beta v''_\gamma c''_I} + \delta_{\beta\gamma} \overline{P' c''_I} - \delta_{\gamma\beta} \overline{\tau_{\gamma\beta} c''_I}) \quad .$$

For the triple correlation the assumption concerning the diffusive mass weighted Reynolds stress transport is extended to concentration correlations. The following first order differential equations results:

$$- \bar{\rho}\, \widetilde{v''_\gamma v''_\delta} \frac{\partial}{\partial x_\delta} (\widetilde{v''_\beta c''_I}) - \bar{\rho}\, \widetilde{v''_\beta v''_\delta} \frac{\partial}{\partial x_\delta} \widetilde{v''_\gamma c''_I} - \bar{\rho}\, \widetilde{v''_\delta c_I} \frac{\partial}{\partial x_\delta} \cdot (\widetilde{v''_\beta v''_\gamma})$$

$$+ \frac{\partial}{\partial x_\delta} \left[\bar{\rho}\, k_\beta \widetilde{v''_\gamma v''_\delta c''_I}) + \bar{\rho}\, k_\gamma \widetilde{v''_\beta v''_\delta c''_I} + \bar{\rho}\, k_c \widetilde{v''_\beta v''_\delta v''_\gamma} + \bar{\rho}\, k_\delta \widetilde{v''_\beta v''_\gamma c''_I} \right]$$

$$= - c' \frac{\epsilon}{k} \bar{\rho}\, \widetilde{v''_\beta v''_\gamma c''_I} \tag{20}$$

in which

$$k_\alpha = \frac{\overline{\rho' v''_\alpha}}{\bar{\rho}} \quad ; \quad k_c = \frac{\overline{\rho' c''_I}}{\bar{\rho}} \quad .$$

The other terms in the diffusive transport contribution correspond to diffusion of concentration by pressure fluctuations and by molecular motion. The pressure effects are neglected here, whereas the molecular effects are represented by a laminar gradient flux law similarly to the Reynolds stress laminar diffusion.

– Redistribution due to pressure fluctuations. Again we use a procedure which is similar to the redistributive term of the Reynolds stress equation, for which we get two different contributions: a fully turbulent and a mean gradient part. The turbulent term may be related to the linear return-to-isotropy approximation of *Rotta* [13]

$$\phi_{\beta c,1} = -c_c \frac{\epsilon}{k} \bar{\rho} \widetilde{v''_\beta c''_I} \tag{21}$$

in which case it must be expected that the time scale (k/ϵ) used here contains also information about the time scale of the turbulent concentration field $\langle c_i''^2\rangle/\epsilon_c$, $\langle c_i''^2\rangle$ being the concentration covariance and ϵ_c its rate of dissipation. *Launder* [14] and *Janicka* [15] pointed out that these two time scales are different. In local equilibrium turbulence they may be assumed proportional. Nevertheless an anisotropy term is added to the "Rotta" term, which carries Reynolds stress anisotropy information. This yields to a form similar to *Lumley's* proposal [7]

$$\phi_{\beta c,1} = -c_{c1} \bar{\rho} \frac{\epsilon}{k} \widetilde{v''_\beta c''_I} - c_{c2} \bar{\rho} \frac{\epsilon}{k} \left(\frac{\widetilde{v''_\beta v''_\alpha}}{k} - \frac{2}{3} \delta_{\alpha\beta} \right) \widetilde{v''_\alpha c''_I} \;. \tag{22}$$

The mean gradient part of the pressure term is transposed from the leading term of the pressure rate of strain term.

$$\phi_{\beta c,2} = c_c \cdot \bar{\rho} \cdot \widetilde{v''_\alpha c''_I} \frac{\partial \tilde{v}_\beta}{\partial x_\alpha} \;.$$

– Dissipation. In the framework of high Reynolds number flows, the fine scale motions are nearly isotropic and the correlation

$$\overline{\left(\tau_{\gamma\beta} \frac{\partial}{\partial x_\gamma} c''_I \right)}$$

which is strictly zero in isotropic turbulence is negligible.

– A mean pressure gradient term exists also in the scalar flux equation. Nevertheless, the isothermal flow assumption avoids the need for a solution of a PDE for the density concentration correlation. It can be expressed in terms of density fluctuations [6, 11]

$$\overline{c''_I} = -\frac{\rho_J \rho_I}{(\rho_I - \rho_J)} \frac{\overline{\rho'^2}}{\bar{\rho}^3} \;. \tag{23}$$

This relationship holds only under the assumption of isothermal flow. If heat transfer problems were to be solved with the dynamic field equations, this part would require some modifications for the description of thermal turbulence.

The modelled scalar flux equation is:

$$\frac{\partial}{\partial t}(\bar{\rho}\,\widetilde{v''_\beta c''_I}) + \frac{\partial}{\partial x_\alpha}(\bar{\rho}\,\tilde{v}_\alpha \widetilde{v''_\beta c''_I}) = -\bar{\rho}\,\widetilde{v''_\alpha c''_I}\,\frac{\partial \tilde{v}_\beta}{\partial x_\alpha} - \bar{\rho}\,\widetilde{v''_\alpha v''_\beta}\,\frac{\partial \tilde{c}_I}{\partial x_\alpha}$$

$$+ \frac{\partial}{\partial x_\alpha}\left(\mu_L \frac{\partial}{\partial x_\alpha}(\widetilde{v''_\beta c''_I}) - \bar{\rho}\,\widetilde{v''_\alpha v''_\beta c''_I}\right) + c_c\,\bar{\rho}\cdot\widetilde{v''_\alpha c''_I}\,\frac{\partial \tilde{v}_\beta}{\partial x_\alpha}$$

$$+ \frac{\rho_I \rho_J}{\rho_I - \rho_J}\,\frac{\overline{\rho'^2}}{\bar{\rho}^3}\,\frac{\partial \bar{P}}{\partial x_\beta} - c_{c1}\,\bar{\rho}\,\frac{\epsilon}{k}\,\widetilde{v''_\beta c''_I} - c_{c2}\,\bar{\rho}\,\frac{\epsilon}{k}\left(\frac{\widetilde{v''_\alpha v''_\beta}}{k} - \frac{2}{3}\,\delta_{\alpha\beta}\right)\widetilde{v''_\alpha c''_I}\,. \tag{24}$$

Density Fluctuations

The derivation of the transport equations for the partial density correlations with the involved modelling efforts has been published in a previous work [12] and thus will not be reproduced here.

Comparison with Experiments

Flow Conditions

Two different sets of experimental data were considered to provide a valuable comparison of computational results against physics. The first set of experiments is that of *Rebollo* [9]. This work emphasizes the influence of non-zero longitudinal mean pressure gradient on a two-dimensional mixing layer. Both adjacent free stream flows have the same dynamic pressure. The gases are helium and nitrogen which yield a density ratio equal to 7. The velocities are 10.9 m/s for the light gas and 4.12 m/s for the heavy one. The respective Reynolds numbers are

helium → Re = 3600/cm
nitrogen → Re = 12000/cm .

From his similarity analysis Rebollo has shown that, together with the dynamic pressure equality, a second equilibrium condition concerns the definition of the longitudinal mean pressure gradient, through the assignment of a constant Falkner-Skan parameter:

$$\alpha = \frac{x}{u}\,\frac{dU}{dx} \quad . \tag{25}$$

The two values considered here are 0. and −0.18.

The second set of experiments is provided by the thesis of *Brown* [10]. This set emphasizes the influence of the free stream velocity ratio, and considers air-air and freon 12-air mixings. The density ratio is either 1 or 4 and the velocity ratio 0, 0.3 or 0.6 in the first case and 0.3 and 0.6 in the second case. In both cases the high speed air stream has a velocity equal to 14.0 m/s.

All the boundary layers at the splitter plate trailing edge are laminar and free stream turbulence level is very low ($Urms/UE < 1\%$). Various experimental techniques have been used in these experiments to provide a wide range of information about turbulence quantities such as Reynolds stress, density fluctuations, concentration fluctuations, density velocity correlations, turbulent fluxes, etc... .

Numerical Method

The numerical method which is used for this calculation is a modified version of the *Patankar* and *Spalding* method [18], described in detail in [11]. All the transport equations are used in their boundary layer form and are solved in a parabolic marching step manner. To start the calculations we must provide the initial values for each transported variable. The given velocity profiles are Blasius-like and agree with the boundary layer thickness values reported by the experiments. The normal Reynolds stresses were assigned to a constant value (isotropic turbulence) in such a way the turbulent kinetic energy remains bounded by the experimentally reported limit. The turbulent shear stress is set to zero and so are all the other turbulent correlations. The dissipation rate is determined to produce a turbulent viscosity equivalent to the laminar one, according to the well known k-ϵ formula [19].

$$\mu_t = 0.09\ \bar{\rho} k^2/\epsilon \quad . \tag{26}$$

The boundary conditions are Dirichlet type for the primitive variables such as velocity or concentration and Neumann type (zero gradient) for all the second order moments. When different from zero, the longitudinal mean pressure gradient must be prescribed analytically as the values of the outer boundaries velocities. The transverse pressure gradient cannot be prescribed but only approximated by a simplified version of the mean transverse momentum equation yielding to

$$\frac{\partial \bar{P}}{\partial y} = - \frac{\partial}{\partial y} (\bar{\rho}\ \widetilde{v''^2}) \quad . \tag{27}$$

The set of ordinary differential equations for the triple correlations is solved once at each x-step by a simple centered difference scheme rather than algebraically. The upper bound of the marching step of the main procedure is determined at each forward position with respect to the value of a characteristic thickness of the mixing layer. To solve a typical problem with this 15 equation turbulence model over a distance of 1 m with 80 node points in the cross direction, the time is about 15 min CPU on a DEC-VAX 11/780 or about 90 seconds CPU on a CDC 7600 computer.

Discussion of the Results

Free stream velocity ratio influence. The main results of the comparison against the *Brown* experiments [10] are summarized in Table 1 for the three different velocity ratios (0.0, 0.3, 0.6) and the two gas compositions (air-air and F12-air). The agreement of the calculations with Brown's numbers is satisfactory except for the following items:

– Virtual origin position. Two different mechanisms establish the virtual origin. First is properly described by the calculation – and second is the transition process which takes place between the attached laminar boundary layers on the splitter plate and the free turbulent mixing layer developing downstream of it. The second mechanism cannot be predicted by the calculation. The calculations develop a transition-like region but there is no rigorous way to relate that phenomenon to the physical laminar to turbulent transition. The spreading rates are compared through the use of the parameter (σ) which is defined from the velocity profile

$$\sigma = 1.19 \ \frac{\Delta x}{\Delta y} \quad . \tag{28}$$

This defintition is similar to the one given in [20] except the characteristic thickness is determined between the bounds (0.2, 0.8) instead of the usual (0.1, 0.9).

– Another point of disagreement concerns the spreading of the F12-air mixing layer with 0.6 velocity ratio. As it is shown on Fig. 1, the experimental point is high above the theoretical homogeneous flow curve defined by

$$\sigma = 11.5 \ \frac{1+\phi_0}{1-\phi_0} \quad ; \quad \phi_0 = \frac{U_0}{U_1} \tag{29}$$

whereas the calculation matches the curve well. Also for this flow case (see Table 1), the measurements show a higher level of *V'rms* compared to *U'rms*. This feature does not appear in the calculations for which the ratio *U'rms*/*V'rms* remains nearly the same value as for the 0.3 velocity ratio case (= 1.6).

Table 1. Velocity ratio influence. (Exp. from *Brown* [10])

	ρ_0	σ	$\frac{U'rms}{\Delta U}$	$\frac{V'rms}{\Delta V}$	$\frac{-10\langle u'v'\rangle}{\Delta U^2}$	Kuv
	air-air					
exp	0.00	11.97	0.150	–	–	–
calc		11.30	0.192	0.125	0.131	0.548
exp	0.30	23.34	0.165	0.125	0.105	0.509
calc		21.22	0.184	0.120	0.120	0.545
exp	0.60	39.27	0.195	0.165	0.162	0.503
calc		40.37	0.184	0.120	0.120	0.542
	F12-air					
exp	0.30	21.86	0.175	0.125	0.112	0.512
calc		25.67	0.181	0.113	0.112	0.548
exp	0.60	23.76	0.168	0.195	0.180	0.549
calc		54.95	0.198	0.126	0.132	0.532

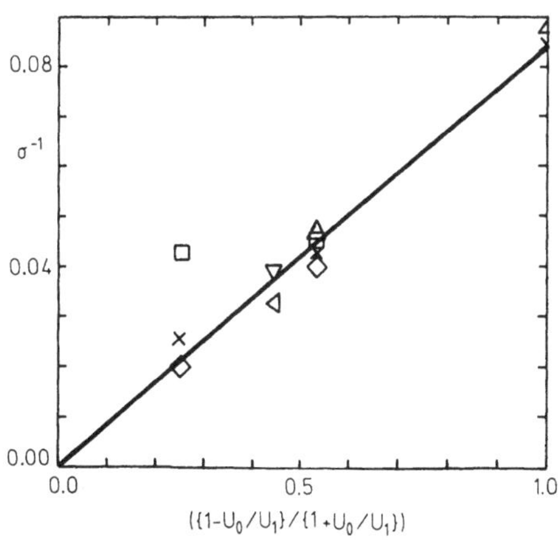

Fig. 1. Spreading rate parameter (line: theory)
× Air-air mixing [10]
□ Freon 12-air mixing [10]
△ Air-air mixing } calculation
◇ Freon 12-air mixing
◁ He-nitrogen $dp/dx = 0$ } calculation
▽ He-nitrogen $dp/dx = 0$

The last point of disagreement concerns the turbulence levels for which Brown reports values increasing with the velocity ratio, where one could expect that turbulence increases when the shear increases (or ϕ_0 decreases) as occurs in the calculations.

The main result of this calculation however, is consistent with the observation that density variations do not have significant influence on the spreading rate of a mixing layer if these fluctuations are due only to inhomogeneities in gas composition.

Pressure gradient influence. More extensive calculations were done with the helium-nitrogen flow [11] than with the previous F12-air mixture, but some conclusions are identical.

The virtual origin cannot be determined correctly, because the transitional mechanism which influences it does not exist in this turbulence model. Thus the only way to compare the spreading of the mixing zone is to look at the value of the downstream spraeding rate parameter. [Figure 1 shows that for this flow ($dP/dX = 0.$) the agreement is as good as for the previous one.] The most interesting results are shown in Table 2 for the two cases which are considered here: $\alpha = 0.$ and $\alpha = -0.18$. In both cases the velocity ratio is $U_0/U_1 = 0.378$. Globally, the agreement is good for the zero pressure gradient case and weaker for the case with positive pressure gradient, for which the turbulence quantities are underestimated by 25–40%.

No direct comparison of the spreading rate parameter is reported here because it has been found in the calculations that the self-similar region begins only after 7–8 cm from the splitter plate, and no experimental value is given beyond this abscissa. Nevertheless the calculated spreading rate parameter agrees with the results of the previous calculations and with the theoretical curve of Fig. 1, when it is calculated further downstream ($8 < X < 30$ cm), where the self-similarity is ensured. For the near wake region, two different thicknesses [9] are plotted against experiments and for which the agreement is good (Fig. 2). Figures 3, 4, and 5 show some typical results for the shear stress, the density velocity correlation and the density fluctuations, for which the experiments provide information. The results of Table 2 point out that the results are good for all the mean quantities but do not agree with the pressure gradient influence. The experiments show

Table 2. Pressure gradient effects. (Exp. from *Rebollo* [9])

$\alpha = \frac{X}{U}\frac{dU}{dX}$	$\alpha = 0.00$		$\alpha = -0.18$	
	exp.	calc.	exp.	calc.
$\delta_1/x-x_0$	0.10	0.10	0.16	0.15
σ	–	31.34	–	25.97
U/U_1	0.70	0.67	0.59	0.59
$\overline{\rho U^2}/\rho_1 U_1^2$	0.87	0.81	0.60	0.59
$\bar{\rho}/\rho_1$	1.78	1.74	1.70	1.62
$\overline{\rho U^2}/\rho_1 U_1^2$	0.52	0.54	0.32	0.31
$\overline{\rho U^2}/\rho_1 U_1^2$	1.22	1.19	1.20	1.22
$\overline{\rho'^2}/\Delta\rho$	1.19	1.18	1.20	1.15
$\overline{\rho u'v'}/\rho_1 U_1^2$	0.012	0.012	0.0205	0.0147
$\rho' v'/\rho_1 U_1$	0.068	0.044	0.082	0.040
Sc_t	0.16	0.22	0.33	0.16

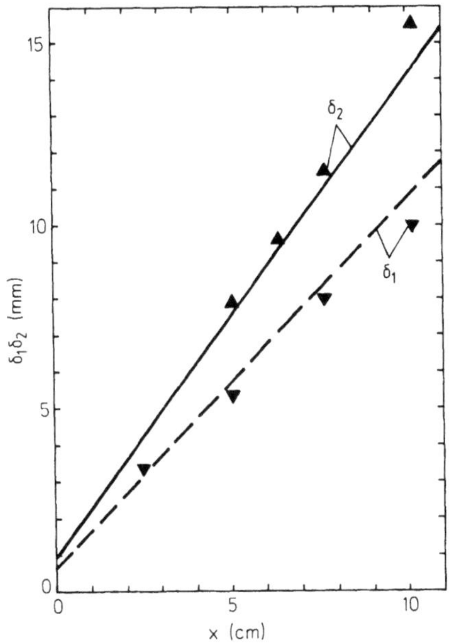

Fig. 2. Spreading of He-N mixing layer (symbols: [9], lines: calculations)

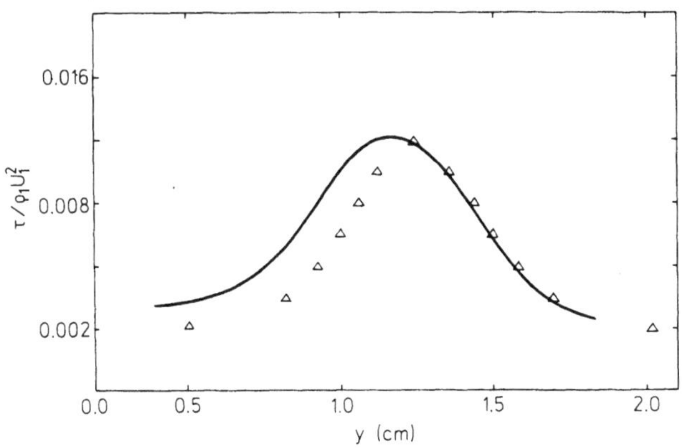

Fig. 3. Shear stress in H-N mixing layer at x = 5.08 cm for $\alpha = 0$ (symbols: [9], lines: calculations)

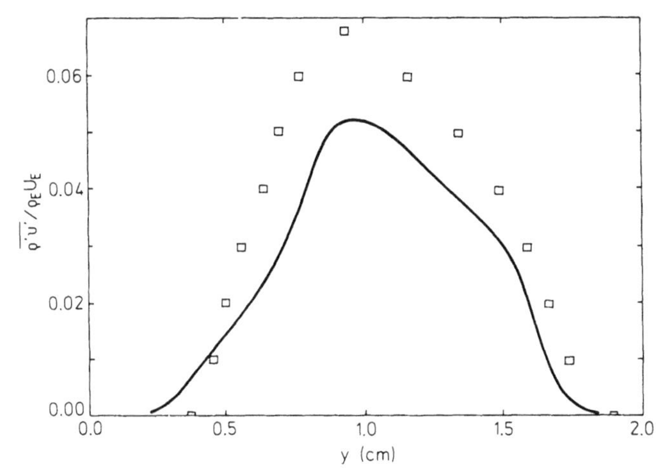

Fig. 4. Density velocity correlations in He-N mixing layer at x = 5.08 cm for $\alpha = 0$ (symbols: [9], lines: calculations)

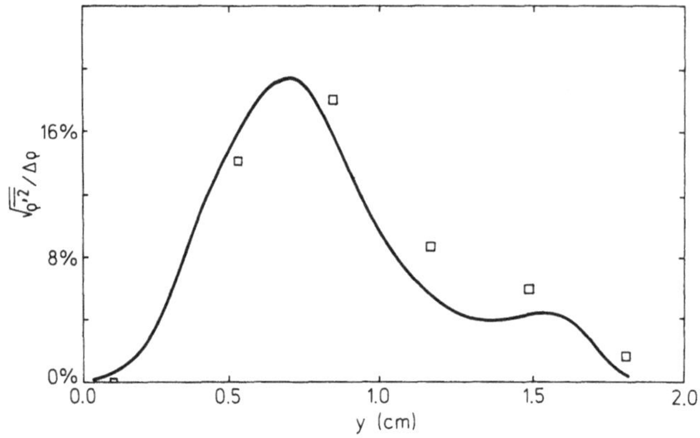

Fig. 5. Density fluctuation intensity in He-N mixing layer at $x = 5.08$ cm for $\alpha = 0$ (symbols: [9], lines: calculations)

Table 3. Values of modelling constants

c	0.11	c_1	1.5	c_2	0.4				
c_ϵ	0.13	$c_{\epsilon 1}$	1.4	$c_{\epsilon 2}$	2.1	$c_{\epsilon 3}$	-1.0	$c_{\epsilon 4}$	2.0
c_s	0.11	$c_{\phi 1}$	4.0	$c_{\phi 2}$	4.0	c_{UDP}	1.0	$c_{\rho 2}$	0.5
c_{Ds}	0.014	c_{cl}	4.0	c_{c2}	-4.8	c_{c}	0.5		
c_{Q}	0.6	c_{Q1}	3.0	c_{Q2}	0.6				

that the pressure gradient tends to increase the spreading rate, the shear stress and the density velocity correlations, but not the density fluctuations whose maximum level remains nearly the same. The calculations show a smaller increase of the shear stress, whereas the density correlations are decreased and do not match at all the results of Rebollo. This weakness is possibly due to the pressure gradient terms in the density correlations equations or in the scalar flux equations. One of these terms is a density con-

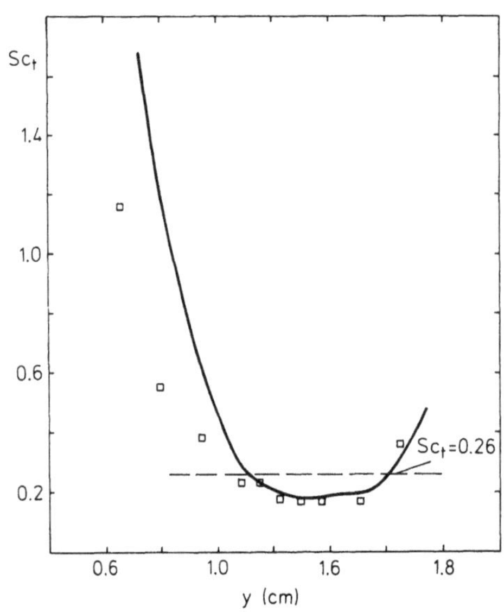

Fig. 6. Turbulent Schmidt-number in He-N mixing layer at $x = 5.08$ cm for $\alpha = 0$ (symbols: [9], lines: calculations)

centration equation, for which it would be better to solve a PDE rather than to approximate it from the density fluctuations. Figure 6 gives an estimate of the Schmidt number value across the mixing layer. The straight line represents the constant value which should be used with the two equation model calculations. This shows clearly that even if such value is realistic in the center part of the mixing zone, it becomes completely irrelevant near the edges.

Conclusions

A full second order closure turbulence model has been set up to describe the low speed mixing of gases with highly different density. Two mixing cases have been studied. In both calculations, the flow predictions are reasonable and promising for futere work in that direction. Nevertheless, this turbulence model version fails to predict correctly the pressure gradient influence on the turbulence (Reynolds stress, density correlations). Some further work would be done in this direction, especially the estimate of $\overline{c''_I}$ through the density fluctuations should be removed.

The main objective of this paper was to demonstrate the value of a second order closure for the prediction of mixing flows where the use of an eddy viscosity with a constant turbulent Schmidt number should fail. Nevertheless, it would be premature to draw any conclusion about universality of such a closure and the values of the constants used here. As some modelling assumptions are based on an analogy with the incompressible Reynolds stress closure [2], future experiments should emphasize measurements on quantities such as turbulent scalar fluxes and density correlations in order to support or to improve these assumptions.

References

1. Boussinesq, J.: "Théorie de l'écoulement tourbillonnant et tumultueux des liquides dans les lits rectilignes a grande section", I-II Gauthier-Villars, Paris
2. Hanjalić, K., Launder, B.E.: A Reynolds stress model of turbulence and its application to thin shear flows" J. Fluid Mech. **52–4**, 609–638 (1972)
3. Launder, B.E., Reece, G.J., Rodi, W.: Progress in the development of a Reynolds stress turbulence closure. J. Fluid Mech. **68–3**, 537–566 (1975)
4. Wolfshtein, M., Naot, D., Lin, A.: "Models of Turbulence", in *Topics in Transport Phenomena*, ed. by C. Gutfinger (Hemisphere, 1975)
5. duPont Donaldson, C.: "Atmospheric Turbulence and the Dispersal of Atmospheric Pollutants", EPA-R4-73-016a, Office of Research and Monitoring, U.S. Environmental Protection Agency, Washington, D.C., March 1973
6. Chassaing, P.: "Mélange Turbulent de Gas Inertes Dans un Jet de Tube Libre"; Thèse de Doctorat d'Etat, Toulouse, 1979
7. Lumley, J.L.: "Prediction Methods for Turbulent Flows. Introduction", VKI Lecture Series 76, Rhode St. Genese, Belgium, 1975
8. Lumley, J.L., Janicka, J.: "Second-Order Modelling in Non-constant Density Flows", FDA 81–01, Cornell University, 1981
9. Rebollo, M.: "Analytical and Experimental Investigation of a Turbulent Mixing Layer of Different Gases in a Pressure Gradient"; Ph.D. Thesis, California Institute of Technology, Pasadena (1973)
10. Brown, J.L.: "Heterogeneous Turbulent Mixing Layer Investigations Utilizing a 2-D 2-color Laser Doppler Anemometer and a Concentration Probe", Ph.D. Thesis, University of Missouri-Columbia (1978)
11. Vandromme, D.: "Turbulence Modelling for Avariable Density Flows"; Ph.D. Thesis, University of Brussels (1980)

12. Kollmann, W., Vandromme, D.: The calculation of free turbulent shear flows with strong density fluctuations. Int. J. Heat Mass Transfer **22**, 1557–1565 (1979)
13. Rotta, J.C.: Statistische Theorie nichthomogener Turbulenz. Z. Phys. **129**, 547–572; **132**, 51–77 (1951)
14. Launder, B.E.: "Heat and Mass Transport", in *Turbulence*, ed. by P. Bradshaw, Topics in Applied Physics, Vol. 12 (Springer, Berlin, Heidelberg, New York 1976)
15. Janicka, J.: "Berechnung Turbulenter Wasserstoff-Luft Diffusions-Flammen"; Ph.D. Thesis, RWTH, Aachen (1979)
16. Harlow, F.H., Nakayama, P.I.: "Transport of Turbulent Kinetic Energy Decay Rate", Los Alamos Scientific Lab Report LA-3854 (1968)
17. Irwin, H.P.: "Measurements in Blown Boundary Layers and Their Prediction by Reynolds Stress Modelling"; Ph.D. Thesis, McGill University (1974)
18. Patankar, S.V., Spalding, D.B.: *Heat and Mass Transfer in Boundary Layers* (Intertext, London 1970)
19. Jones, W.P., Launder, B.E.: The prediction of laminarization with a two equation model of turbulence. Int. J. Heat Mass Transfer **15**, 301–314 (1972)
20. Free Turbulent Shear Flows, NASA SP 321, Vol II, p. 14

The Turbulence Modelling of Variable Density Flows – A Mixed-Weighted Decomposition

H. Ha Minh

Institut de Mécanique des Fluides, Institut National Polytechnique, F–31071 Toulouse, France

B.E. Launder and J. MacInnes

Mechanical Engineering Department, University of Manchester, Institute of Science and Technology, Manchester M60 1 QD, England

Abstract

A new decomposition is presented of a variable density turbulent field into mean and fluctuating parts which adopts *both* the mass-weighted and the conventional mean velocity as dependent variables. By so doing, the different roles played by the velocity vector in the equations of motion can be properly accommodated. For the mean field conservation equations of momentum and species, the effects of turbulence in each equation are confined to a single correlation between fluctuating quantities; indeed the form is identical to that for a uniform density flow. The transport equations for the turbulent correlations are more complex than when mass-weighted averages are adopted but, arguably, the physical processes are easier to approximate. Various extrapolations of uniform-density second-moment closure methodology are discussed in connection with the behaviour observed by *Rebollo* [1] in the variable density mixing layer. The effective turbulent Prandtl-Schmidt number in a variable density medium is shown to be crucially sensitive to the specific forms of the approximation for correlations containing the fluctuating pressures.

Nomenclature

d	dimensionless density gradient (Fig. 2)
$\tilde{G}_i$	instantaneous mass velocity ($\equiv \tilde{\rho}\tilde{U}_i$)
G_i, g_i	mean, fluctuating part of $\tilde{G}_i$ ($\overline{g_i} = 0$)
$\tilde{M}$	instantaneous mass fraction
M, m	mean, fluctuating part of $\tilde{M}$
$\tilde{N}$	instantaneous mass-weighted mass fraction ($\equiv \tilde{\rho}\tilde{M}$)
N, n	mean, fluctuating part of $\tilde{N}$
$\tilde{U}_i$	instantaneous velocity
U_i, u_i	mean, fluctuating part of $\tilde{U}_i$
$\tilde{P}$	instantaneous static pressure
P, p	mean, fluctuating part of $\tilde{P}$
x_i	cartesian coordinate vector
ϵ	dissipation rate of $\overline{g_l^2/2}$
ϵ_{ij}^{ab}	dissipation rate of $\overline{a_i b_j}$
$\tilde{\rho}$	instantaneous density
ρ, ρ'	mean, fluctuating part of $\tilde{\rho}$
ϕ_{ij}^{ab}	fluctuating pressure-strain correlation term in transport equation for $\overline{a_i b_j}$
ϕ_j^{am}	fluctuating pressure-strain correlation term in transport equation for $\overline{a_j m}$
$\tilde{\tau}_{ij}$	instantaneous Stokes' stress tensor
τ_{ij}, τ_{ij}'	mean and instantaneous parts of $\tilde{\tau}_{ij}$

Introduction

In problems of turbulent combustion, in the mixing of combustion products or in the accidental release of heavy gases into the atmosphere, density fluctuations at a point may be of a magnitude comparable with that of the mean density. In such circumstances any credible turbulence closure must specifically address the influence that the large density excursions will have on the transport processes themselves. Here we consider only those situations where inertial effects entirely swamp contributions due to buoyancy.[1] Until about 1973 nearly all computational studies of variable-density mixing neglected any contribution of density fluctuations to the transport processes. More recently, turbulence closure studies have tended to follow one of two routes. In the first, the instantaneous density is represented as the sum of mean and fluctuating parts so, on averaging the convective transport term, correlations between fluctuating density and velocity appear in addition to the conventional Reynolds stress. This approach, in the context of second-moment closure, has been followed by *Donaldson* and co-workers [4], *Janicka* and *Kollmann* [5] and others. Those following this path have, implicitly, invoked the principle of receding influence since, in the transport equations for the second-moment quantities ($\overline{u_i u_j}$, etc.), uniform density forms have been adopted.

The second and increasingly followed approach, originally proposed by *Favre* [6], adopts a mass weighting of velocity and other dependent variables except pressure. With this arrangement the mean momentum equation, with molecular transport neglected, is analogous in appearance to that for a uniform density flow if the Reynolds stresses, like the velocities, are obtained by taking a density-weighted average. The simplicity of appearance of the resultant equation when mass-weighted averaging is used (a simplicity that also carries through to the exact form of the transport equations for turbulence correlations such as the turbulence energy) has encouraged the hope that closure ideas evolved for the invariant density case would be directly applicable to variable density mixing, e.g. *Bilger* and *Kent* [7] and *Jones* [8]. *Libby* [9], while more cautious about the prospect of direct extrapolation, recommend that closure models be evolved within the framework of mass-weighted averages to keep the computational burden to manageable proportions.

The most decisive set of experimental data for distinguishing between different approaches to closure in variable density flows is that of *Rebollo* [1] (see also the paper by *Brown* and *Roshko* [10]) who in two experiments examined the mixing of helium and nitrogen in a plane mixing layer with velocity ratios adjusted to give uniform dynamic pressure in the two streams. In one case, with zero streamwise pressure gradient, the turbulent Schmidt number in the central part of the layer fell below 0.2, i.e. to about one third the value found in the uniform density case. Up to September 1981 none of the reported attempts at computing this flow using either conventional or mass-weighted averaging had, so far as we could judge, been successful in reproducing the measured behaviour unless empirical coefficients were adjusted to values that would give unacceptable predictions in the uniform density case. The contribution by *Vandromme* and *Kollmann* [11] which appears in the present volume achieves a more impressive measure of success though, notwithstanding the values of their chosen coefficients appear to imply a local-equilibrium value for the turbulent Prandtl/Schmidt number in a uniform density flow of only about 0.45 which is some 30% less than reported from experiments.

In principle it is possible to extend and generalize the uniform-density model in order that the model give satisfactory agreement with measurements for both uniform and variable density mixing. Yet the very simplicity of the exact equations in mass-weighted coordinates gives little guidance as to the direction in which to proceed. In the present

---

1 In many problems gravitational forces *will* be influential though, to first order, they can be treated by the same approaches adopted when density fluctuations are weak (i.e. when the Boussinesq approximation applies); see, for example, [2, 3] for a discussion.

contribution we have adopted an alternative decomposition which we believe may display more transparently the physical processes at work. We refer to it as "mixed weighting" since, in the momentum equations both conventional velocity and the density weighted velocity appear. The basic ideas were developed in the "conference version" of this paper [12]. Since then, however, there has been an opportunity to re-examine the closure model and, as a result, numerous improvements in consistency have been made and several errors removed. The numerical computations have also been entirely reworked. The present paper should thus be regarded as superseding the version presented at the 3rd Turbulent Shear Flow Symposium.

Before turning to the analysis it remains only to emphasize that there are *many* possible approaches to decomposing variables into mean and fluctuating parts. To underline this fact we note that *Janicka* and *Lumley* [13] have very recently proposed a further, attractive decomposition no less different from established approaches than that presented herein. It is not yet clear what will eventually represent the best way forward. It could well be the case that the best arrangement for working out the physical modelling is not the best one, from a computational point of view, for calculating inhomogeneous shear flows. Once the physical representation in one decomposition has been settled, however, it should be possible to devise a nearly equivalent form in some other decomposition.

Analysis

The Mean Flow Transport Equations

The Navier Stokes equations for turbulent mixing of a variable density, uniform viscosity[2] stream may be written:

$$\frac{\partial}{\partial t}\,\tilde{\rho}\tilde{U}_i + \frac{\partial}{\partial x_j}\,\tilde{\rho}\tilde{U}_j\tilde{U}_i = -\frac{\partial \tilde{P}}{\partial x_i} + \frac{\partial}{\partial x_j}\left[\mu\left(\frac{\partial \tilde{U}_i}{\partial x_j} + \frac{\partial \tilde{U}_j}{\partial x_i} - \frac{2}{3}\,\delta_{ij}\,\frac{\partial \tilde{U}_k}{\partial x_k}\right)\right] \tag{1}$$

where tildes denote the instantaneous value (mean plus fluctuating) of the dependent variable in question. We note that the velocity vector in the second term on the left has two roles: $\tilde{\rho}\,\tilde{U}_j$ represents the mass flow rate per unit area while $\tilde{U}_i$ is the specific x_i-mometum flux. Here these separate roles are brought into prominence by defining

$$\tilde{\rho}\,\tilde{U}_j \equiv \tilde{G}_j \quad . \tag{2}$$

Let us introduce (2) into (1) and define the mean value Φ of a dependent variable whose instantaneous and fluctuating parts are $\tilde{\Phi}$ and ϕ by:

$$\Phi \equiv \frac{1}{2T}\int_{-T}^{T} \tilde{\Phi}\;dt\,; \qquad \tilde{\Phi} \equiv \Phi + \phi \quad . \tag{3}$$

It follows from (2) and (3) that

2 The unrealistic limitation of a stream of uniform viscosity is of no practical consequence since, in a free shear flow at high Reynolds numbers, viscous terms only indirectly affect the turbulence structure.

$$G_j = \rho U_j + \overline{\rho' u_j} \quad \text{and}$$

$$g_j = \rho u_j + \rho' U_j + (\rho' u_j - \overline{\rho' u_j}) \quad .$$

On averaging (1) in the conventional way implied by (3) we obtain for a stationary turbulent free shear flow at high Reynolds number:

$$\frac{\partial}{\partial x_j} G_j U_i = -\frac{\partial P}{\partial x_i} - \frac{\partial}{\partial x_j} \overline{g_j u_i} \quad . \tag{4a}$$

The asymmetric, mixed-weighted Reynolds stress, $\overline{g_j u_i}$, is now the quantity for which a turbulence model must be devised and that topic forms the main preoccupation of the remainder of the article. Before considering $\overline{g_j u_i}$ more closely, however, let us note that we could equally have linked $\tilde{\rho}$ with the velocity component $\tilde{U}_i$. In that case, Eq. (4a) becomes:

$$\frac{\partial}{\partial x_j} U_j G_i = -\frac{\partial P}{\partial x_i} - \frac{\partial}{\partial x_j} \overline{u_j g_i} \quad . \tag{4b}$$

Thus, again the mixed-weighted second-rank tensor appears but note that, in a thin shear flow, a different component of it will make the dominant contribution: $\overline{g_1 u_2}$ instead of $\overline{g_2 u_1}$. In a uniform density field the correlation is symmetric so the terms are equal but in a strongly stratified medium this may no longer be the case.

We suppose that the density variations arise from the inert mixing of two different gases and that the instantaneous mass fraction $\tilde{M}$ of one of these is governed by

$$\frac{\partial \tilde{\rho}\tilde{M}}{\partial t} + \frac{\partial}{\partial x_j} \tilde{\rho}\tilde{U}_j\tilde{M} = \lambda \frac{\partial^2 \tilde{M}}{\partial x_j^2} \quad . \tag{5}$$

Introduction of the mass velocity G_j and subsequent averaging, according to Eq. (3) leads, on neglect of molecular diffusion, to:

$$\frac{\partial}{\partial x_j} G_j M = -\frac{\partial}{\partial x_j} \overline{g_j m} \tag{6a}$$

for a stationary flow. Alternatively, defining $\tilde{\rho}\tilde{M} \equiv \tilde{N} \equiv (N + n)$ allows us to write (6a) as

$$\frac{\partial}{\partial x_j} U_j N = -\frac{\partial}{\partial x_j} \overline{u_j n} \quad . \tag{6b}$$

The use of (6b) in place of (6a) has certain attractions since N is linearly proportional to density:

$$\tilde{\rho} = a\tilde{N} + b \tag{7}$$

where a and b are constants. The difference between the 'a' and 'b' versions of Eq. (4, 6) is only significant at a higher level of closure than adopted in this paper.

It is evident that the averaged forms of the transport equations governing the velocity and scalar fields which arise from the present decompositions are just as compact as when density-weighted averaging is used for all terms. This compactness does not carry over to the transport equations for turbulence correlations, however. This, we would argue, is a *helpful* feature for, if we are fortunate, the intrinsic physical interactions giving rise to the strong alteration of the turbulent Schmidt number discussed above will become transparent in the additional terms.

Continuity and Its Implications for Closure

On averaging the instantaneous continuity equation

$$\frac{\partial \tilde{\rho}}{\partial t} + \frac{\partial}{\partial x_j} \tilde{\rho} \tilde{U}_j = 0 \quad , \tag{8}$$

there follows

$$\frac{\partial G_j}{\partial x_j} = -\frac{\partial \rho}{\partial t} \quad . \tag{9a}$$

Now, from (7) and the definition of $\tilde{N}$, (5) may be rewritten:

$$\frac{\partial}{\partial t} \frac{(\tilde{\rho} - b)}{a} + \frac{\partial}{\partial x_j} \frac{(\tilde{\rho} - b)}{a} \tilde{U}_j = \lambda \frac{\partial^2 \tilde{M}}{\partial x_j^2}$$

or, with the help of (8)

$$\frac{\partial \tilde{U}_j}{\partial x_j} = -\frac{\lambda a}{b} \frac{\partial^2 \tilde{M}}{\partial x_j^2} \tag{10}$$

whose mean part is evidently

$$\frac{\partial U_j}{\partial x_j} = -\frac{\lambda a}{b} \frac{\partial^2 M}{\partial x_j^2} \quad . \tag{9b}$$

Comparing (9a, b), we observe that the divergence of G_j is exactly zero in a stationary flow but not otherwise while, at high Reynolds numbers, the divergence of U_j is always essentially zero since the molecular diffusion on the right side of (9b) is then negligible.

The corresponding equations for the divergence of the fluctuating motion are obtained by subracting (9a) from (8) and (9b) from (10) respectively

$$\frac{\partial g_j}{\partial x_j} = -\frac{\partial \rho'}{\partial t} \tag{11a}$$

$$\frac{\partial u_j}{\partial x_j} = -\frac{\lambda a}{b} \frac{\partial^2 m}{\partial x_j^2} \quad . \tag{11b}$$

Clearly the divergence of neither g_j nor u_j vanishes. We note, however, that the contribution to $\partial u_j/\partial x_j$ arises overwhelmingly from the fine scale motion since only there are concentration gradients steep enough for molecular diffusion to be significant. Thus, for the energy containing motions, comprising eddies two or more orders of magnitude larger in dimension than the dissipative eddies, the divergence of the fluctuating velocity field is zero to a good degree of approximation (this certainly represents a better approximation than the closure hypotheses introduced in later sections). The same measure of simplification does not arise with g_j however. This difference between the spectral character of the divergence of u_j and g_j has far reaching consequences for modelling, particularly in respect of the correlations containing fluctuating pressures. For these, we shall find it advantageous to re-express correlations containing $\partial g_j/\partial x_k$ in terms of $\rho\partial u_j/\partial x_k$ (plus other quantities) in order to take advantage of the fact that the correlations are centered in the energy containing range where $\partial u_j/\partial x_j$ is negligible. The development by *Janicka* and *Lumley* [13] has independently drawn analogous conclusions.

The Second-Moment Correlation Equations and Closure Rationale

Exact transport equations for the second-moment correlations $\overline{g_j u_i}$ and $\overline{g_j m}$ (or $\overline{u_j n}$) may be formed in the conventional way. For example that for the first correlation is obtained by multiplying the equation of motion for $\tilde{U}_i$ by g_j and averaging and adding it to the corresponding equation for $\tilde{G}_j$ multiplied by u_i and averaged. In this connection the "corresponding" $\tilde{G}_j$ equation is formed by multiplying and dividing the $\tilde{U}_i$ equation by density:

$$\frac{\partial \tilde{G}_k \tilde{U}_j}{\partial x_k} = \frac{\partial \tilde{G}_k \tilde{G}_j/\tilde{\rho}}{\partial x_k} = \frac{1}{\tilde{\rho}}\left(\frac{\partial \tilde{G}_k \tilde{G}_j}{\partial x_j} - \frac{\tilde{G}_k\, \tilde{U}_j \partial\tilde{\rho}}{\partial x_k}\right) .$$

Then by straightforward manipulation we find that the transport of $\overline{g_i u_i}$ is described by:

$$\begin{aligned}
\frac{\partial G_k \overline{g_j u_i}}{\partial x_k} = & -\overline{g_j g_k}\,\frac{\partial U_i}{\partial x_k} - \overline{g_k u_i}\,\frac{\partial G_j}{\partial x_k} - G_j \overline{u_i \frac{\partial g_k}{\partial x_k}} - \overline{g_j u_i \frac{\partial g_k}{\partial x_k}} \\
& + \left\{ G_k U_j \overline{u_i \frac{\partial \rho'}{\partial x_k}} + G_k\, \overline{u_i u_j}\,\frac{\partial \rho}{\partial x_k} + \overline{g_k u_i}\, U_j \frac{\partial \rho}{\partial x_k} + G_k \overline{u_i u_j \frac{\partial \rho'}{\partial x_k}} \right. \\
& \left. + \overline{U_j u_i g_k \frac{\partial \rho'}{\partial x_k}} + \overline{g_k u_i u_j}\,\frac{\partial \rho}{\partial x_k} + \overline{g_k u_i u_j \frac{\partial \rho'}{\partial x_k}} \right\} \\
& - \overline{u_i \rho'}\,\frac{\partial P}{\partial x_j} + \overline{p\left(\frac{\partial \rho u_i}{\partial x_j} + \rho' \frac{\partial u_i}{\partial x_j} + u_i \frac{\partial \rho'}{\partial x_j} + \frac{\partial g_j}{\partial x_i}\right)} \\
& - \frac{\partial}{\partial x_k}\left(\overline{g_j g_k u_i} + \left[\rho \overline{u_i p} + \overline{\rho' u_i p}\right]\delta_{jk} + \overline{g_j p}\;\;\delta_{ik}\right) \\
& + \overline{g_j \frac{\partial \tau'_{ik}}{\partial x_k}} + \overline{\rho' u_i}\,\frac{\partial \tau_{jk}}{\partial x_k} + \rho\, \overline{u_i \frac{\partial \tau'_{jk}}{\partial x_k}} + \overline{\rho' u_i \frac{\partial \tau'_{jk}}{\partial x_k}} \quad .
\end{aligned} \qquad (12)$$

The corresponding equation for $\overline{g_j m}$ emerges as:

$$\begin{aligned}\frac{\partial G_k \overline{g_j m}}{\partial x_k} = & -\overline{g_k m}\,\frac{\partial G_j}{\partial x_k} - \overline{g_j g_k}\,\frac{\partial M}{\partial x_k} - G_j\,\overline{m\,\frac{\partial g_k}{\partial x_k}} - \overline{m\,g_j\,\frac{\partial g_k}{\partial x_k}} \\ & + \left[G_k U_j\,\overline{m\,\frac{\partial \rho'}{\partial x_k}} + G_k\,\overline{u_j m}\,\frac{\partial \rho}{\partial x_k} + \overline{g_k m}\,U_j\,\frac{\partial \rho}{\partial x_k} + G_k\,\overline{u_j m\,\frac{\partial \rho'}{\partial x_k}}\right. \\ & \left. + \overline{U_j\,g_k m\,\frac{\partial \rho'}{\partial x_k}} + \overline{g_k u_j m}\,\frac{\partial \rho}{\partial x_k} + \overline{g_k u_j m\,\frac{\partial \rho'}{\partial x_k}}\right] \\ & - \overline{\rho' m}\,\frac{\partial P}{\partial x_j} + \overline{p\left(\frac{\partial \rho m}{\partial x_j} + \rho'\,\frac{\partial m}{\partial x_j} + m\,\frac{\partial \rho'}{\partial x_k}\right)} \\ & - \frac{\partial}{\partial x_k}\left(\overline{g_j g_k m} + \overline{p\,(\rho m + \rho' m)}\;\delta_{jk}\right) \\ & + \overline{\rho' m}\,\frac{\partial \tau_{jk}}{\partial x_k} + \overline{\rho\, m\,\frac{\partial \tau'_{jk}}{\partial x_k}} + \overline{\rho' m\,\frac{\partial \tau'_{jk}}{\partial x_k}} + \overline{g_j\,\frac{\partial}{\partial x_k}\left(\lambda_m\,\frac{\partial m}{\partial x_k}\right)}\ . \qquad (13)\end{aligned}$$

These equations are clearly more cumbersome in appearance than their uniform-density counterparts. As when a fully mass-weighted treatment is employed (see, for example, *Rubesin* and *Rose* [14]), mean pressure gradients appear in (12) and (13). Moreover, the principal generation terms (involving mean velocity or mass-fraction gradients) introduce a further unknown second-moment tensor, $\overline{g_j g_k}$.

Against this somewhat daunting complexity one must set the fact that the observed effects of density stratification are large (as noted above, one experiment showed a threefold modification in effective Schmidt number). One should therefore be able to make fairly sweeping simplifications to the transport equations on the grounds that certain terms, while not negligible, cannot foreseeably be responsible for changes of such major proportions. We shall therefore assume that the *essential* physics are retained in the local-equilibrium form of the equations and, at the same time, discard all correlations involving terms of similar structure to those associated with the neglected transport terms. In particular, triple correlations involving density fluctuations are discarded in the dynamic equations. When, in correlations, we wish to eliminate the fluctuating mass velocity g_j, we likewise adopt the approximation:

$$g_j \approx \rho u_j + \rho' U_j \quad .$$

Consistently we take $G_1 = \rho U_1$; the neglected term $\overline{\rho' u_1}$ is at most 4% of G_1. (Source terms in the second moment equations involving G_2 are ones neglected in thin shear flows, though of course $\overline{\rho' u_2}$ may be of the same magnitude as ρU_2). With these simplifications, (12) can be expressed:

$$0 = -\left(\overline{\rho g_k u_i}\frac{\partial U_j}{\partial x_k} + \overline{g_j g_k}\frac{\partial U_i}{\partial x_k} + G_j\overline{u_i u_k}\frac{\partial \rho}{\partial x_k}\right)$$

$$- \overline{u_i \rho'}\frac{\partial P}{\partial x_j} + \overline{p\left(\frac{\partial \rho u_i}{\partial x_j} + \frac{\partial g_j}{\partial x_i}\right)} - \rho\epsilon_{ji}^{gu} \quad . \tag{14}$$

The last term in (14) denotes the rate of destruction of $\overline{g_j u_i}$ by viscous action. The corresponding truncated form for the equation for $\overline{g_i g_j}$ may, under like simplification, be written:

$$0 = -\rho\left(\overline{g_j g_k}\frac{\partial U_i}{\partial x_k} + \overline{g_i g_k}\frac{\partial U_j}{\partial x_k}\right) - \left(G_i\overline{g_j u_k}\frac{\partial \rho}{\partial x_k} + G_j\overline{g_i u_k}\frac{\partial \rho}{\partial x_k}\right)$$

$$+ \overline{p\left(\frac{\partial \rho g_j}{\partial x_i} + \frac{\partial \rho g_i}{\partial x_j}\right)} - \rho\epsilon_{ij}^{gg} \quad . \tag{15}$$

Alternatively, by following the 'b' decomposition and applying analogous local equilibrium approximations we obtain:

$$0 = -\rho\left(\overline{g_j u_k}\frac{\partial G_i}{\partial x_k} + \overline{g_i u_k}\frac{\partial G_j}{\partial x_k}\right) + \rho\overline{p\left(\frac{\partial g_j}{\partial x_i} + \frac{\partial g_i}{\partial x_j}\right)} - \rho\epsilon_{ij}^{gg} \tag{16}$$

To the present order of approximation Eqs. (15) and (16) may be readily shown to be equivalent. The latter form brings out more clearly, however, the fact that generation of $\overline{g_k g_k}$ by the 'conventional' terms vanishes when the mass-velocity gradient is zero. As we see from Fig. 1, stationary values of G_1 occur at several positions on the low-density side of Rebollo's mixing layer. Although the disappearance of these terms must make one wary of applying local-equilibrium approximations in such regions, they are not *necessarily* inadequate because the correlation $\overline{p\partial g_k/\partial x_k}$ (which vanishes in a uniform density flow) then takes on positive values – at least, according to the present closure approximation. The corresponding local-equilibrium equations for the other relevant second-moment equations are:

$\overline{g_j m}$:

$$0 = -\left(\overline{g_j g_k}\frac{\partial M}{\partial x_k} + \overline{\rho g_k m}\frac{\partial U_j}{\partial x_k} + G_j\overline{\rho' u_k}\frac{\partial \rho}{\partial x_k}\right) - \overline{\rho' m}\frac{\partial P}{\partial x_j}$$

$$+ \frac{\overline{p\partial \rho m}}{\partial x_j} - \rho\epsilon_j^{gm} \tag{17}$$

$\overline{u_j m}$:

$$0 = -\overline{u_j g_k}\frac{\partial M}{\partial x_k} - \overline{g_k m}\frac{\partial U_j}{\partial x_k} + \frac{\overline{p\partial m}}{\partial x_j} - \rho\epsilon_j^{um} \quad .$$

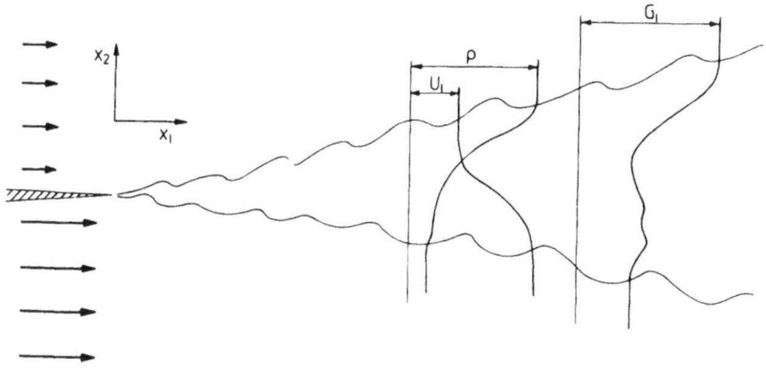

Fig. 1. The plane mixing layer in variable density flow; profiles of ρ, U_1, and G_1 in Rebollo's experiment

The last of these equations will make an appearance in the next section where the pressure-containing correlations are considered. For completeness the generation terms in all the local-equilibrium equations are grouped in the left hand column of Table 3.

The Role of Pressure Fluctuations in Variable Density Turbulence

It will be seen in what follows that the whole character of predictions of turbulent mixing in strong density gradients depends crucially on the approximation of the pressure-containing correlations.

An equation describing the instantaneous level of pressure above its mean level is obtained by taking the divergence of the equation of motion and subtracting its mean part. The result may be written:

$$-\frac{\partial^2 p}{\partial x_n^2} = \underbrace{\frac{\partial^2}{\partial x_n \partial x_m}(u_n g_m - \overline{u_n g_m}) + \frac{\partial G_m}{\partial x_n}\frac{\partial u_n}{\partial x_m} + \frac{\partial U_n}{\partial x_m}\frac{\partial g_m}{\partial x_n}}_{\text{A}}$$

$$+ \underbrace{\frac{\partial \rho}{\partial x_n}\frac{\partial u_n}{\partial t} + G_m \frac{\partial^2 u_n}{\partial x_n \partial x_m} + \frac{\partial^2 \tau'_{nm}}{\partial x_n \partial x_m}}_{\text{B}} \quad . \tag{19}$$

In the above arrangement those terms grouped as 'A' are clearly recognizable generalizations of terms present in uniform density flow; those in group 'B' are non-zero only because of density fluctuations. Following the discussion of continuity for the turbulent motion above, we neglect the second and third of these terms on the grounds that though the terms may be individually large they are associated with such fine scale fluctuations that they make an insignificant contribution to the volume integrals that go to make up the pressure-strain and similar correlations. The first term of group B can be expected to dampen somewhat the contribution of density fluctuations implied by group A.[3] Accordingly, we retain only group A with the understanding that contributions from the tempo-

3 Taylor's hypothesis cannot be applied to remove the time derivative because the pressure-fluctuation equation essentially examines *departures* from Taylor's hypothesis. If we were to make the substitution $\partial u_n/\partial t = -U_k(\partial u_n/\partial x_k)$, however, the effect would be to cancel the principal variable density contribution of the second term in group A. From this misapplication we may infer that neglecting the time derivative will give somewhat too great a sensitivity of the pressure equation to density gradients.

ral derivative term of group B are at least partly accounted for in the choice of coefficients in modelling the correlations arising from group A.

After integrating (19) the various correlations of interest may be expressed in the forms shown in Table 1. For brevity D_{mn} is used to denote $1/2(\partial G_m/\partial x_n + \rho\partial U_m/\partial x_n)$. Spatial homogeneity of the flow has been assumed so the (uniform) mean velocity and density gradients have been moved outside the integral signs. Correlations $\overline{pu_i}$, $\overline{pm}$ etc. which will vanish in a strictly homogenous flow are also dropped. In line with our earlier discussion, terms with g_i have been expanded using the first-order approximation while G_i is replaced by ρU_i.

Concerning the "mean strain" or linear parts of the pressure correlations, ϕ_2, the task of providing suitable approximations is reduced to that of representing, in closed form, the basic integrals[4]:

$$I_{mj}^{ni} \equiv \int_v \overline{\left(\frac{\partial u_n}{\partial x_m}\right)'' \left(\frac{\partial u_i}{\partial x_j}\right)} \frac{dV}{r}$$

$$I_{mj}^{i} \equiv \int_v \overline{\left(\frac{\partial \rho_i'}{\partial x_m}\right)'' \left(\frac{\partial u_i}{\partial x_j}\right)} \frac{dV}{r}$$

$$I_{mj} \equiv \int_v \overline{\left(\frac{\partial \rho'}{\partial x_m}\right)'' \left(\frac{\partial \rho'}{\partial x_j}\right)} \frac{dV}{r} \quad .$$

In fact, the terms containing the third integral (see Table 1) are, by virtue of the thinness of the shear flow, roughly an order of magnitude smaller than those associated with the second. Thus, in line with our policy of retaining only the principal effects of density fluctuations, these terms are discarded. They could have been retained by adopting an entirely consistent closure proposal with that assumed for the other two-point correlation integrals; indeed, for consistency, they would *need* to be retained in any closure treatment higher than the local-equilibrium truncation adopted here.

We explore below the consequences of applying uniform-density representations for I_{mj}^{ni} and I_{mj}^{i}. Two approaches to approximating ϕ_2 have received extensive attention: the "quasi-isotropic" model (QIM) [15, 16] and the "redistribution of production" model (RPM) [15, 17]. The first provides an exact solution when the two-point correlation functions, such as appear in the integrals of Table 1, depend on radius only. The second is simpler and more intuitive: the idea being that a fixed fraction of the shear-stress generation rate is obliterated while, for the normal stresses, the same fraction of any generation in excess of the mean is redistributed among the other normal stresses. While the former model has the clearer pedigree, the RPM has, in practice, been at least as successful; indeed it has seemed to offer a more secure basis for extrapolation to new phenomena, for then it has consistently given the correct trends while the quasi-isotropic model has sometimes given spectacularly wrong results (e.g. the effect of swirl on the spread of a jet [18]).

In a uniform density shear, the RPM can alternatively be regarded as a simplified QIM in which the principal term (only) is retained with a numerical coefficient adjusted

4 For the purpose of evaluating the volume integral, the assumption of a linear relationship between ρ' and m should suffice. Integrals involving $\overline{(\partial\rho'/\partial x_m)''\,\partial m/\partial x_n}$ thus reduce to the third of the basic types.

Table 1. Fluctuating pressure correlations, expressed using a Poisson solution for p, simplified for the case of high Re, homogeneous flow

Eqn.	Correlation	Approximation ϕ_1		Approximation ϕ_2
$\overline{u_i u_j}$	$\overline{p\left(\frac{\partial u_i}{\partial x_j}+\frac{\partial u_j}{\partial x_i}\right)}$	$\underbrace{\frac{1}{4\pi}\int_V \overline{\left(\frac{\partial^2 g_m u_n}{\partial x_n \partial x_m}\right)''\left(\frac{\partial u_i}{\partial x_j}+\frac{\partial u_j}{\partial x_i}\right)}\frac{dV}{r}}_{\phi_{1ij}^{uu}}$	$+$	$\underbrace{\frac{1}{2\pi}D_{mn}\int_V \overline{\left(\frac{\partial u_n}{\partial x_m}\right)''\left(\frac{\partial u_i}{\partial x_j}+\frac{\partial u_j}{\partial x_i}\right)}\frac{dV}{r}+\frac{1}{4\pi}U_n\frac{\partial U_m}{\partial x_n}\int_V \overline{\left(\frac{\partial \varrho'}{\partial x_m}\right)''\left(\frac{\partial u_i}{\partial x_j}+\frac{\partial u_j}{\partial x_i}\right)}\frac{dV}{r}}_{\phi_{2ij}^{uu}}$
$\overline{g_i u_j}$	$\overline{p\left(\frac{\partial g_i}{\partial x_j}+\frac{\partial \varrho u_j}{\partial x_i}\right)}$	$\varrho\phi_{1ij}^{uu}+\frac{1}{4\pi}U_i\int_V \overline{\left(\frac{\partial g_m u_n}{\partial x_n \partial x_m}\right)''\frac{\partial \varrho'}{\partial x_j}}\frac{dV}{r}$	$+$	$\varrho\phi_{2ij}^{uu}+\frac{1}{2\pi}D_{mn}U_i\int_V \overline{\left(\frac{\partial u_n}{\partial x_m}\right)''\frac{\partial \varrho'}{\partial x_j}}\frac{dV}{r}+\frac{1}{4\pi}U_n\frac{\partial U_m}{\partial x_n}U_i\int_V \overline{\left(\frac{\partial \varrho'}{\partial x_m}\right)''\frac{\partial \varrho'}{\partial x_j}}\frac{dV}{r}$
$\overline{g_i g_j}$	$\overline{p\left(\frac{\partial \varrho g_i}{\partial x_j}+\frac{\partial \varrho g_j}{\partial x_i}\right)}$	$\varrho^2\phi_{1ij}^{uu}+\frac{\varrho}{4\pi}\int_V \overline{\left(\frac{\partial^2 g_m u_n}{\partial x_n \partial x_m}\right)''\left(U_i\frac{\partial \varrho'}{\partial x_j}+U_j\frac{\partial \varrho'}{\partial x_i}\right)}\frac{dV}{r}$	$+$	$\varrho^2\phi_{2ij}^{uu}+\frac{\varrho}{2\pi}D_{mn}\int_V \overline{\left(\frac{\partial u_n}{\partial x_m}\right)''\left(U_i\frac{\partial \varrho'}{\partial x_j}+U_j\frac{\partial \varrho'}{\partial x_i}\right)}\frac{dV}{r}$ $+\frac{\varrho}{4\pi}U_n\frac{\partial U_m}{\partial x_n}\int_V \overline{\left(\frac{\partial \varrho'}{\partial x_m}\right)''\left(U_i\frac{\partial \varrho'}{\partial x_j}+U_j\frac{\partial \varrho'}{\partial x_i}\right)}\frac{dV}{r}$
$\overline{u_i m}$	$\overline{p\frac{\partial m}{\partial x_i}}$	$\underbrace{\frac{1}{4\pi}\int_V \overline{\left(\frac{\partial^2 g_m u_n}{\partial x_n \partial x_m}\right)''\frac{\partial m}{\partial x_i}}\frac{dV}{r}}_{\phi_{1i}^{um}}$	$+$	$\underbrace{\frac{1}{2\pi}D_{mn}\int_V \overline{\left(\frac{\partial u_n}{\partial x_m}\right)''\frac{\partial m}{\partial x_i}}\frac{dV}{r}+\frac{1}{4\pi}U_n\frac{\partial U_m}{\partial x_n}\int_V \overline{\left(\frac{\partial \varrho'}{\partial x_m}\right)''\frac{\partial m}{\partial x_i}}\frac{dV}{r}}_{\phi_{2i}^{um}}$
$\overline{g_i m}$	$\overline{p\frac{\partial \varrho m}{\partial x_i}}$	$\varrho\phi_{1i}^{um}$	$+$	$\varrho\phi_{2i}^{um}$

Table 2. The various ϕ_2 pressure terms

	ϕ_2 pressure terms resulting from modeling hypothesis		
	QIM	SQIM	RPM
ϕ_{2ij}^{uu}	$-\frac{(c_2+8)}{11}[-(\overline{u_i u_k}D_{jk}+\overline{u_j u_k}D_{ik})+\frac{2}{3}\overline{u_l u_k}D_{lk}\delta_{ij}]$ $-\frac{(15c_2-1)}{55}\overline{u_k u_k}[D_{ij}+D_{ji}]$ $-\frac{(8c_2-2)}{11}[-(\overline{u_i u_k}D_{kj}+\overline{u_j u_k}D_{ki})+\frac{2}{3}\overline{u_l u_k}D_{lk}\delta_{ij}]$	$-\alpha[-(\overline{u_i u_k}D_{jk}+\overline{u_j u_k}D_{ik})+\frac{2}{3}\overline{u_l u_k}D_{lk}\delta_{ij}]$	$-\alpha'[P_{ij}^{uu}-\frac{1}{3}P_{ll}^{uu}\delta_{ij}]$
ϕ_{2ij}^{gu}	$\varrho\phi_{2ij}^{uu}+0.8\,U_i\overline{u_k\varrho'}D_{jk}-0.2\,U_i\overline{u_k\varrho'}D_{kj}$	$\varrho\phi_{2ij}^{uu}+C_{2m}U_i\overline{u_k\varrho'}D_{jk}$	$-\alpha'[P_{ij}^{gu}-\frac{1}{3}P_{ll}^{gu}\delta_{ij}]$
ϕ_{2ij}^{gg}	$\varrho^2\phi_{2ij}^{uu}+0.8\,\varrho(U_i\overline{u_k\varrho'}D_{jk}+U_j\overline{u_k\varrho'}D_{ik})$ $-0.2\,\varrho(U_i\overline{u_k\varrho'}D_{kj}+U_j\overline{u_k\varrho'}D_{ki})$	$\varrho^2\phi_{2ij}^{uu}+C_{2m}\varrho(U_i\overline{u_k\varrho'}D_{jk}+U_j\overline{u_k\varrho'}D_{ik})$	$-\alpha'[P_{ij}^{gg}-\frac{1}{3}P_{ll}^{gg}\delta_{ij}]$
ϕ_{2i}^{um}	$0.8\,\overline{u_k m}D_{ik}-0.2\,\overline{u_k m}D_{ki}$	$C_{2m}\overline{u_k m}D_{ik}$	$C'_{2m}\overline{g_k m}\frac{\partial U_i}{\partial x_k}$
ϕ_{2i}^{gm}	$\varrho\phi_{2i}^{um}$	$\varrho\phi_{2i}^{um}$	$\varrho\phi_{2i}^{um}$

Table 3. The production terms and the forms used for the dissipation rate and 'return' part of the pressure correlations (ϕ_1)

Production	Dissipation	Return terms
$P_{ij}^{uu} = -\overline{u_j g_k}\dfrac{\partial U_i}{\partial x_k} - \overline{u_i g_k}\dfrac{\partial U_j}{\partial x_k}$	$\varepsilon_{ij}^{uu} = \dfrac{\varepsilon_{kk}^{uu}}{3}\delta_{ij}$	$\phi_{1ij}^{uu} = -\dfrac{\varrho\overline{u_l u_l}}{T_{uu}}\left[C_1 a_{ij} + C_3 b_{ij} + C_1'\left(a_{ik}a_{jk} - \dfrac{a_{lk}^2}{3}\delta_{ij}\right)\right]$
$P_{ij}^{gu} = -\varrho\overline{g_k u_j}\dfrac{\partial U_i}{\partial x_k} - \overline{g_i g_k}\dfrac{\partial U_j}{\partial x_k} - G_i\overline{u_j u_k}\dfrac{\partial \varrho}{\partial x_k}$	$\varepsilon_{ij}^{gu} = \varrho\varepsilon_{ij}^{uu}$	$\phi_{1ij}^{gu} = \varrho\phi_{1ij}^{uu} + U_i\phi_{1i}^{u\varrho}$
$P_{ij}^{gg} = -\varrho\left(\overline{g_j g_k}\dfrac{\partial U_i}{\partial x_k} + \overline{g_i g_k}\dfrac{\partial U_j}{\partial x_k}\right)$ $-(G_i\overline{g_j u_k} + G_j\overline{g_i u_k})\dfrac{\partial \varrho}{\partial x_k}$	$\varepsilon_{ij}^{gg} = \varrho^2\varepsilon_{ij}^{uu}$ $+U_i U_j\varepsilon^{mm}$	$\phi_{1ij}^{gg} = \varrho^2\phi_{1ij}^{uu} + \varrho U_i\phi_{1j}^{u\varrho} + \varrho U_j\phi_{1i}^{u\varrho}$
$P_i^{um} = -\overline{u_i g_k}\dfrac{\partial M}{\partial x_k} - \overline{g_k m}\dfrac{\partial U_i}{\partial x_k}$	$\varepsilon_i^{um} = 0$	$\phi_{1i}^{um} = -\dfrac{\varrho}{T_{um}}\left[C_{1m}\overline{u_i m} + C_{3m}\dfrac{U_i}{\varrho}\overline{\varrho' m} + C_{1m}' a_{ik}\overline{u_k m}\right]$
$P_i^{gm} = -\overline{g_i g_k}\dfrac{\partial M}{\partial x_k} - \varrho\overline{g_k m}\dfrac{\partial U_i}{\partial x_k} - G_i\overline{m u_k}\dfrac{\partial \varrho}{\partial x_k}$	$\varepsilon_i^{gm} = 0$	$\phi_{1i}^{gm} = \varrho\phi_{1i}^{um}$
$P^{mm} = -2\overline{g_k m}\dfrac{\partial M}{\partial x_k}$	$\varepsilon^{mm} = \dfrac{\overline{mm}}{\overline{u_l u_l}}\dfrac{\varepsilon_{kk}^{uu}}{R}$	(no pressure terms in $\overline{m^2}$ equation)

where $a_{ij} \equiv (\overline{u_i u_j}/\overline{u_k u_k} - \delta_{ij}/3)$

$b_{ij} \equiv [(U_i\overline{u_j\varrho'} + U_j\overline{u_i\varrho'})/2 - U_k\overline{u_k\varrho'}\,\delta_{ij}/3]/\varrho\overline{u_k u_k}$

somewhat to compensate for the discarded terms. In a variable density flow, the corresponding simplification of the QIM (the SQIM) does not reduce to the RPM.

These three basic closure concepts for ϕ_2 have all been examined in our study. The specific forms of the models for the different second-moment correlations are set out in Table 2. The QIM forms are 'exact' in the same sense as they are for uniform density flows; of course, it does not necessarily follow that they are adequate in approximating anisotropic turbulence.

The representation of ϕ_1, the non-linear parts of the pressure correlations, adopts the forms commonly employed for flows of uniform density and adds to them density flux contributions suggested by the exact integrals in Table 1. The approximations adopted are listed in the right-hand column of Table 3. A linear connection between density and mass fraction is again assumed so $\phi_{1j}^{u\rho}$ is identical to ϕ_{1j}^{um} provided m is replaced by ρ'. Three levels of complexity have been examined. The simplest is where a linear 'return-to-isotropy' approximation is made, i.e., where only c_1 and c_{1m} are taken as non-zero; at the next level, the non-linear contributions with coefficients c_1' and c_{1m}' are added while, finally, terms arising from the "$\rho' U_j$" part of g_j are included by assigning non-zero values to c_3 and c_{3m}. Only linear forms have been included for the latter processes. We doubt the wisdom of attempting a more elaborate approximation until more detailed experimental data become available.

Dissipation Rates and Time Scales

The application of local isotropy coupled with the discarding of triple correlations involving viscous terms in the second-moment equations gives the simple relationships between the dissipation rates for second-moment correlations shown in the second column of Table 3. The actual level of the kinetic energy dissipation rate ϵ_{kk}^{uu} which serves to define the others is not brought into question in this local-equilibrium study as we shall be concerned only with time-scale *ratios* rather than absolute levels. The dissipation rate of $\overline{m^2}$ is obtained via the dimensionless time-scale-ratio R ($\equiv T_{mm}/T_{uu}$) which, in simple free shear flows, has been found to be approximately equal to 0.5 [19]. We have retained this value here assuming that even severe density fluctuations have but negligible effect on this dimensionless quantity. The assumption is open to question but it must be said that no other choice appears to involve less uncertainty at present. The time scale T_{um} associated with the destruction of the scalar flux $\overline{u_i m}$ by pressure fluctuations is taken as $(0.5)^{1/2}\ T_u$, reflecting the combined influence of the fluctuating velocity and scalar fields.

Discussion of Results

We have examined, for conceptually homogeneous plane flow in local equilibrium, the effect of density gradients on the various dimensionless quantities characterizing the second moments and their interaction with the mean field. The strength of the density gradient has been expressed in terms of the dimensionless parameter d defined as $-U_1(\partial\rho/\partial x_2)/\rho(\partial U_1/\partial x_2)$, x_1 being the flow direction and x_2 the direction of velocity and density gradients. Prime attention has been given to the dependence of the turbulent Schmidt number on d, for it is this parameter that, as we have discussed, offers the most startling experimental evidence of density gradients.

No experimental data are available for a homogeneous shear flow with appreciable density gradients (yet unaffected by buoyancy). The underlying rationale of our closure however, is that the local-equilibrium approximation would still be useful in calculating *in*homogeneous flows not strictly in local equilibrium. We here apply the reverse of that idea by plotting, for the available data of plane mixing layers in zero pressure gradient, the local measured values of Sc_t versus d. The data for mass or heat diffusion at essentially uniform density (e.g. [20]) indicate a mean turbulent Prandtl-Schmidt number of about 0.65. Experiments from Caltech of helium-nitrogen mixing provide the other guidance: Rebollo's data for positive d [1] are shaded in on the right side of Figs. 2 and 3 while those of *Brown* and *Roshko* [10], obtained in experiments where the density and velocity gradients are in the same sense, appear on the left of the figures. In the latter case, turbulent Schmidt numbers have been evaluated as part of the present study from the reported mean profiles of velocity and density assuming self-similar flow. When the data are plotted they formed patches rather than single lines underlining the fact that the flows are in fact appreciably influenced by transport effects. This feature limits our ability to draw firm conclusions but does not prevent us distinguishing broad trends.

Results are reported for five different treatments of the pressure correlations. Our computations have, in fact, spanned the full range of combinations of model but the limited comparison shown in Figs. 2 and 3 allows all the main effects to be discerned. Before considering these results it is first noted that, despite its record in uniform-density flows, the RPM model produces a turbulent Schmidt number, Sc_t, entirely *independent* of d. It is seen that the various curves shown in Figs. 2 and 3 all pass through the common point $d = 0$, $Sc_t = 0.65$. This agreement with the data for the uniform-density limit was insisted upon. Indeed in choosing different sets of coefficients, we have required in addition (for $d = 0$):

$$-\overline{u_1\rho'}/\overline{u_2\rho'} = 1.2\ ;\quad \overline{u_1^2}/\overline{u_k^2} \cong 0.46\ ;\quad \overline{u_2^2}/\overline{u_k^2} \cong 0.24 \quad \text{and} \quad \overline{u_3^2}/\overline{u_k^2} \cong 0.30\ .$$

The different coefficients associated with the curves are given in Table 4. Model 'A' on Fig. 2, obtained with the SQIM scheme for ϕ_2 and the simplest return-to-isotropy model for ϕ_1, displays substantial sensitivity of Sc_t to d; but unfortunately the depen-

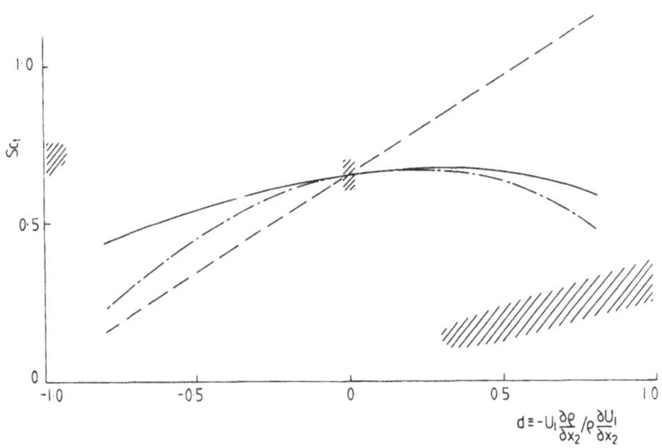

Fig. 2. Comparison of different forms of ϕ_1 and ϕ_2 models. (---) Model A, (—) Model B, (—·—·) Model C, (see Table 4); (////) Expt. data [1] and [10]

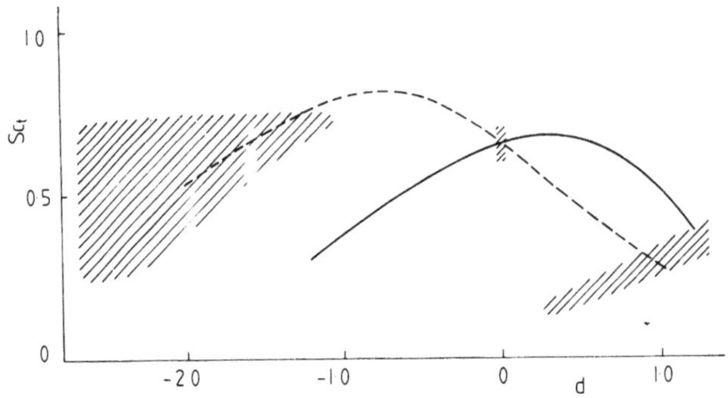

Fig. 3. Influence of density-flux correction. (—) Model B, (---) Model D; (////) Expt. data [1] and [10]

Table 4. Coefficients for the various models

	c_1	c_3	c_1'	c_{1m}	c_{3m}	c_{1m}'	c_2	α	c_{2m}
Model A	1.80	0	0	2.42	0	0	–	0.66	0.33
Model B	1.82	0	−0.78	3.31	0	−6.91	0.47	–	–
Model C	1.80	0	0	4.32	0	−10.57	–	0.66	–
Model D	1.82	−1.0	−0.78	3.31	1.0	−6.91	0.47	–	–

dence is in the opposite sense from experiment, at least for positive d. The other two curves on Fig. 2 show a variation that is less strikingly contrary to experiment.

Although, for case B, the quasi-isotropic model has been used for ϕ_2, the principal agency for the improvement is the non-linear model for the 'return-to-isotropy' processes involving scalar fluxes. It is known [21, 22] that a non-linear form for ϕ_{1j}^{um} must be used in conjunction with the QIM or else the target values for turbulent Schmidt number and scalar flux ratio quoted above cannot both be achieved. Model C is a hybrid, adopting SQIM for the dynamic field (in conjunction with a linear ϕ_1) while retaining QIM for the scalar fluxes (with a non-linear ϕ_1 contribution). Both B and C show a reduction in Sc_t for sufficiently large d though not as markedly as do the data. The reduction of Sc_t for negative d is much reduced compared with curve A though it is still steeper than experimental data indicate (see the replot of the curve for case B in Fig. 3).

Finally, model D in Fig. 3 shows the effects of adding terms designed to account for the 'density-fluctuation' part of g_j in ϕ_1. The extra terms are found to have a very pronounced influence on the behaviour; the choice $c_3 = -1.0$ and $c_{3m} = +1.0$ appears to be close to the optimum. The effect, in simple terms is to shift the predictions to the left as experiment requires. An interesting suggestion of Model D is that, for *small* negative d, turbulent Schmidt numbers occur which are appreciably higher than those found in the uniform-density mixing layer or for steeper density stratification. Fig. 4a, b compare the dependence of the normalized second-moment correlations on d for models B and D of Fig. 3. The "leftwards" shift is again evident in Fig. 4b. There are, however, marked variations in the shape of some of the correlations, perhaps most notably for the shear stress $\overline{u_1 u_2}$ and streamwise normal stress $\overline{u_1^2}$.

It may be of interest to note that the paper by Van Dromme and Kollmann appearing in this volume also uses a non-linear model of the ϕ_1 process in the density-velocity correlation equation [Eq. (22) of their paper]; this presumably accounts in part for the

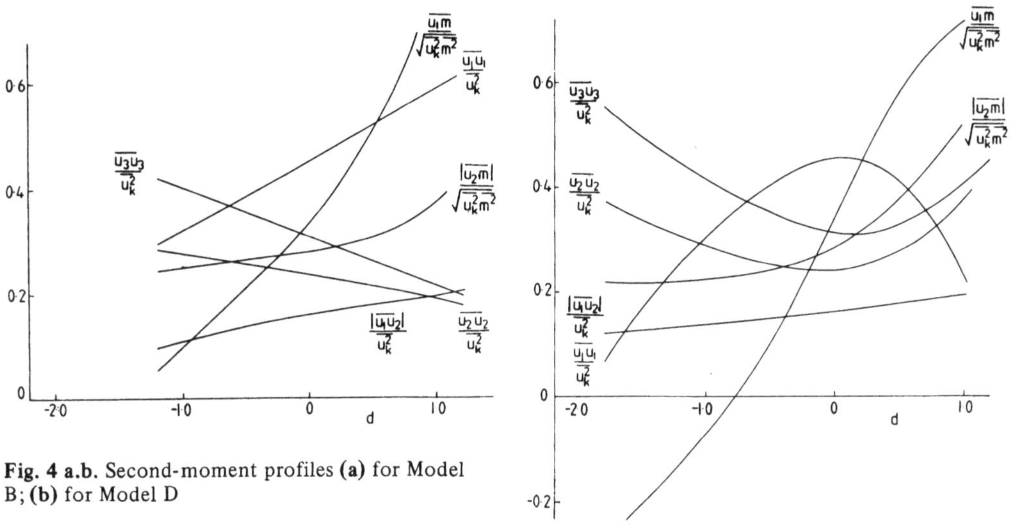

Fig. 4 a.b. Second-moment profiles **(a)** for Model B; **(b)** for Model D

improvement these workers achieve in relation to an earlier attempt [11]. Their use of the non-linear form is, however, coupled with the SQIM (or RPM) version for ϕ_2, a combination which our experience suggests is not well adapted to the calculation of uniform density flows.

Concluding Remarks

The use of the mixed weighting of velocities presented in this contribution helps illuminate the different roles that mean velocity and mass velocity gradients play in determining the level of the turbulent stresses and mass fluxes. Our studies suggest that the key to accounting for the observed phenomena is the approximation of the pressure-strain and other pressure-containing correlations. The results obtained for a notionally homogeneous shear flow exhibit strong sensitivity to the nature of these approximations.

Of the variants tested, version D appears to mimic reasonably well the effects of substantial density gradient for both positive and negative d (corresponding to density gradients in the opposite and the same sense as the velocity gradient). It also reduces to a well-tested form in the uniform density limit. The paper has, of course, taken as its point of departure the view that the experimental phenomena can be usefully thought about in local-equilibrium terms. While not withdrawing that proposition, it must firmly be said that that is not the *whole* story nor, possibly, is it enough of the story even for engineering purposes. The ideas need now to be tested by undertaking numerical simulations that recognize the *in*homogeneities that are actually present, a task on which the authors are embarked.

An inescapable conclusion of this study is that there is an urgent need for more experimental data of a fundamental kind that would allow a far more complete guide to the dependence of Sc_t on d (and possibly other factors) than is presently available.

Acknowledgements. Besides the assistance provided by our respective institutions, the research has been supported by the US Department of Energy, (Division of Basic Energy Sciences) through Contract No. AS03–76SFO 0034 and through the UK Science and Engineering Research Council Grant No. GR/B70056.

Authors' names appear alphabetically

References

1. Rebollo, R.M.: "Analytical and Experimental Investigation of a Turbulent Mixing-Layer of Different Gases in a Pressure-Gradient"; Ph.D. Thesis, California Inst. of Technology, Pasadena (1973)
2. Gibson, M.M., Launder, B.E.: ASME J. Heat Transfer **C 98**, 81 (1976)
3. Launder, B.E.: J. Fluid Mech. **67**, 569 (1975)
4. DuPont-Donaldson, C., Sullivan, R.C., Rosebaum, H.: AIAA J. **10**, 162 (1972)
5. Janicka, J., Kollmann, W.: "A Prediction Model for Turbulent Diffusion Flames Including NO-Formation", AGARD-PEP 54th Meeting, Cologne, 1979
6. Favre, A.: "Statistical Equations of Turbulent Gases", in *Problems of Hydrodynamics and Continuum Mechanics* (Society for Industrial and Applied Mathematics, Philadelphia, PA 1969) pp. 237–266
7. Kent, J.H., Bilger, R.W.: "The Prediction of Turbulent Flame Fields and Nitric Oxide Formation" Proc. 16th Int. Symp. of Combustion (The Combustion Inst. 1976) p. 1643
8. Jones, W.P.: "Models for Turbulent Flows with Variable Density and Combustion", VKI Lecture Series, 1979–2, Jan. 15–19, 1979
9. Libby, P.A.: "Studies in Variable-Density and Reacting Turbulent Shear Flows", in *Studies in Convection*, Vol. 2, ed. by B.E. Launder (Academic, New York 1977)
10. Brown, G.L., Roshko, A.: J. Fluid Mech. **64**, 775 (1974)
11. Kollmann, W., Vandromme, D.: "Turbulent Shear Flow with Fluctuating Density", AIAA 12th Fluid and Plasma Dynamics Conference, Williamsburg, VA, July 23–25, 1979
12. Ha Minh, H., Launder, B.E., MacInnes, J.: "A New Approach to the Analysis of Turbulent Mixing in Variable Density Flows", 3rd Turbulent Shear Flow Symp. p. 19.19, Davis, CA, Sept. 1981
13. Janicka, J., Lumley, J.L.: "Second-Order Modelling in Non-Constant Density Flows", Fluid Dynamics and Aerodynamics, Sibley School of Mechanical and Aerospace Engineering, Cornell University, Ithaca, NY, Jan. 1981
14. Rubesin, M.W., Rose, W.C.: "The Turbulent Mean Flow, Reynolds Stress and Heat-Flux Equations in Mass-Averaged Dependent Variables", NASA TNX 62248, 1973
15. Launder, B.E., Reece, G.J., Rodi, W.: J. Fluid Mech. **68**, 537 (1975)
16. Naot, D., Shavit, A., Wolfshtein, M.: Phys. Fluids **16**, 738 (1973)
17. Naot, D., Shavit, A., Wolfstein, M.: Interactions between components of the turbulent velocity correlation tensor due to pressure fluctuations. Israel J. Technol. **8**, 259 (1970)
18. Launder, B.E., Morse, A.: "Numerical Prediction of Axisymmetric Free Shear Flows with a Reynolds Stress Closure", in *Turbulent Shear Flows I*, ed. by F. Durst, B.E. Launder, F.W. Schmidt, J.H. Whitelaw (Springer, Berlin, Heidelberg, New York 1979)
19. Béguier, C., Dekeyser, I., Launder, B.E.: Ratio of scalar and velocity dissipation time scales in shear flow turbulence. Phys. Fluids **21**, 307 (1978)
20. Watt, W.E.: "The Velocity-Temperature Mixing Layer", University of Toronto, Dept. of Mech. Eng. Technical Publication, TP 6705, 1967
21. Launder, B.E.: "Heat and Mass Transport", in *Turbulence*, ed. by P. Bradshaw, Topics in Applied Physics, Vol. 12 (Springer, Berlin, Heidelberg, New York 1976) Chap. 6
22. Samaraweera, D.S.A.: "Turbulent Heat Transport in 2- and 3-Dimensional Temperature Fields"; Ph.D. Thesis, Faculty of Engineering, University of London (1978)

Direct Simulation of Homogeneous Turbulent Shear Flows on the Illiac IV Computer: Applications to Compressible and Incompressible Modelling

W.J. Feiereisen[1], E. Shirani[2], J.H. Ferziger, and W.C. Reynolds
Department of Mechanical Engineering, Stanford University, Stanford, CA 94305, USA

Abstract

Full simulations of sheared homogeneous turbulence were made on the ILLIAC IV computer using 64 x 64 x 64 mesh points; both incompressible and compressible cases were run. The results are validated by comparison with experimental data and are used as the basis for a study of phenomenological turbulence models. A flaw in eddy viscosity models is pointed out. The Rotta pressure-strain model is shown to be reasonably good when one includes the anisotropy of the dissipation in the model. On the other hand, the commonly used model for the rapid pressure-strain term is shown to be not very accurate. Compressibility effects are fairly small, but the acoustic pressure fluctuations in these flows can be quite large. Despite this, little change in the modeling is required.

1. Objectives

The objectives of this paper are to provide guidance to the development of simple phenomenological models of turbulence through comparisons of models against exact numerical simulation of simple turbulent flows. The flow of interest is a homogeneous shear flow in which the mean velocity varies linearly in one direction.

Homogeneous shear flow is of complexity intermediate between unsheared homogeneous turbulence and inhomogeneous turbulent shear flows. Its study can provide better understanding of turbulence and can be used in testing turbulence theories and statistical turbulence models.

2. Review of Previous Work

Reis (1952) studied homogeneous shear flows mathematically. *Rose* (1966) studied the behavior of the turbulence intensities experimentally at low Reynolds numbers. *Champagne* et al. (1970), *Harris* (1977), *Rose* (1970), *Hwang* (1971), and *Mulhearn* and *Luxton* (1975) studied this flow at higher Reynolds numbers. *Tavoularis* (1978) studied this flow and included a passive scalar. They all measured the shear stress correlations various length scales, one- and two-point velocity correlations, spectra, and the

---

1 *Present address:* Brown, Boveri et Cie., CH–5401 Baden, Switzerland
2 *Present address:* Universität Karlsruhe, D–7500, Karlsruhe, Fed. Rep. of Germany

Table 1. Description of various experiments on homogeneous shear flows

	Shear Rate S [s^{-1}]	Reynolds No. Re_λ [a]	Shear No. S [b]	Total Shear St
Rose (1966)	13.68	133.0	9.53	3.2
Champagne et al. (1970)	12.90	204.8	7.79	3.2
Harris et al. (1977)	44.0	396.4	8.28	10.81
Tavoularis (1978)	46.8	397.9	14.76	12.65

[a] The values of Re_λ and S are given at $x_1/h = 7.5$ for *Rose* and Harris et al., at $x_1/h = 8.5$ for *Champagne* et al., and at $x_1/h = 11$ for Tavoularis.
[b] The shear number is defined as $S = SL/q$, where L is integral length scale and $q = \langle u_i u_i \rangle^{1/2}$.

kinetic energy of turbulence. Table 1 summarizes some of the basic parameters of these flows.

Numerical simulations of homogeneous shear flow were made by Shaanan (1975) using large eddy simulation. *Rogallo* (1977) made full simulations of low Reynolds number homogeneous turbulent shear flows. This paper reports further simulations of this kind.

We also look at the effect of compressibility on homogeneous turbulent shear flow. Early studies of compressible turbulence introduced decomposions of the flow variables into modes and studied their interaction. *Kovasznay* (1953) and *Chu* and *Kovasznay* (1958) derived equations for the vorticity mode, which is responsible for convective and vortical motions, and for the pressure (acoustic) and entropy (temperature) modes. The analytically studied interactions among these modes. Moyal divided the kinetic energy spectra into "eddy turbulence" and "random noise" parts. He did this by Fourier-transforming the velocity vector and decomposing it into vectors that are perpendicular and parallel to the wave-number vector. His analysis, like that of Chu and Kovasznay, is limited to low turbulence levels and therefore not applicable in technologically useful flows. *Morkovin* (1962) showed that the Reynolds stresses in supersonic boundary layers were structurally similar to those in incompressible flows. *Laufer* (1969) used Favre averages and came to the same conclusion. *Bradshaw* (1977) agreed with Morkovin and showed that incompressible models should not be applied in boundary layers with external Mach numbers greater than five, in shock-boundary layer interactions, nor in free-shear layers with Mach numbers greater than 1.5. The Free Shear Layers Conference (1972) concluded that existing turbulence models were inadequate to predict the decrease in mixing-layer spreading rate with Mach number.

These failures have been attributed to variations in density. However, *Brown* and *Roshko* (1974) concluded that spreading-rate variations must be a "compressibility effect." *Oh* (1974) simulated a two-dimensional, high-speed mixing layer and proposed a kinetic-energy equation model that provided for a nonzero pressure-dilatation interaction.

3. Review of Turbulence Modeling

Turbulent flows contain a large range of length scales. This causes problems for the flow simulator because the smallest and largest scales of motion in the flow cannot be represented at the same time on a relatively coarse mesh. Only simple flows at low Reynolds numbers can now be simulated.

Most of the fundamental work on turbulence modeling has been based on the incompressible Navier-Stokes equations. For many years, models developed from the incompressible equations have been applied to compressible flows, frequently with great success. At higher Mach numbers, existing turbulence models become increasingly inadequate.

The most common approach to the prediction of turbulent flows is based on Reynolds averaging. The mean equations are not closed, and some terms (principally the Reynolds stresses, $<u_i'u_j'>$) need to be modeled. The simplest models for the turbulent Reynolds stress assumes a Newtonian-like constitutive equation – the Boussineq eddy-viscosity concept:

$$<u_i'u_j'> = -2\nu_T S_{ij} + \frac{1}{3}\delta_{ij}q^2 \tag{1}$$

where $<>$ denotes volume average, u_i' is the velocity fluctuation defined as the difference between the total velocity, u_i, and the ensemble average velocity, $\bar{u}_i$

$$u_i' = u_i - \bar{u}_i \tag{2}$$

ν_T is eddy viscosity and itself needs to be modeled. S_{ij} and q are mean strain and r.m.s. turbulent velocity, respectively.

The simplest model represents ν_T in terms of the velocity field. In higher-level models, partial differential equations for the turbulent kinetic energy equation, dissipation and/or turbulent Reynolds stresses are solved. To close these equations, various terms need to be modeled. In Sect. 5, we present and test some of the models for these terms.

4. Problem Description and Numerical Methods

In this paper, we consider homogeneous shear flows with the following mean values of velocity components and pressure:

$$\bar{u}_1 = SX_2, \quad \bar{u}_2 = \bar{u}_3 = 0, \quad \bar{p} = \text{const} \tag{4}$$

where S is the shear rate. The velocity field is shown in Fig.1. We choose to simulate a portion of the flow field and impose periodic boundary conditions on the fluctuating quantities. In order to do this, it is necessary to transform to a coordinate system which shears with the mean flow. The physical region being simulated becomes distorted with time, and, without "remeshing," the simulation, if continued, would become invalid. We "remesh" the coordinate system when St is a half-integer; the sheared coordinate system is replaced by one sheared in the other direction, and the velocity field is placed on the new grid. The mesh points of the "before" and "after" systems are equivalent, but the velocity field needs to be truncated in Fourier space to maintain continuity.

In the compressible flow, the initial condition represents isotropic turbulence with a given spectral shape and intensity, and is divergence-free, i.e., it is an incompressible initial field. Compressibility effects arise from the high shear rate and the fact that the ratio of the turbulence intensity, q, to the mean sound speed is as high as 0.3. The full compressible equations are advanced by means of a fourth-order Runge-Kutta method. Spectral methods are used to evaluate all spatial derivates; both this and the program described below used 64 mesh points in each spatial direction.

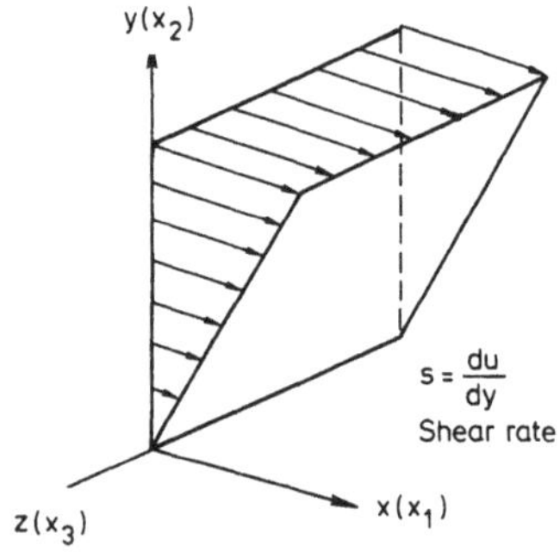

Fig. 1. The mean flow

Incompressible flows were simulated both by running the compressible code at low Mach number and by a separate imcompressible code. The latter solves a Poisson equation for the pressure and uses a second-order Adams-Bashforth method for the time advancement. Spectral methods are used for the spatial derivatives. This program also has the ability to include a passive scalar. Further details of these methods can be found in reports by the authors [*Feiereisen* et al. (1981) and *Shirani* et al. (1981)].

5. The Simulations

Several runs were made with each of the codes described above. In each run, the earliest part of the simulated flow field is not representative of a real flow, because the initial conditions are not sufficiently realistic; at $St = 4$, the flow attains the characteristics observed in the laboratory flow. The length scales grow throughout the simulation; eventually they become long with respect to the size of the region simulated, and the results are no longer realistic. This happens in most cases at approximately $St = 7$. Thus, only the period $4 \leqslant St \leqslant 7$ is useful for model testing.

Analysis of the independent variables in these flows shows that there are only two independent, nondimensional parameters (other than the nondimensional time St) in the imcompressible flows. We have chosen these to be the Reynolds number $R_\lambda = q\lambda/\nu$ and a quantity we have called the shear number, which is defined as $\mathcal{S} = SL/q$. The range of these parameters is approximately $15 \leqslant R_\lambda \leqslant 120$ and $3 < \mathcal{S} < 70$. Here L is the integral scale and λ, the microscale. Due to the restriction on the Reynolds number, the ratio of length scales L/λ is nearly constant in these flows and cannot be considered as an independent, nondimensional variable. In the compressible cases, the Mach number $M = q/c$ is another independent parameter; we have considered the range $0 \leqslant M < 0.3$.

Detailed comparison of the results with experiments is not possible, because it is impossible to duplicate the initial conditions of the experiments with sufficient precision. We have compared several results with those of experiments, and they appear to agree quite well on a qualitative basis. Some of these will be shown later. For further details, the reader may wish to consult the reports cited earlier.

6. Models for Incompressible Turbulence

This and the following section are devoted to studying turbulence models using the computed results. Incompressible flows and their modeling are considered here, and compressibility effects are considered in the next section.

First, let us consider eddy viscosity models. The usual assumptions are:

$$\tau_{ij} = \frac{R_{kk}}{3}\,\delta_{ij} = \langle u_i' u_j' \rangle - \frac{q^2}{3}\,\delta_{ij} = -\nu_T S_{ij}\ ;\quad \nu_T = cqL \tag{5}$$

where L is a length scale which we shall take to be the integral scale (the choice is not critical in view of the above) and c is assumed constant. However, c may be a function of the dimensionless parameters of the problem, so we attempted to fit it as a function of the Reynolds and shear numbers. A power-law fitting function was selected:

$$F = C\,R_\lambda^{\alpha}\,S^{\beta} \tag{6}$$

This function will be used for other parameters as well. It may not be the optimum choice, but there is little theoretical guidance for the selection of a more appropriate function. The intent of this fit is not to develop a model for the term but instead to use the fit to discover which parameter most sensitively affects the quantity under study (e.g., c). The three constants in Eq. (6) were determined by least-squares fitting, and Fig. 2 gives the constants and shows that the results are well fit; three experimental points are also shown, and they agree quite well with the fit. The results show the eddy viscosity constant c to be nearly inverse to the shear number. This result can be written:

$$\frac{\langle u_1' u_2' \rangle}{q^2} = 0.265\,R_\lambda^{-0.08}\,S^{-0.07} \tag{7}$$

which agrees quite well with the experimental observation that in many flows $\langle u'v' \rangle / q^2$ is approximately constant. This means that the assumption that c is constant in Eq. (5) is inconsistent with the experimental observation of constant $\langle u_1' u_2' \rangle / q^2$. Furthermore, one can show from these results that the mixing length is given by

$$l^2 \equiv \frac{\langle u_1' u_2' \rangle}{S^2} = \frac{\langle u_1' u_2' \rangle}{q^2}\,\frac{q^2}{S^2} \cong \text{const}\,\lambda^2\ . \tag{8}$$

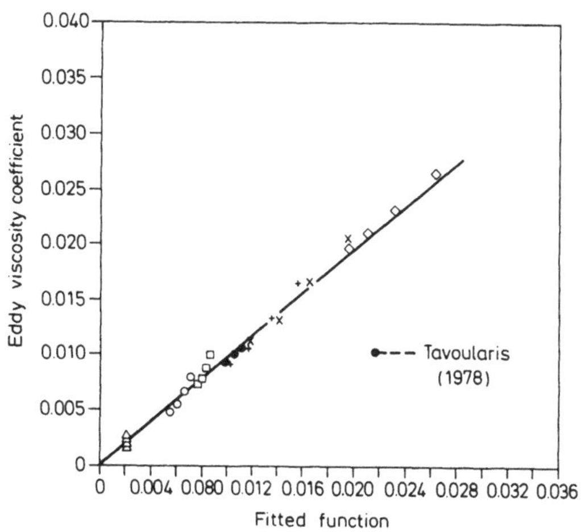

Fig. 2. ν_T/qL vs the fitting function $0.265\,\mathrm{Re}_\lambda^{-0.08}\,S^{-1.07}$

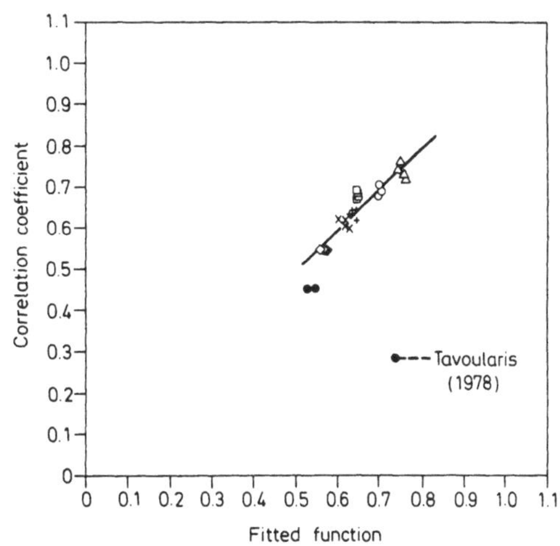

Fig. 3. The Reynolds shear stress correlation vs the fitting function $0.515\,\mathrm{Re}_\lambda^{-0.043}\,S^{0.134}$

Thus, the mixing length appears to grow in proportion to the microscale.

Figure 3 gives a related result. The shear stress correlation $\langle u_1' u_2' \rangle (\langle u_1'^2 \rangle \langle u_2'^2 \rangle)^{1/2}$ is fitted with the function (6). The fit is not as good as in Fig. 2, but we see that there is little variation of the correlation with Reynolds or shear number.

Now let us turn our attention to higher-order models. One common assumption in turbulence modeling is that the dissipation is isotropic. We have looked at the anisotropy of the dissipation defined by:

$$d_{ij} = \left(\epsilon_{ij} - \frac{\epsilon_{kk}}{3} \delta_{ij} \right) \Big/ \epsilon \tag{9}$$

where ϵ_{ij} is the rate of viscous dissipation of $\langle u_i' u_j' \rangle$ and $\epsilon = \epsilon_{kk}$. In order to condense the presentation of the results, we plot the dissipation anisotropy vs anisotropy of the Reynolds stress b_{ij}. Figure 4 shows that the relationship is nearly linear; the slope is approximately 0.85, indicating strong anisotropy of d_{ij}. Although it might be thought that this is attributable to the relatively low Reynolds number of the simulation, we find no systematic variation of the dissipation anisotropy with Reynolds number over the range covered. This result suggests that the assumption of small d_{ij} be reexamined.

Finally, we have studied the pressure-strain models that are commonly used in a full Reynolds stress model. The pressure obeys the Poisson equation in which the source terms are velocity gradients; the latter can be divided into two groups, depending on whether they contain the mean velocity or not. Following Lumley, we call the terms containing the mean velocity the "rapid" terms and the remainder, the "Rotta" terms.

First, we consider the Rotta terms. The common model is:

$$\tau_{ij}^1 \equiv \left\langle p' \left(\frac{\partial u_i'}{\partial x_j} + \frac{\partial u_j'}{\partial x_i} \right) \right\rangle_{\text{Rotta}} = c\epsilon b_{ij}/q^2 \quad . \tag{10}$$

The "constant" was evaluated from the computational results and found to be smaller than 1; since the constant must be at least 2 to give relaxation to isotropy, this is unacceptable. The "constant" is also strongly dependent on Reynolds number, as shown in Table 2. When the anisotropy of the dissipation [a redistribution term, *Lumley* (1978)] is added to the Rotta term, we get the results shown in Table 3. The best estimate of a single value for the constant is 2.7, and the dependence on Reynolds and shear number

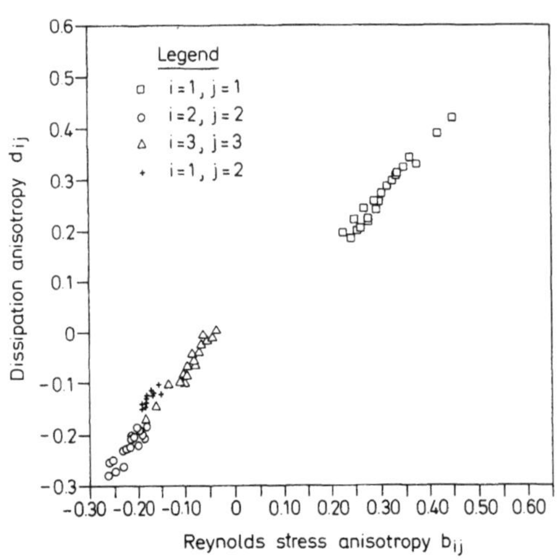

Fig. 4. Dissipation anisotropy vs Reynolds stress anisotropy. All four nonzero components are shown

Table 2. "Constant in the Rotta model Eq. (10). The "constant" was fit by $c = C\,\mathrm{Re}_\lambda^\alpha\, S^\beta$. The last column is the r.m.s. difference between the „constant" and the fit

Equation	A	C	α	β	r.m.s. error
$i = 1,\ j = 1$	0.673 ± 0.599	1.143	0.678	-1.195	0.15
$i = 2,\ j = 2$	0.673 ± 0.570	0.734	0.715	-1.075	0.13
$i = 1\ \ j = 2$	0.679 ± 0.452	0.238	0.813	-0.776	0.15

Table 3. "Constant" in the Rotta model the anisotropy of the dissipation is included with the pressure-strain term. The meaning of the last four columns is the same as in Table 2

Equation	A_3	C	α	β	r.m.s. error
$i = 1,\ j = 1$	2.862 ± 0.349	3.342	0.144	-0.225	0.045
$i = 2,\ j = 2$	3.080 ± 0.428	4.200	0.119	-0.245	0.048
$i = 1,\ j = 2$	2.472 ± 0.426	1.633	0.330	-0.266	0.085

is reasonably small. In model calculations, assuming the dissipation isotropic is equivalent to lumping its anisotropy with the Rotta term, and we conclude that, although the arguments used in constructing the model are not correct, the resulting model is reasonable. This provides a reinterpretation of the model and an estimate for the constant that is in good agreement with the ones in common use. We have also tested a correlation suggested by *Lumley* and *Newman* (1977) for the Rotta "constant." The results showed large scatter, and some of the coefficients were very large; we cannot recommend this correlation on the basis of our results.

Now let us turn to consideration of the rapid term. A model for this term has been given by several authors. They are essentially equivalent, so we shall use the one given by *Reynolds* (1976):

$$T_{ij}^2 = \tfrac{2}{5}(1 + A_1)\,q^2 S_{ij} - \tfrac{3}{5}A_1\,(R_{ik}S_{kj} + R_{jk}S_{ki} + \tfrac{2}{3}\,P\,\delta_{ij})$$
$$- \tfrac{4}{5}\left(\tfrac{5}{3} + \tfrac{7}{12}A_1\right)(R_{ik}\Omega_{kj} + R_{jk}\Omega_{ki}) \qquad (11)$$

where $P = R_{ik}S_{ki}$ is the production,

$$S_{ij} = \frac{1}{2}\left(\frac{\partial \bar{u}_i}{\partial x_j} + \frac{\partial \bar{u}_j}{\partial x_i}\right)$$

is the strain rate, and

$$\Omega_{ij} = \frac{1}{2}\left(\frac{\partial \bar{u}_i}{\partial x_j} - \frac{\partial \bar{u}_j}{\partial x_i}\right)$$

is the rotation. The "constant" A_1 was computed for each set of tensor indices in each flow for each case, and a remarkable result was found. The "constant" is indeed nearly independent of the Reynolds and shear numbers, but it varies with the tensor index. The results are given in Table 4. We cannot offer an explanation of this result, but people do-

Table 4. "Constant" for the rapid pressure-strain term. The meaning of the last four columns is the same as in the previous tables

Equation	A_1	C	α	β	r.m.s. error
$i = 1, \quad j = 1$	-6.94 ± 0.519	9.75	-0.028	-0.082	0.038
$i = 2, \quad j = 2$	-1.83 ± 0.400	1.55	0.038	0.008	0.060
$i = 3, \quad j = 3$	-1.57 ± 0.380	6.82	-0.185	-0.282	0.093
$i = 1, \quad j = 2$	-1.52 ± 0.077	2.01	-0.082	0.010	0.032

ing model calculations might want to try an ad hoc modification of the model implied by these results, even though it is not in proper tensorial form.

A simpler model used by *Launder* et al. (1975) was also tested but did not fare so well as the model represented by Eq. (10).

Finally, fitting the full pressure-strain term to the combined models using a least-squares fit to determine the two constants gives results that are significantly better than either model alone.

7. Compressible Flow

As stated earlier, we also ran several cases in which compressibility is important. The effect of compressibility is indicated by the turbulence Mach number $M = q/c$ differing from zero; we considered values as high as $M = 0.3$, approximately the value found in supersonic shear layers. An alternative measure of the importance of compressibility is the parameter SL/c, the difference in speed across a large eddy divided by the Mach number. This parameter is, of course, the product of the shear and Mach numbers defined above.

In this section, we shall concentrate on the effects produced by the introduction of compressibility into the flow. On the whole , these effects are quite small, but it is instructive to study them in some detail.

One tool for examining the effect of compressibility is the Moyal decomposition. Moyal considers the velocity field to be a combination of a field which contains vorticity but no dilatation and one which has dilation but no vorticity; these two components are called u_i^s (solenoidal) and u_i^d (dilatational), respectively. It turns out that the energy

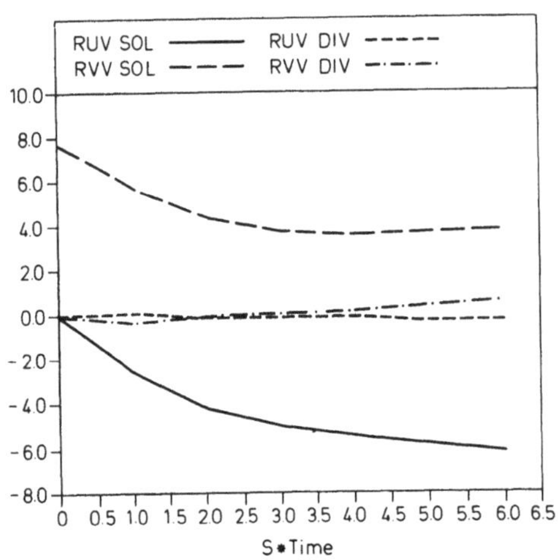

Fig. 5. The solenoidal and dilatational components of the Reynolds stresses $\langle u_1' u_2' \rangle$ and $\langle u_1' u_2' \rangle$

in the delatational part of the field is always small compared to the energy in the solenoidal component; at the highest Mach number, the former is approximately two orders of magnitude larger than the former. The only exception is the 2–2 component of the dilatation field, which is less than an order of magnitude smaller than the corresponding solenoidal component. This is the gradient direction, and we believe that the relatively large size of $<u_2^d u_2^d>$ is associated with sound-wave propagation in the direction normal to the flow. There is also a possibility that this result is connected with the decreased rate of growth of shear layers at high Mach numbers, but it is difficult to offer convincing evidence for this. A typical time history of the flow is given in Fig. 5.

We now turn our attention to the pressure field and the pressure-strain terms. In a compressible flow, the pressure obeys the equation:

$$\nabla^2 p = -(\rho u_i u_j)_{,ij} - (\rho u_i)_{,j} u_{j,i} + \frac{\partial^2 \rho}{\partial t^2} \quad . \tag{12}$$

In an incompressible flow, the first two terms give rise to the Rotta and rapid components of the pressure field. In view of what we discovered above, we do not expect these terms to change very much when compressibility is considered. When we note that $\partial^2 \rho / \partial t^2 = c^{-2} \, \partial^2 p / \partial t^2$, the last term changes the equation from a Poisson equation to a wave equation. Alternatively, we can regard the last term as a new source term due to the compressibility, and the pressure field is then regarded as being composed of three parts – the two used in the incompressible flow and a new compressible term. (There is no need to use the Moyal decomposition here.)

It was expected that the new term would be small, but it turned out that the compressible component of the pressure is surprisingly large. As it is the pressure-strain term that we are primarily concerned with, we present in Fig. 6 the ratio of the mean-square pressure-strain term (the second invariant of the pressure-strain tensor as de-

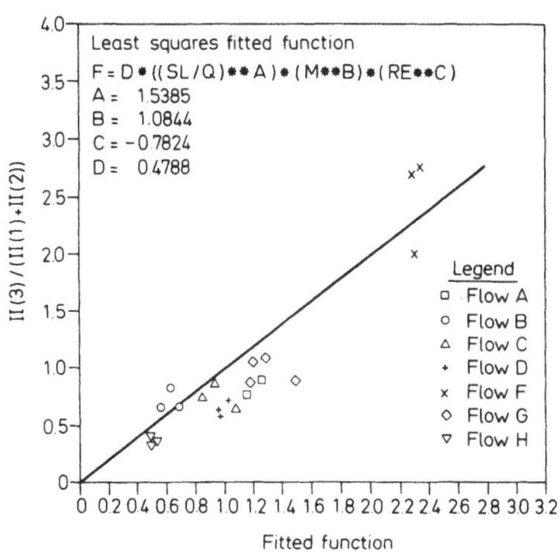

Fig. 6. Ratio of the second invariant of the compressible pressure-strain term to the sum of the second invariants of the rapid and Rotta terms vs a fitting function

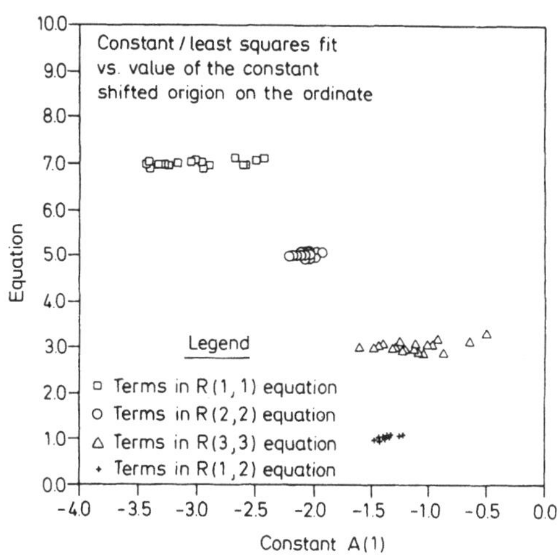

Fig. 7. Test of the model of Eq. (11) in compressible flow. The model is applied to the sum of the rapid and compressible terms. The abscissa is the constant obtained; the ordinate is the ratio of the constant to a correlation fit. The curves are shifted for clarity of presentation

fined by Lumley) for the compressible and incompressible components of the field. In this figure, the results have been correlated using an extension of Eq. (7) which includes a power of the Mach number. The results are given in the figure, and the size of the new term is indeed much larger than one might have expected. This term was computed in two different ways which differed by 10%, so we have confidence in the results. We suspect that the results are due to sound waves propagating through the flow. It is worth noting that the ratio of the r.m.s fluctuations of the components of the pressure field is less than 1%, so there is apparently some mechanism operating which strongly correlates the sound waves with the strain in the field. The origin of the compressible portion of the pressure field was found to be dilatation and not the simple existence of density differences.

Finally, we considered the modeling of the compressible term. It was found to behave like the rapid term, and it is preferable to model the sum of the two terms as a unit. When this is done, the result is qualitatively similar to what was found in the incompressible flow; however, the constants are changed somewhat (compare the results in Fig. 7 with those in Table 4). Thus it appears that, the compressible terms have a measuable effect on the pressure-strain term and this may be responsible for the change in $<u'_2 u'_2>$ noted above.

We have also investigated models in which one assumes structural similarity or constancy of the anisotropic component of the Reynolds stresses. We were unable to find a "constant" for these models which fit the results well.

8. Conclusions

We have shown that it is possible to fully simulate sheared homogeneous turbulence. The results compare quite well with the experiments, at least on a qualitative basis. It is possible to use the results of the simulations to produce information that cannot be obtained directly from experiments. In particular, we have computed the pressure-strain terms and compared them with model predictions. For the Rotta term, good results are obtained, provided that one includes the anisotropy of the dissipation along with the pressure-strain term. For the rapid term, we are unable to produce a good fit between the model and the computational results. Finally, it is shown that the effects of compressibility are small, for the most part. The only effect which was not found to be small is a modification of the pressure term, and this seems to be responsible for producing a change in the normal component of the Reynolds stress.

References

Bradshaw, P.: Compressible turbulent shear layers. Ann. Rev. Fluid Mech. 33–54 (1977)

Brown, G.L., Roshko, A.: On density effects and large structure in turbulent mixing layers. J. Fluid Mech. **64**, 775 (1974)

Champagne, F.H., Harris, V.G., Corrsin, S.: Experiments on nearly homogeneous turbulent shear flow. J. Fluid Mech. **41**, 81 (1970)

Chu, B-T, Kovasznay, L.S.G.: Nonlinear interactions in a viscous heat-conducting compressible flow. J. Fluid Mech. **3**, 494 (1958)

Feiereisen, W.J.: "Numerical Simulation of a Compressible, Homogeneous, Turbulent Shear Flow"; Ph.D. Thesis, Mech. Eng. Dept., Stanford University (1981) (Also published as Rpt. TF-13)

Harris, V.G., Graham, J.A., Corrsin, S.: Further experiments in nearly homogeneous turbulent shear flow. J. Fluid Mech. **81**, 657 (1977)

Hwang, W.S.: "Experimental Investigation of Turbulent Shear Flows"; Ph.D. Thesis, University of Virginia (1971)

Kovasznay, L.S.G.: Turbulence in supersonic flow. J. Aero. Sci. **20**, 657 (1953)
Laufer, J.: "Thoughts on Compressible Turbulent Boundary Layers", Memorandum RM-5946-PR, RAND Corp., March 1969
Launder, B.E., Reece, G.J., Rodi, W.: Progress in the development of a Reynolds-stress turbulence closure. J. Fluid Mech. **68**, 537 (1975)
Lumley, J.L., Newman, G.R.: The return to isotropy of homogeneous turbulence. J. Fluid Mech. **82**, 161 (1977)
Lumley, J.L.: Computational modeling of turbulent flows. Adv. Appl. Mech. **18** (1978)
Morkovin, M.V.: "Effects of Compressibility on Turbulence", in *Mechanique de la Turbulence* (C.N.R.S., Paris 1962)
Mulhearn, P.J., Luxton, R.E.: The development of turbulence structure in a uniform shear flow. J. Fluid Mech. **68**, 577 (1975)
Oh, Y.H.: "Analysis of Two-Dimensional Free Turbulent Mixing," AIAA 7th Fluid & Plasma Dynamics Conf., AIAA Paper 74–594 (1974)
Reis, F.B.: "Studies of Correlation and Spectra in Homogeneous Turbulence"; Ph.D. Thesis, MIT, Cambridge, MA (1952)
Reynolds, W.C.: Computation of turbulent flows. Annu. Rev. Fluid Mech. **8**, 183–208 (1976)
Rogallo, R.S.: An ILLIAC program for the numerical simulation of homogeneous incompressible turbulence. NASA Tech. Memo. **73**, 203 (1977)
Rose, W.G.: Results of an attempt to generate a homogeneous turbulent shear flow. J. Fluid Mech. **25**, 97 (1966)
Rose, W.G.: Interaction of grid turbulence with a uniform mean shear. J. Fluid Mech. **44**, 767 (1970)
Shaanan, S., Ferziger, J.H., Reynolds, W.C.: "Numerical Simulation of Turbulence in the Presence of Shear", Rpt. No. TF-6, Mech. Eng. Dept., Stanford University (1975)
Shirani, C.E., Ferziger, J.H., Reynolds, W.C.: "Mixing of a Passive Scalar in Homogeneous Turbulent Shear Flows"; Ph.D. Thesis, Eng. Dept., Stanford University (1981) (Also published as Rpt. TF-15)
Tavoularis, S.: "Experiments in Turbulent Transport and Mixing"; Ph.D. Thesis, John Hopkins University (1978)

Index of Contributors

Turbulent Shear Flows 1

Selected Papers from the First International Symposium on Turbulent Shear Flows, The Pennsylvania State University, University Park, Pennsylvania, USA, April 18–20, 1977

Editors: **F. Durst, B.E. Launder, F.W. Schmidt, J.H. Whitelaw**

1979. 256 figures, 4 tables. VI, 415 Seiten
ISBN 3-540-09041-X

Contents: Free Flows. - Wall Flows. - Recirculating Flows. - Developments in Reynolds Stress Closures. - New Directions in Modeling.

Here is a survey of the latest developments in the calculation of turbulent shear flows, emphasizing their flow and heat transfer properties. The improvement of physical understanding and related measurements is considered essential throughout. Skillfully edited and carefully selected, a third of the papers presented at the symposium have been included in this volume.

There are five distinct themes:
- Free shear flows
- Wall flows
- Recirculating flows
- Stress transport models
- Modelling developments

Each section is preceded by an introductory review article contributed by the editors so that the volume provides a major and unified statement on the field. Researchers in most fields of turbulence will find the book a valuable aid in assessing likely trends of future work. It is a volume destined to have a lasting impact on the field.

Springer-Verlag
Berlin
Heidelberg
New York

MIX
Papier aus verantwortungsvollen Quellen
Paper from responsible sources
FSC® C105338

If you have any concerns about our products,
you can contact us on
ProductSafety@springernature.com

In case Publisher is established outside the EU,
the EU authorized representative is:
Springer Nature Customer Service Center GmbH
Europaplatz 3, 69115 Heidelberg, Germany

Printed by Libri Plureos GmbH
in Hamburg, Germany